national
STATiSTiCS

Cancer trends in England and Wales 1950–1999

Studies on Medical and Population Subjects No. 66

Mike Quinn, Penny Babb, Anita Brock, Liz Kirby, Jenny Jones

London: The Stationery Office

ISBN 0 11 621393 0

Applications for reproduction:
Copyright Manager, Office for National Statistics
Zone B1/09, 1 Drummond Gate,
London SW1V 2QQ
Fax: 020 7533 5685
E-mail: copyright@ons.gov.uk

Contact points
For enquiries about this publication,
contact Penny Babb
Tel: 020 7533 5266
E-mail: penny.babb@ons.gov.uk

To order this publication, call The Stationery Office
on **0870 600 5522.** See also back cover.

For general enquiries, contact the National Statistics
Public Enquiry Service on **0845 601 3034**
(minicom: 01633 812399)
E-mail: info@statistics.gov.uk
Fax: 01633 652747
Letters: Zone DG/18, 1 Drummond Gate, London SW1V 2QQ

You can also find National Statistics on the internet –
go to **www.statistics.gov.uk**

A National Statistics publication
Official statistics bearing the National Statistics logo are produced
to high professional standards set out in the National Statistics
Code of Practice. They undergo regular quality assurance reviews
to ensure that they meet customer needs. They are produced free
from any political interference.

About the Office for National Statistics
The Office for National Statistics (ONS) is the government
agency responsible for compiling, analysing and disseminating
many of the United Kingdom's economic, social and
demographic statistics, including the retail prices index, trade
figures and labour market data, as well as the periodic census of
the population and health statistics. The Director of ONS is also
the National Statistician and the Registrar General for England
and Wales, and the agency administers the statutory registration
of births, marriages and deaths there.

LIST OF TABLES

LIST OF FIGURES

ABBREVIATIONS

ACCR	Advisory Committee on Cancer Registration
ALL	acute lymphoid leukaemia
AML	acute myeloid leukaemia
ARV	(ONS) annual reference volume (eg Cancer registrations, series MB1)
ASR	age-standardised rate
CIR	comparative incidence ratio
CLL	chronic lymphoid leukaemia
CML	chronic myeloid leukaemia
CMR	comparative mortality ratio
CRCG	Cancer Registries Consultative Group
CRITG	Cancer Registries' Information Technology Group
CSG	Cancer Surveillance Group
CSO	Central Statistical Office
ESR	directly age-standardised rate using the European standard population
DCO	death certificate only
DH	Department of Health
DHSS	Department of Health and Social Security
GRO	General Register Office
HAA	hospital activity analysis
HSQ	Health Statistics Quarterly
IACR	International Association of Cancer Registries
IARC	International Agency for Research on Cancer
ICD	International Classification of Diseases
ISD	Information and Statistics Division (Scotland)
LSHTM	London School of Hygiene and Tropical Medicine
MAC	Medical Advisory Committee
MOTNAC	Manual of Tumour Nomenclature and Coding
NCIC	National Cancer Intelligence Centre
NCREG	National Cancer Registration Executive Group
NHL	non-Hodgkin's lymphoma
NHS	National Health Service
NHSCR	National Health Service Central Register
NMSC	non-melanoma skin cancer
ONS	Office for National Statistics
OPCS	Office of Population Censuses and Surveys
RHA	regional health authority
SLA	service level agreement
UICC	International Union against Cancer
UK	United Kingdom
UKACR	United Kingdom Association of Cancer Registries
USA	United States of America
WHO	World Health Organisation
WSR	directly age-standardised rate using the World standard population

Symbols and conventions

–	nil
..	not available
:	not appropriate
nos	not otherwise specified

In the UK, as in most developed countries, cancer is a major public health problem. At least one in three people will be diagnosed with cancer during their lifetime, and cancer is the major cause of death in both men and women. A wealth of information on cancer trends has been made available in the past, but in separate publications. This book brings together for the first time not only the long term trends in cancer incidence, mortality, prevalence and survival for all the major cancers for England and Wales but also some recent data for the whole of the UK, accompanied by brief notes on aetiology and risk factors.

There are wide inequalities within the UK in terms of who gets cancer, and what happens to them when they do. People from deprived backgrounds are more likely to get some types of cancer and overall are more likely to die from it once they have been diagnosed. For several reasons, cancer patients in the UK have poorer survival than in other European countries. Although many cancer patients receive excellent treatment, services are patchy – patients in different parts of the country receive varying quality and types of treatment. Experience of cancer care also varies, and long waits and uncertainty add to the inevitable anxiety of patients. The recently published NHS Cancer Plan sets out in detail what needs to be done to address these problems and how progress will be monitored. The government has set a target (for England) of reducing cancer deaths in people aged under 75 by 20% by the year 2010.

The Office for National Statistics and its predecessors have for many years published annual statistics on both cancer incidence and mortality (for England and Wales). In collaboration with the London School of Hygiene and Tropical Medicine it has also published a comprehensive analysis of cancer survival trends over time by region and socio-economic deprivation. Annual data on incidence and mortality and analyses of survival trends have also been published for Scotland. In addition, individual cancer registries publish a vast amount of information relating to the areas they cover, as well as conducting research which is published in papers in peer reviewed scientific and medical journals.

We know so much more about cancer than many other diseases because population based data have been collected and collated through the cancer registration system. The report in 1995 by the Expert Advisory Group on Cancer to the Chief Medical Officers of England and Wales (the Calman-Hine report) recognised the key role of cancer registries, which are essential for monitoring incidence, the effectiveness of screening programmes, and outcomes – particularly survival rates in relation to treatment. The NHS Cancer Plan recognises that these public health benefits depend on the completeness of cancer registration in the population and on its quality and timeliness. The government will publish an action plan to improve the organisation and effectiveness of cancer registries in Autumn 2000.

I welcome the publication of this book which paints the broad picture of the cancer burden in this country and illustrates the baselines against which progress in implementing the NHS Cancer Plan will be measured.

Professor M A Richards, National Cancer Director

Chapter 1

INTRODUCTION

Cancer, together with heart disease, has been given high priority by the government[1,2]. The cost of cancer treatment amounts to over £1 billion each year, accounting for about 6% of all NHS expenditure. Support for research into cancer is currently about £260 million each year[3]. More than one in three people in England and Wales will develop a cancer* some time in their life, and cancer causes one in four deaths. There are 220,000 new cases of cancer registered every year, and 140,000 deaths from cancer. People diagnosed with cancer in the past 10 years who are still alive represent just over 1% of the total population[4]. Since 1950, deaths from major causes such as heart disease, stroke and infectious diseases have all declined dramatically – but cancer mortality has remained fairly stable in both men and women. Cancer became the most common cause of death in females in 1969 and in males in 1995. Since the 1970s, five-year survival has improved for some cancers – particularly for those of the bladder, breast, large bowel, melanoma of skin, and testicle, and for Hodgkin's disease, non-Hodgkin's lymphoma and childhood leukaemia – but not for some of the most common and highly fatal ones such as cancers of the liver, lung, oesophagus and pancreas which represent about one third of cancers in men and one fifth in women. There is overwhelming evidence that cancer survival is generally lower among patients in the most deprived groups than in the affluent[5]. And for the vast majority of cancers, survival in England and Wales, and in Scotland, is well below the rates in other western European countries[5,6,7] and in the USA[8].

The government has set a target (for England) of reducing the cancer death rate in people aged under 75 by 20% over the years to 2010[1]. A National Cancer Director, charged with overseeing the achievement of this target, has been appointed and a National Cancer Plan has been published[9]. Over the next ten years, trends in cancer incidence, survival and mortality will be closely monitored at the national, regional and local levels. Most cancers take many years to develop before being diagnosed, and cancer incidence and mortality generally exhibit only slowly increasing or decreasing trends. For some cancers, the trends for many years to come will be determined to a great extent by people's earlier exposure to various risk factors – for example, lung cancer rates depend on smoking habits 20 or more years previously. So with the exception of cancers for which there have been major public health interventions – such as screening for breast cancer – past trends are a good guide to the future.

Contents

This volume brings together for the first time a large amount of information on cancer incidence (the number of newly diagnosed cases in a particular period), deaths from cancer, prevalence (the number of people who have been diagnosed with cancer in a particular period and are still alive at the end of it) and survival from cancer. A very large number of publications on cancer have been produced by ONS (and its predecessor, the Office of Population Censuses and Surveys – OPCS) – both separately and in collaboration with others – and by the regional cancer registries in England and the registries in Wales, Scotland and Northern Ireland. A bibliography is given in Appendix I.

Following this introduction, a summary chapter draws together information on trends in the incidence of and mortality from all malignant cancers and for nine of the most important individual cancer sites; the prevalence of major cancers; and an outline of survival trends over time and variations by age, sex and socio-economic deprivation. More detailed results on incidence, mortality, prevalence and survival are then presented in 20 separate chapters for each of the main cancer sites which together constitute almost 90% of all malignancies in both males and females. There is also a chapter on several types of childhood cancer which together constitute 82% of all childhood cancers (which in total account for about ¹/₂% of all cancers).

Each of chapters 3-22 starts with a brief description of the epidemiology of the particular cancer, including its global significance, known or suspected causes and risk factors, histological sub-types, and treatment. The trends in incidence and mortality are then described together. Information on incidence covers the period 1971-1997; the results for the latest three years are provisional, but are reliable, being based on data from half of the regional cancer registries in England and Wales. Mortality data cover the period 1950-1999. For both incidence and mortality there are charts showing the rates by (usually ten year) age-group, and an overall (directly) age-standardised rate which allows for differences in the size and age structure of the population of males and females, and for changes in the age structure of the population over time.

Trends in incidence and mortality are presented both by year of diagnosis or death, and by year of birth. The variation of incidence and mortality by region in England and Wales is illustrated; and the overall rates for England and Wales are compared with those for selected countries in all continents of the world. The prevalence of cancer is illustrated in charts showing the numbers and proportions of people alive at the beginning of 1993 who had been diagnosed with cancer in the preceding three years and in the preceding ten years. The information presented on survival comprises trends for cases diagnosed over the period 1971 to 1993 and followed up to the end of 1998; five-year relative survival by sex and age-group for patients diagnosed in 1991-93; one- and five-year survival by socio-economic deprivation (based on Carstairs' deprivation measure); and comparisons with other countries in Europe and the USA.

The appendices contain key statistics about the UK; detailed tables of cancer incidence for England and Wales in 1994 and provisional results for 1995 to 1997; trends in incidence from 1971 to 1997; detailed tables of cancer mortality for 1999; trends in mortality from 1950 to 1999; tables of cancer registration data quality indicators; and population estimates for 1971, 1981, 1991 and 1998, together with the World and European standard populations[10]. Other appendices, in addition to the bibliography mentioned above, include a history of the cancer registration system, which gives an outline of the role of ONS, and details of the United Kingdom Association of Cancer Registries (UKACR);

---

* All malignant neoplasms excluding non-melanoma skin cancer (see Appendix H)

full descriptions of the data; a survey of potential problems in the interpretation of cancer registration data; and descriptions of the methods used for calculating incidence rates, mortality, prevalence and survival. There is also a glossary of terms used. Lists of the references, which are numbered within each chapter, are given at the end of the book.

Data and Methods

We know so much more about cancer than other diseases because in the UK there is a population based registration system which includes follow up of cases. The history of the cancer registration system in England and Wales is described in detail in Appendix G. There are some 150 cancer registries around the world, following guidelines established by the International Union Against Cancer (UICC), the International Agency for Research on Cancer (IARC), the International Association of Cancer Registries (IACR), and the World Health Organisation (WHO)[11,12].

Some aspects of the cancer registration system in England and Wales which are relevant to the interpretation of incidence and survival data have been discussed by Swerdlow[13]. A detailed assessment of the factors which affect the quality of cancer registration data – including the ascertainment of cases; completeness of data on each record; validity; accuracy; geographic coverage; late registrations, deletions and amendments; duplicate and true multiple registrations; use of death certificates; diagnostic practice, clinical and pathological (and registry) definitions of cancer; coding systems (for cancer site and histology); flagging of cancers for follow up at the National Health Service Central Register (NHSCR); population denominators; and error – is given in Appendix H.

Details of the system of registration of deaths and advice on the interpretation of mortality data have been published in *Mortality statistics – Cause*[14]; and Rooney and Devis have described the impact of the introduction in 1993 of automated coding of cause of death, and related changes[15].

Only one chart in this book shows trends over time in the numbers of cancer cases and deaths, because such numbers depend on the size and age structure of the underlying population. Crude rates – the total number of cancer cases divided by the total number of residents – are also not suitable for use in making comparisons between different areas or between different periods of time in the same area. This is because the incidence of (and the mortality from) cancer depends heavily on age. Therefore the crude rates will be highest in those areas with the highest proportions of elderly people, and would rise if the age distribution of the population shifted towards the more elderly, even if there were no change in the rate in any age-group.

For comparison of rates between areas and over time, standardised rates are used, based on a hypothetical standard population. Age-standardised rates using both the World standard and the European standard populations have been presented in this book. As the European standard population is closely similar in age structure to that of the UK, this measure is used in the text (except for the world-wide comparisons of incidence and mortality for which the World standard population was used – Figures 2.32 and 2.33 in Chapter 2 and Figures 13 and 14 in each of chapters 3-22). Further details of the methods used to compile and present the data on incidence, mortality and prevalence are given in Appendix H.

Statistics on cancer survival in England and Wales are compiled by ONS from information on individual cases of cancer collected by the regional cancer registries. These data are subsequently linked with national death registrations to produce information on the survival of all patients diagnosed with cancer. The national cancer system is person-based, and second or subsequent tumours in the same person are linked. The results given below relate to survival following a first malignant neoplasm. Crude survival is the probability of survival irrespective of the cause of death. Cancer patients do not all die from their disease, however; the relative survival rate allows for this by comparing the crude survival among cancer patients with the expected survival in the general population (the background mortality). The relative survival rate can be interpreted as survival from cancer in the absence of other causes of death. Further details of the survival methods are given in Appendix H; an in-depth description is given in the *Cancer Survival Trends* volume[5].

Care is required in the interpretation of apparent trends over time and of geographical differences (particularly between countries) in cancer incidence, mortality and survival[16,17] because of

- changes in the definition of cancer, the level of anatomical detail in coding, the criteria for malignancy and coding rules – and diagnostic drift;
- variation in the likelihood of diagnosis, which may change with age and with screening;
- improvements in the efficiency of the cancer registration process; and
- the inadequacy of population denominators (used to calculate rates) for years between censuses.

Mortality is sometimes preferred to incidence as an indicator of trends because the data are more timely and cancer registration may not be complete. The quality and utility of incidence and mortality data in the USA up to the late 1970s was compared by Doll and Peto[16]. They showed that while mortality data were largely reliable and stable over time, there were considerable problems with the incidence data and that mortality data were generally more trustworthy. Their conclusions, published in 1981, do not apply to the data for the UK described in this volume. Mortality trends are not immune to bias or criticism. Death is not always correctly certified, nor is the underlying cause always correctly coded, even for cancer. For cancers with good survival, such as breast cancer, those dying in a given year may have been diagnosed and treated 10 or more years earlier. Cancer mortality trends are therefore a fuzzy indicator of trends in the efficacy of treatment – they reflect earlier trends in incidence and survival, and cannot be interpreted sensibly without them. Incidence and survival trends from the national cancer registry, based on data from the regional cancer registries, provide additional insight into the complex problems of cancer control. None of these indicators is perfect, and none is adequate on its own[18].

Acknowledgements

We are grateful for all the work of the regional cancer registries over the years that the national scheme has been in operation, and their close co-operation with the national registry. The full addresses, telephone and fax numbers of the regional registries in England, and the registries in Wales, Scotland and Northern Ireland, are given in Appendix G. The current (September 2000) directors of the registries in the UK are:

Northern & Yorkshire   Professor R Haward
Medical Director
Professor D Forman
Director of Information and Research

Trent   Dr J Botha

East Anglian   Dr T W Davies
General Director
Dr C H Brown
Medical Director

Thames   Ms E Davis

South & West   Dr J A E Smith

Oxford   Dr M Roche

West Midlands   Dr G Lawrence

Merseyside & Cheshire   Dr E M I Williams

North Western   Professor C B J Woodman

Wales   Dr J Steward

Scotland   Dr D Brewster

Northern Ireland   Dr A Gavin

We are also grateful to Dr Sunjai Gupta at the Department of Health for his original suggestion of a "compendium" of cancer statistics.

Finally we thank all our colleagues at the Office for National Statistics and its predecessors, the Office of Population Censuses and Surveys and the General Register Office, for all their painstaking work over the years on which the accuracy of the data – and the conclusions drawn from them – so heavily depend. We particularly thank those involved in the processing of the cancer and death registration data in our Titchfield office and in the flagging of cancers at the National Health Service Central Register in Southport.

Chapter 2

SUMMARY OF CANCER TRENDS IN ENGLAND AND WALES

Introduction

This chapter summarises a large amount of information on trends in cancer incidence, mortality, prevalence and survival at the national and regional levels during the second half of the last century.

Following brief comparisons of cancer mortality with the other major causes of death – heart disease, stroke and infectious diseases – the variation in cancer incidence and mortality with age in the late 1990s is described, together with comparisons of rates for some of the individual major cancers such as those of the lung and breast.

Trends in the incidence of all cancers over the period 1971–1997, and mortality from all cancers over the period 1950–1999, are compared – both with and without allowance for changes in the age structure of the population. Trends for individual cancer sites vary widely – those for cancers of the lung, colon and rectum, breast (in females), prostate, stomach, bladder, uterus, ovary and cervix are described. Estimates of the cumulative risk of a person being diagnosed during their lifetime with one of the major cancers are presented. The subsequent three sections describe variations in cancer incidence and mortality by socio–economic deprivation, by geographical area within the UK, and internationally.

These are followed by estimates of the prevalence of cancer – the numbers of people who had been diagnosed with a malignant cancer in the three-year period 1990–92, and in the ten year period 1983–92, and were still alive at the beginning of 1993.

The information presented on survival includes the rates for the major cancers by sex and age-group; the trends for cases diagnosed in the early 1970s up to those diagnosed in the early 1990s; the variation by socio–economic deprivation; and comparisons with survival rates elsewhere in Europe and in the USA.

Brief explanations of some of the technical terms used here are given in Chapter 1; further details can be found in Appendix H.

Background

Over the 50 year period 1950–1999, age-standardised mortality from all causes of death fell by around 45% in both males and females (Figure 2.1).

There were large declines in mortality from heart disease, stroke and infectious diseases (Figure 2.2). From 1950 onwards, heart disease in males declined slowly for about 30 years, but in the 1980s the fall accelerated and by 1999 mortality was only around half the rate in 1950. In females, mortality from heart disease fell much more steeply than in males in the 1950s and 1960s, levelled off in the 1970s, and began to fall again in the 1980s; by 1999, the rate was only about one third of that in 1950. There was little change in mortality from stroke during the 1950s, but a steady decline began in the mid–1960s in both males and females. Mortality in both sexes in 1999 was only around 40% of the rate in 1950. Mortality from infectious diseases fell rapidly during the 1950s and by the mid–1960s had declined to around 10% of the rates in 1950.

In contrast, age-standardised mortality from cancer (malignant and non–malignant neoplasms) in both males and females changed relatively little during the 50 year period (Figure 2.2). There was a slight increase for males during the 1950s and 1960s to a plateau in the 1970s, followed by a slight decline, but the rate remained mostly between 250 and 280 per 100,000. The rate in females also rose slightly to a plateau – in the late 1980s – followed by a slight decline, but remained mostly in the range 170 to 180 per 100,000 throughout the period. With the large reductions in mortality from the other major causes described above, the proportion of deaths due to cancer rose over the period from 15% to 27% in males and from 16% to 23% in females. Cancer became the most common cause of death in females in 1969 and in males in 1995.

Cancer incidence and mortality in the late 1990s

About 133,000 new cases of cancer (malignant and non–malignant neoplasms) in males and 157,000 in females were diagnosed in 1997 (provisional figures[1]; the latest available complete data are for 1994[2]). Of these, about 6,000 cases in males and 28,000 in females

Figure 2.1 **Age-standardised mortality, all causes of deaths by sex, England and Wales, 1950-1999**

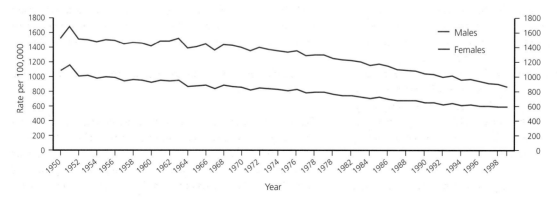

Figure 2.2 **Age-standardised mortality by cause of death, England and Wales, 1950-1999**

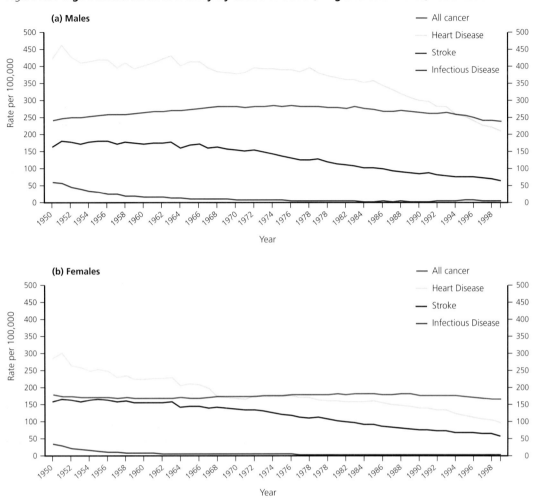

were non–malignant. Around two thirds of the non–malignant cases in females were carcinoma in situ of the cervix. Excluding cases of non–melanoma skin cancer (which for technical reasons is under–registered, and is rarely fatal[2], there were about 109,000 malignant cancers diagnosed in males, and 113,000 in females. [All further discussion of results for 'all cancers' relates to all malignant neoplasms excluding non–melanoma skin cancer.]

Cancer is predominantly a disease of the elderly. The overall crude annual rates of cancer incidence in 1997, 423 per 100,000 population for males and 426 for females, conceal wide differences between the sexes and across the age-groups, as illustrated in Figure 2.3. For adults, rates increased continuously across the age range from around the age of 30 for both males and females. Rates of cancer rose more quickly with age in females than in males. In the 40–44 age–group, the rate in females was double that for males. Subsequently, the overall rates rose more rapidly for males and were broadly similar to those for females in the 60–64 age–group. After this, the rates rose much more rapidly for males – they were about 45% higher than those for females in the 65–69 age–group and almost double in those aged 80–84.

The age distribution of people diagnosed with a malignant cancer

in 1997 is shown in Figure 2.4[1] . Of the total of around 222,000 registrations, only 1,250 (0.6 per cent) occurred in children aged under 15; of these, just over 400 (one third) were leukaemias. The percentages of cancers in the five–year age–groups tended to rise earlier in females than in males, owing largely to the influence of the incidence of cancers of the breast and cervix. Cancers in those aged under 45 amounted to just under 6% of the total for males and 9% for females. The peaks in the age distributions for both males and females occurred in the 70–79 age-group.

In 1999 there were 69,000 deaths from all cancers combined (ie excluding non–melanoma skin cancer and non–malignant neoplasms) in males, and 64,000 in females. These represented 26% and 22% of all deaths in males and females, respectively.

Age-specific mortality and the age distribution of deaths are shown in Figures 2.5 and 2.6.

For both males and females just three cancer sites (different ones for each sex) constituted around half of the total registrations in 1997 (Table 2.1). [In the ICD 10th Revision, there are 88 three–character site codes relating to malignant neoplasms; of these, four relate to males only and eight to females only[3].]

Figure 2.3 **Age-specific incidence of all cancers†, England and Wales, 1997***

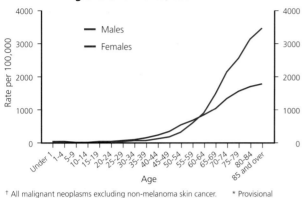

† All malignant neoplasms excluding non-melanoma skin cancer.   * Provisional

Figure 2.4 **Incidence of all cancers†: frequency distribution by age-group, England and Wales, 1997***

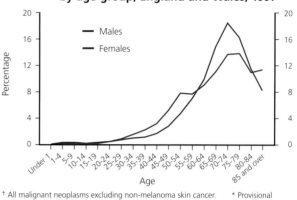

† All malignant neoplasms excluding non-melanoma skin cancer.   * Provisional

Figure 2.5 **Age-specific mortality from all cancers†, England and Wales, 1999***

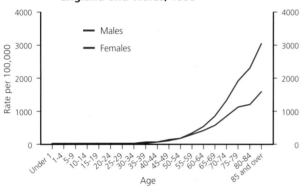

† All malignant neoplasms excluding non-melanoma skin cancer.   * Provisional

Figure 2.6 **Mortality from all cancers†: frequency distribution by age-group, England and Wales, 1999***

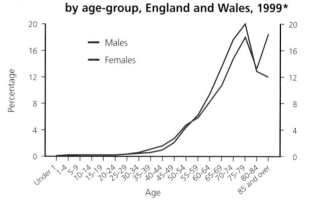

† All malignant neoplasms excluding non-melanoma skin cancer.   * Provisional

Table 2.1 Incidence of the major cancers, England and Wales, 1997*

Males

Rank	Site	Number of registrations	% all cancers
1	Lung	21,000	19
2	Prostate	18,300	17
3	Colorectal	14,900	14
Sub-total 1-3		54,200	50
4	Bladder	8,500	8
5	Stomach	5,800	5
6	NHL	4,000	4
7	Oesophagus	3,600	3
8	Kidney	3,000	3
9	Leukaemia	2,900	3
10	Pancreas	2,700	2
11	Lip, mouth and pharynx	2,300	2
12	Brain	2,100	2
13	Melanoma of skin	2,000	2
14	Larynx	1,500	1
15	Multiple myeloma	1,500	1
16	Testis	1,400	1
Sub-total 4-16		41,300	38
	Other	13,100	12
	All cancers†	**108,600**	**100**

Females

Rank	Site	Number of registrations	% all cancers
1	Breast	33,100	29
2	Colorectal	14,000	12
3	Lung	12,300	11
Sub-total 1-3		59,400	53
4	Ovary	6,100	5
5	Uterus	4,000	4
6	Bladder	3,600	3
7	NHL	3,400	3
8	Stomach	3,300	3
9	Pancreas	3,000	3
10	Cervix	2,800	2
11	Melanoma of skin	2,700	2
12	Oesophagus	2,500	2
13	Leukaemia	2,300	2
14	Kidney	1,800	2
15	Brain	1,500	1
16	Lip, mouth and pharynx	1,400	1
17	Multiple myeloma	1,400	1
Sub-total 4-17		39,800	35
	Other	13,900	12
	All cancers†	**112,900**	**100**

* Provisional
† All malignant neoplasms excluding non-melanoma skin cancer

Figure 2.7 **Age-standardised incidence, major cancers, England and Wales, 1997***

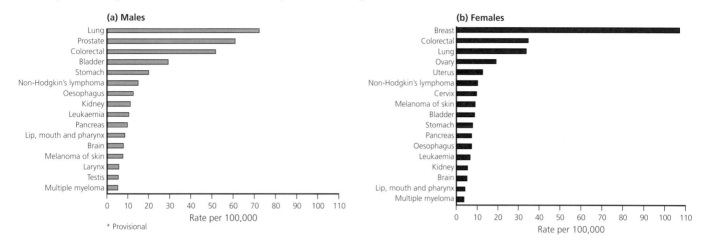

Figure 2.8 **Age-standardised mortality, major cancers, England and Wales, 1999***

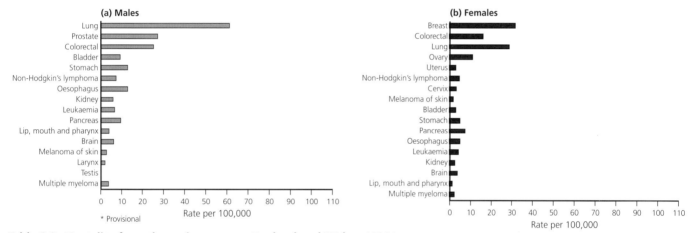

Table 2.2 Mortality from the major cancers, England and Wales, 1999*

Males				Females			
Rank	Site	Number of deaths	% all cancers	Rank	Site	Number of deaths	% all cancers
1	Lung	18,300	27	1	Breast	11,500	18
2	Prostate	8,500	12	2	Lung	11,100	17
3	Colorectal	7,500	11	3	Colorectal	7,100	11
Sub-total 1-3		**34,300**	**50**	**Sub-total 1-3**		**29,700**	**46**
4	Stomach	3,800	6	4	Ovary	3,900	6
5	Oesophagus	3,700	5	5	Pancreas	3,100	5
6	Bladder	2,800	4	6	Stomach	2,300	4
7	Pancreas	2,800	4	7	Oesophagus	2,300	4
8	NHL	2,100	3	8	NHL	1,900	3
9	Leukaemia	1,900	3	9	Leukaemia	1,700	3
10	Kidney	1,700	2	10	Bladder	1,500	2
11	Brain	1,600	2	11	Uterus	1,200	2
12	Multiple myeloma	1,100	2	12	Brain	1,200	2
13	Lip, mouth and pharynx	1,100	2	13	Cervix	1,100	2
14	Melanoma of skin	800	1	14	Multiple myeloma	1,100	2
15	Larynx	600	1	15	Kidney	1,000	2
16	Testis	100	0	16	Melanoma of skin	700	1
Sub-total 4-16		**24,100**	**35**	17	Lip, mouth and pharynx	600	1
				Sub-total 4-17		**23,600**	**37**
	Other	10,500	15		Other	10,600	17
	All cancers†	**68,800**	**100**		**All cancers†**	**63,900**	**100**

* Provisional

† All malignant neoplasms excluding non-melanoma skin cancer

7

The age-standardised incidence rates for these and the other major sites for which there are separate descriptive chapters in this book are illustrated in Figure 2.7 (ranked on incidence rate, by sex). The numbers of registrations for the 16 sites (counting lip, mouth and pharynx; colorectal; non–Hodgkin's lymphoma; and leukaemia each as one) for males represent 88% of the total; those for the 17 sites for females also represent 88%.

Deaths from the three most common cancers in 1999 constituted just under half of all cancer deaths (Table 2.2). Mortality from the 16 major cancers in males and 17 in females accounted for 85% and 83% of the totals, respectively (Table 2.2); age standardised mortality for these cancers is illustrated in Figure 2.8 (the sites are presented in the same order as in Figure 2.7).

Figure 2.9 **Age-specific incidence of all cancers†, England and Wales, 1971-1997***

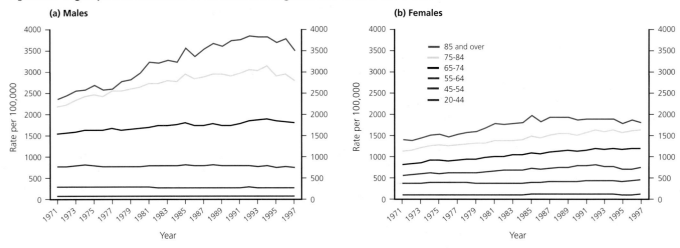

† All malignant neoplasms excluding non-melanoma skin cancer.   * Figures for 1995-1997 for incidence are provisional.

Figure 2.10 **Age-specific mortality from all cancers†, England and Wales, 1950-1999***

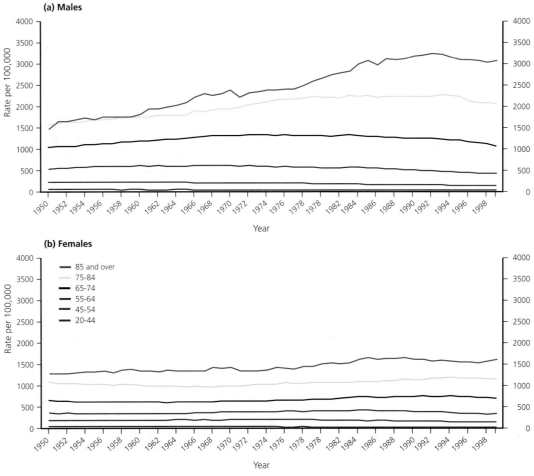

† All malignant neoplasms excluding non-melanoma skin cancer.   * Figures for 1999 for mortality are provisional.

Trends in cancer incidence and mortality – all cancers

Over the period 1971–1997, the incidence of all cancers (all malignant neoplasms excluding non–melanoma skin cancer) remained fairly stable in men aged under 65 (Figure 2.9). There have, however, been steady increases in the rates in older men, particularly those aged 85 and over, in whom the rise was about 50%. Overall incidence has also remained fairly stable in women aged under 55. Rates in women aged 55–64, however, rose gradually to peak in the early 1990s. This is largely as a result of the general increase in the incidence of breast cancer, and of the effect of the introduction of the NHS breast screening programme after 1988[4] which detected a large number of cases earlier than they would have been otherwise. Incidence in women in both the age-groups 65–74 and 75–84 increased steadily, and by around 45% in total, over the period. In very elderly women, by the mid–1980s rates had increased by about one third, but then stabilised.

Age-specific trends in mortality from cancer have been quite different from those in incidence. Over the period 1950–1999, mortality in males aged under 55 declined by around 40% (Figure 2.10). Mortality increased slightly during the 1950s in men aged 55–64, was stable during the 1960s and early 1970s, and then began to fall. There were similar patterns, but with plateaux in the 1970s for men aged 65–74 and in the 1980s for men aged 75–84. For very elderly men, mortality more than doubled up to 1994 and subsequently declined noticeably. These trends are largely due to changes in smoking habits and consequent changes in the incidence of and mortality from lung and other smoking-related cancers, which make up about 40% of the total[5].

The trends in cancer mortality in females are different from those in males (Figure 2.10). The patterns reflect the lower overall risk for women born towards the end of the 19th century and the relatively high risk for women born in the mid–1920s. This is largely due to the increase in smoking in the latter group during the second world war (and afterwards). Mortality decreased during the 1950s in women aged 55–84, but then rose steadily for the next 20 to 30 years. Mortality began to decline again from the mid–1970s in women aged 45–54, from the mid–1980s in those aged 55–64, and in the mid–1990s in those aged 65–74. Mortality in very elderly women has not risen as fast as in very elderly men.

The numbers of all newly diagnosed cases of cancer increased from 1971 up to the early 1990s by just over 45% in males and around 55% in females (Figure 2.11). Some of the apparent increases in incidence, however, particularly during the 1970s, will have arisen from improved ascertainment of cases by the cancer registries (see below). The corresponding numbers of deaths increased up to the late 1980s by around 15% in males and 25% in females and then declined slowly.

The trends in the "crude" rates – that is the total numbers of cases or deaths divided by the total population, shown in Figure 2.12 – indicate that around 10% points of the increases in numbers resulted from the growth of the population during the 30 year period.

Figure 2.11 **The numbers of new cases of, and deaths from, all cancers[†] by sex, England and Wales, 1971-1999* (1971=100)**

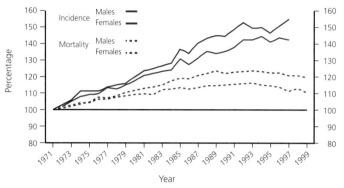

[†] All malignant neoplasms excluding non-melanoma skin cancer.
* Figures for incidence for 1995-1997 and mortality for 1999 are provisional.

Figure 2.12 **Crude rates of incidence of, and mortality from, all cancers[†] by sex, England and Wales, 1971-1999* (1971=100)**

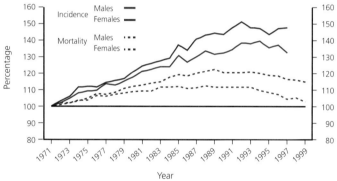

[†] All malignant neoplasms excluding non-melanoma skin cancer.
* Figures for incidence for 1995-1997 and mortality for 1999 are provisional.

Figure 2.13 **Age-standardised incidence of, and mortality from, all cancers[†] by sex, England and Wales, 1971-1999* (1971=100)**

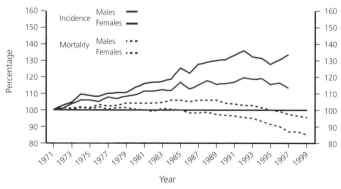

[†] All malignant neoplasms excluding non-melanoma skin cancer.
* Figures for incidence for 1995-1997 and mortality for 1999 are provisional.

During this time there was also a large shift in the age distribution of the population towards older ages – see the population "trees" for 1971, 1981, 1991 and 1998 in Appendix F. The proportions of men and women aged 85 years and over rose from 0.5% and 1.3%, respectively, in 1971 to 1.0% and 2.8%, respectively, in 1998. The age-standardised incidence rates, which adjust for changes in both the size and the age structure of the population over time, show much smaller increases than the numbers and the crude rates – around 20% in males and 30% in females (Figure 2.13). Thus about 25% points of the increases in the number of cases in both males and females, resulted from changes in the population. Age-standardised mortality in males was more or less stable at its 1971 level up to the mid–1980s, subsequently falling by around 15%. Overall cancer mortality in females rose by about 5% from 1971 up to the late 1980s, but subsequently declined to 5% below the 1971 level in 1999.

The overall trends in cancer incidence (allowing for some under–ascertainment in the early 1970s) and mortality are broadly consistent with trends in cancer survival. During the 1970s and 1980s, five–year survival from all cancers rose from 19% to 31% for men and from 32% to 43% for women[6]. [These overall survival trends must be interpreted with caution, as they will be affected by changes in incidence: for example, if a highly fatal cancer became less common, overall survival would improve even if it did not improve at all for any of the individual cancer sites. Also, much of

the apparent difference in survival between men and women arises because cancers of the breast, cervix and uterus have good survival (60–70%) (see below).]

Trends in incidence and mortality – major cancers
The trends in incidence and mortality for all cancers combined conceal very different patterns in the separate cancer sites. As lung cancer is highly fatal – one–year survival is only 20% and five–year survival just 5% – the trends in incidence and mortality are closely similar (Figure 2.14), although there appears to have been some under ascertainment of cases in the early 1970s when the incidence rates were only very slightly higher than mortality. In males, both incidence and mortality had reached a plateau in the early 1970s, but both have since fallen steadily. In contrast, there were increasing trends in both incidence and mortality in females up to the end of the 1980s, since when rates have been fairly stable. These patterns of course reflect the earlier trends in smoking which showed decreases in men but not in women. Cohort analyses indicate that men at highest risk are those born in the 1890s, whereas the risk is highest for women born in the 1920s (Chapter 12).

Figure 2.14 **Age-standardised incidence of, and mortality from, lung cancer by sex, England and Wales, 1950-1999***

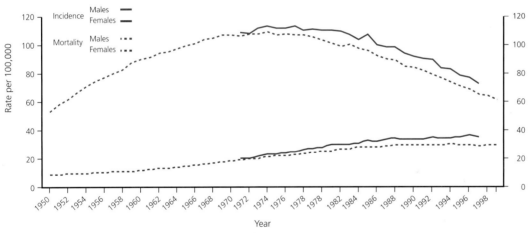

* Figures for incidence for 1995-1997 and mortality for 1999 are provisional.

Figure 2.15 **Age-standardised incidence of, and mortality from, colorectal cancer by sex, England and Wales, 1950-1999***

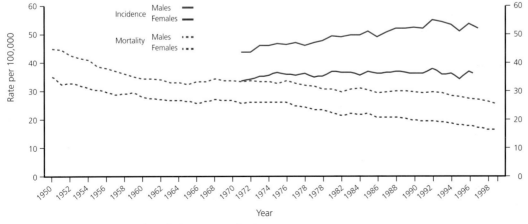

* Figures for incidence for 1995-1997 and mortality for 1999 are provisional.

10

The incidence of cancer of the large bowel (colon and rectum) rose gradually from 1971 in males, but not in females (Figure 2.15). The risks by birth cohort rise in males and (slightly faster) in females up to the 1930s and early 1940s and then decline (Chapter 7). Mortality has fallen more or less continuously since 1950, and in all cohorts since those born in the mid–1860s.

The incidence of breast cancer in females increased throughout the 1980s and peaked in 1992 (Figure 2.16). The rise from 1988 onwards occurred principally in women aged 50–64 years which (as noted above) was associated with the NHS breast screening programme[4,7]. With the bulk of the first or "prevalent" round of screening completed, incidence was 5% lower in 1993, and has remained at around 100 per 100,000 since then. The incidence of carcinoma in situ of the breast rose sharply with the introduction of screening.

Mortality from breast cancer rose to its highest level in the mid–1980s, when the rate was among the highest in the world. By 1999, mortality had fallen by just over 20% since 1989. In addition to cohort effects (Figure 2.17), mortality in the 1990s will have been affected directly by screening, and by the increasingly widespread use of tamoxifen[8], improvements in chemotherapy, earlier presentation of cases outside the screening programme, and

structural changes in the NHS following the Calman–Hine report[9]. It is estimated that by 1998 about one third of the reduction in breast cancer mortality since 1990 was due directly to the screening programme[10]. Although the risk of being diagnosed with breast cancer has increased continuously from women born in the 1880s up to those born in the 1960s, the corresponding mortality has been more stable, with the highest risk being for women born in the mid–1920s (Figure 2.17). This is consistent with the observed long term improvements in five–year survival of 4–5% points every five years since the early 1970s[6] (Chapter 5).

The incidence of prostate cancer first exceeded that of colorectal cancer in men in 1993. The increase in incidence in the 1980s may have resulted partly from the detection of latent carcinoma following operations for enlarged prostate, or at post mortem in men who had died from other causes[11]. In more recent years, the availability of prostate specific antigen (PSA) testing probably accounts for the very sharp rise since 1991 (Figure 2.18). Similar rises have occurred in Scotland[12] and elsewhere; in the USA, incidence rose by 82% between 1986 and 1991[13]. Mortality in England and Wales was fairly stable in the 1960s and 1970s; it fell very slightly in both 1998 and 1999 following a consistent rise during the 1980s and early 1990s[14] (Chapter 19).

Figure 2.16 **Age-standardised incidence of, and mortality from, female breast cancer, England and Wales, 1950-1999***

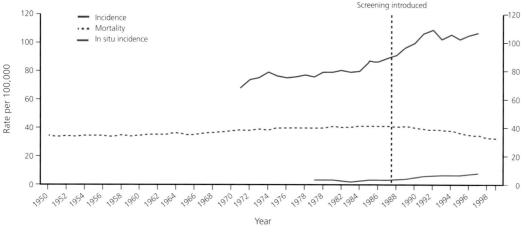

* Figures for incidence for 1995-1997 and mortality for 1999 are provisional.

Figure 2.17 **Female breast cancer mortality* by birth cohort, England and Wales (1916=100)**

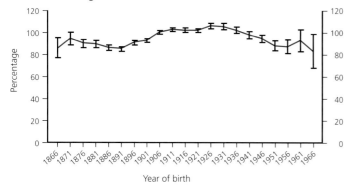

* Bars indicate 95% confidence interval.

Figure 2.18 **Age-standardised incidence of, and mortality from, prostate cancer, England and Wales, 1950-1999***

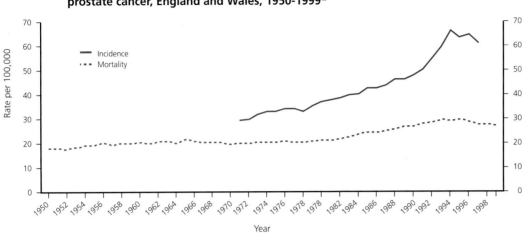

* Figures for incidence for 1995-1997 and mortality for 1999 are provisional.

Figure 2.19 **Age-standardised incidence of, and mortality from, stomach cancer by sex, England and Wales, 1950-1999***

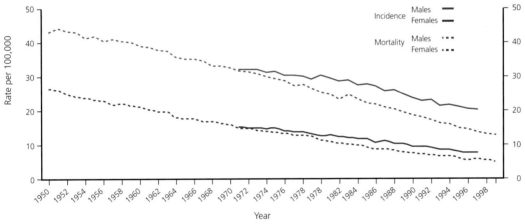

* Figures for incidence for 1995-1997 and mortality for 1999 are provisional.

Both the incidence of and mortality from stomach cancer have declined consistently for a very long time (Figure 2.19). Over the period 1971–1997, incidence fell in all age-groups in both males and females; the age-standardised rates fell by more than 40% in males and 50% in females. As with lung cancer, however, there appears to have been some under–ascertainment of cases in the early 1970s. Mortality has also fallen steadily in all age-groups in both males and females. Age-standardised mortality in males is now only 30%, and that in females only 20%, of the levels 50 years ago (Chapter 20).

The age-standardised incidence of bladder cancer in males is almost four times that in females (Table 3.1) – the highest ratio in any of the major cancers. Incidence rose throughout the 1970s and early 1980s in both males and females and then levelled off (Figure 2.20); the increases occurred mainly in those aged 75 and over. Mortality in males rose in the 1950s and 1960s then declined gradually to almost the same level as in the 1950s. Mortality in females was stable over the whole 50 year period. Cohort analyses show that the patterns of risk for males and females were closely similar, with those born in the 1920s being at highest risk of being diagnosed with bladder cancer, while the peak for mortality in males was in those born 10–20 years earlier (Chapter 3).

The incidence of ovarian cancer in women aged under 55 has been fairly stable since the early 1970s, but rates in older women have increased steadily. By 1997, the overall rate had increased by about 30% compared with the early 1970s (Figure 2.21). Overall mortality rose during the 1950s and 1960s, and has subsequently remained at around 12 per 100,000. Since the early 1970s, mortality in women aged 65 and over has continued to rise, the rate in women aged 55–64 has stabilised, and mortality in younger women (45–54 years) has declined. The risks for both incidence and mortality increased with successive cohorts from the 19th century to those born in the 1920s. Subsequently, although the risk for incidence first stabilised and then rose for women born from the 1950s onwards, the risk for mortality fell (Chapter 17). These patterns are consistent with the observed improvements in five–year survival of around 3% points every five years during the 1970s and 1980s[6].

The relatively stable overall incidence of cancer of the uterus of around 12 per 100,000 conceals diverging trends in women aged 45–54 and the more elderly women (Chapter 22). In contrast, there has been a long term decline in mortality (which has occurred in all age-groups except the very elderly) with the age-standardised rate at the end of the 1990s only just over half that in 1950 (Figure 2.21). The cohorts of women at highest risk of being diagnosed with cancer of the uterus are those born in the 1920s and 1930s. In contrast, mortality has declined more or less continuously in successive cohorts from those born in the 1890s (Chapter 22).

Up to 1990, the trends in the incidence of and mortality from cervical cancer depended on the levels of risk in cohorts of women born in different periods[15,16,17]. Risk is elevated in women born at the end of the 19th century, in the mid–1920s, and after 1950. These women would have been in their late teens and early twenties and hence becoming sexually active at the times of the first world war, the second world war, and the introduction of oral contraceptives, respectively (Chapter 6). Cervical cancer was first identified as a sexually transmitted disease by Beral in 1974[15]. From

1971 to 1990, incidence remained between 14 and 16 per 100,000. Following changes made to the screening programme – the introduction of national call and recall and of incentive payments to GPs – the coverage of the screening programme increased dramatically to around 85%. As a direct result[16,17], incidence fell for seven consecutive years after 1990, falling below 10 per 100,000 in 1996, about 40% lower than in the mid–1980s (Figure 2.22). Mortality, which had been falling continuously since 1950, fell more steeply after 1990 and there were only 1,100 deaths from cervical cancer in England and Wales in 1999.

The changes in the age-standardised incidence between 1971 and 1997 for all the major cancers covered in separate chapters in this volume are illustrated in Figure 2.23. The largest increases were for malignant melanoma of the skin: the latest rates are over four-fold higher in males and almost three-fold higher in females than in 1971. Rates of non–Hodgkin's lymphoma increased almost three fold in both males and females. The incidence of cancer of the kidney almost doubled in both males and females, as did lung

Figure 2.20 **Age-standardised incidence of, and mortality from, bladder cancer by sex, England and Wales, 1950-1999***

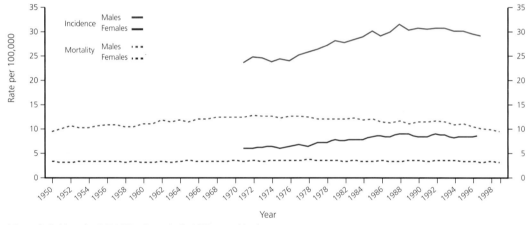

* Figures for incidence for 1995-1997 and mortality for 1999 are provisional.

Figure 2.21 **Age-standardised incidence of, and mortality from, uterine and ovarian cancers, England and Wales, 1950-1999***

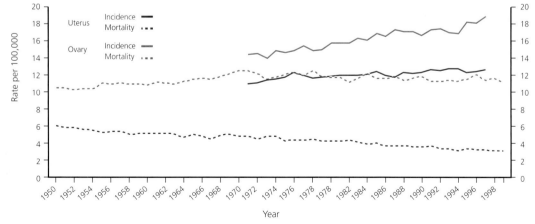

* Figures for incidence for 1995-1997 and mortality for 1999 are provisional.

13

Figure 2.22 **Age-standardised incidence of, and mortality from, cervical cancer, England and Wales, 1950-1999***

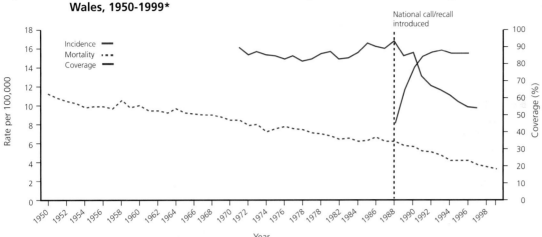

* Figures for incidence for 1995-1997 and mortality for 1999 are provisional.

Figure 2.23 **Percentage change in age-standardised incidence by sex and site, England and Wales, 1997* compared with 1971**

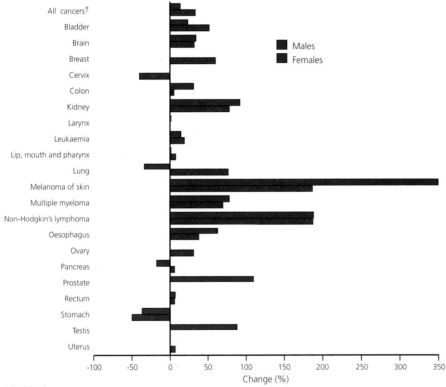

* Provisional
† All malignant neoplasms excluding non-melanoma skin cancer.

cancer in women, multiple myeloma in both males and females, and prostate and testicular cancers. Increases of around 50% occurred in breast cancer in females and in oesophageal cancer in both sexes. As noted above, some part of the apparent increase for some cancers may have arisen as a result of improved ascertainment by the cancer registries. Marked decreases in incidence occurred only for cancers of the cervix, lung in males, and stomach (both sexes).

Mortality in 1999 is compared with that in 1971 in Figure 2.24. The largest increase, in line with that for incidence, was for melanoma

of the skin in males: about two and a half fold. Increases of between 50% and 100% occurred for lung cancer and melanoma of the skin in females, multiple myeloma, non–Hodgkin's lymphoma, and oesophageal cancer in males. There were smaller increases in mortality for brain tumours and cancers of the kidney and prostate. Mortality fell by more than half for cancers of the cervix, stomach, testis and uterus; and by about a third for colon cancer in females, lung cancer in males, and rectal cancer in both sexes. There were only small falls in mortality from cancers of the bladder in males, breast in females, colon in males, larynx in males, leukaemia, lip, mouth and pharynx, ovary, and pancreas in males.

14

Figure 2.24 **Percentage change in age-standardised mortality by sex and site, England and Wales, 1999* compared with 1971**

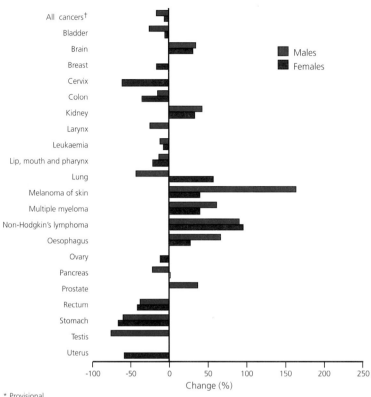

* Provisional
† All malignant neoplasms excluding non-melanoma skin cancer.

Figure 2.25 **Cumulative risk of incidence of all cancer† by age and sex, England and Wales, 1992-94**

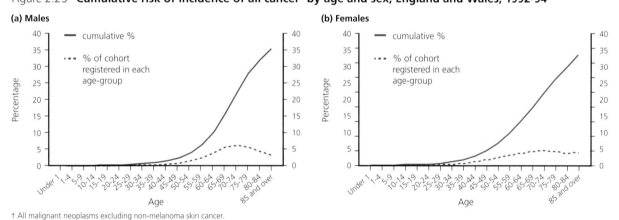

(a) Males

(b) Females

† All malignant neoplasms excluding non-melanoma skin cancer.

The reasons for the trends in some of the major cancers are described above; further details on all the separate cancers are given in chapters 3–22.

Lifetime risk of cancer
Based on incidence rates for 1992–94[2], it is estimated that 35% of males and 33% of females – approximately one person out of every three – would eventually be diagnosed with some form of malignancy (Figure 2.25). Registrations would not, however, be equally spread across age-groups. Only 7% of males (one fifth of the total cases) and 10% of females (just under a third) would be registered before the age of 60.

The cumulative risk of a person being diagnosed during their lifetime with one of the major cancers is summarised in Figure 2.26. For males, the major risks are equivalent to 1 in 13 for lung cancer – although this is of course an average across the whole population, and the risk is higher in smokers (and very much lower in non–smokers); 1 in 14 for prostate cancer; and 1 in 18 for colorectal cancer. The risks for cancers of the kidney and pancreas and for leukaemia are about 1 in 100. For females, the risk is 1 in 9 for breast cancer; 1 in 20 for colorectal cancer; and 1 in 23 for lung cancer (again on average); the risk for cervical cancer, as a result of the improved effectiveness of the national screening programme[16,17], is now less than 1 in 100.

Figure 2.26   **Lifetime risk of being diagnosed with major cancers, England and Wales, 1997***

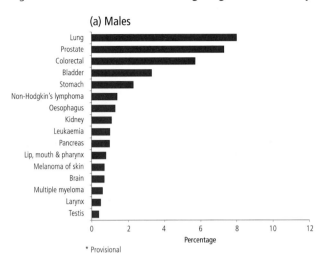

(a) Males

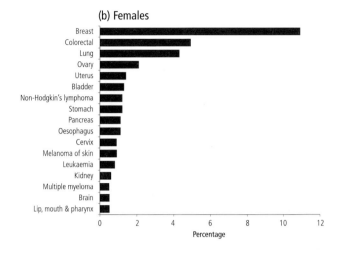

(b) Females

* Provisional

Figure 2.27   **Lung cancer incidence and mortality by deprivation category, males, England and Wales, 1993**

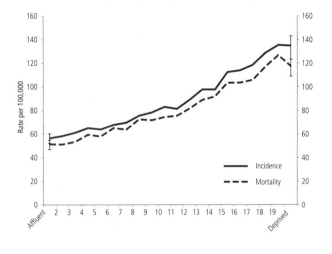

Variation in cancer incidence and mortality with socio–economic deprivation

In the early 1990s, the major cancer sites fell into three main groups with respect to socio–economic deprivation: those with a positive gradient (higher age-standardised incidence and/or mortality in the most deprived groups); little or no gradient; and a negative gradient (higher rates in the affluent groups).

Cancers of the cervix, larynx and lung all show a strong positive gradient in both incidence and mortality. Cancers of the bladder, kidney, lip, mouth and pharynx, oesophagus, stomach, pancreas, and rectum in males also show a positive gradient. For lung cancer, rates in the most deprived groups are around twice those in the most affluent (Figure 2.27 gives the patterns for males – those for females show a similar two–fold range, but at lower overall rates).

If the incidence of the above 10 cancers (counting all lip, mouth and pharyngeal cancers as one) in people in all deprivation categories were as low as in those in the most affluent (categories 1

Table 2.3 Number of preventable cases of, and death from, selected cancers related to socio-economic deprivation*

Site	Preventable cases			Preventable deaths		
	Males	Females	Total	Males	Females	Total
Bladder	490	330	820	390	240	630
Cervix	..	1,410	1,410	..	590	590
Kidney	100	-	100	30	-	30
Larynx	620	-	620	290	-	290
Lip, mouth and pharynx	710	100	810	380	40	420
Lung	7,720	4,070	11,790	7,440	3,560	11,000
Oesophagus	380	350	730	550	270	820
Pancreas	180	230	410	200	190	390
Rectum	790	-	790	520	-	520
Stomach	1,540	940	2,480	1,160	810	1,970
Total	12,520	7,430	19,950	10,950	5,690	16,640

* The number of preventable cases (deaths) is calculated as the difference between the average age specific rate in deprivation categories 1 and 2 and the rate in each of the other deprivation categories, multiplied by the population in each category, summed across age groups and deprivation categories.

- Not calculated (see text)
.. Not applicable.

and 2) then 20,000 fewer cases of these cancers would be diagnosed every year (Table 2.3). These would include about 7,700 lung cancers in males and 4,100 in females. There would be 16,600 fewer deaths from the 10 cancers every year.

For cancers of the colon in males, both colon and rectum in females, multiple myeloma and uterus there is virtually no relationship at all in either incidence or mortality with deprivation (Figure 2.28 colon cancer).

The cancers which exhibit a negative gradient in either incidence or mortality with deprivation include brain, breast (in females), leukaemia, melanoma of the skin, non–Hodgkin's lymphoma, prostate and testis. The incidence of breast cancer is around 25% higher in the most affluent groups than in the most deprived, but there is no gradient in mortality (Figure 2.29). This is consistent with the affluent groups having better survival[6].

On the same principle as used to calculate the results given in Table 2.3, if the incidence of breast and prostate cancers in people in all deprivation categories were as high as in those in the most affluent (categories 1 and 2), there would be about 3,300 more cases of breast cancer and 1,900 more cases of prostate cancer diagnosed each year.

Geographical variations in cancer incidence and mortality in England and Wales

Any regional patterns in overall cancer incidence should be interpreted with caution because it is difficult to separate the effect of variation in levels of ascertainment of cases by the cancer registries from genuine geographical differences in incidence. However, there are generally higher levels of deprivation in the north of England than in the south, and most of the cancers which have a positive gradient with deprivation in incidence also exhibit a north–south gradient. A major contributory factor to the positive gradients is that there are higher levels of smoking in deprived areas than in affluent areas[18]. The geographical pattern for the incidence of all cancers combined is shown in Figure 2.30, and the corresponding pattern in mortality in Figure 2.31. Rates are above average in Northern and Yorkshire and the North West, and below average in Anglia and Oxford, the Thames regions, and, for

Figure 2.28 **Colon cancer incidence and mortality by deprivation category, males, England and Wales, 1992–93**

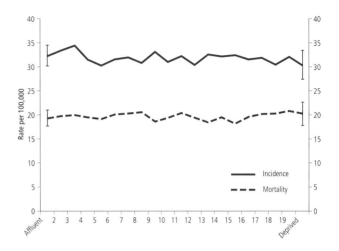

Figure 2.29 **Breast cancer incidence and mortality by deprivation category, females, England and Wales, 1993**

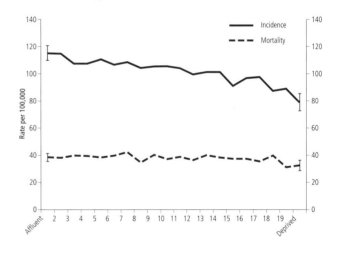

Figure 2.30 **Comparative incidence ratio (CIR) for all cancers,[†] by region and sex, 1997* (England and Wales =100)**

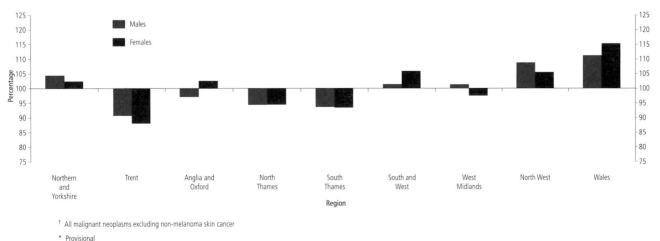

† All malignant neoplasms excluding non-melanoma skin cancer

* Provisional

17

Figure 2.31 **Comparative mortality ratio (CMR) for all cancers,† by region and sex, 1999* (England and Wales =100)**

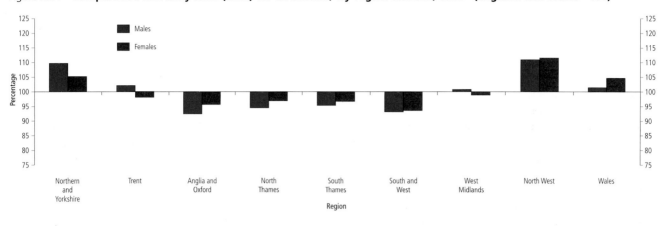

† All malignant neoplasms excluding non-melanoma skin cancer

* Provisional

Figure 2.32 **Age-standardised international incidence* of all cancers†**

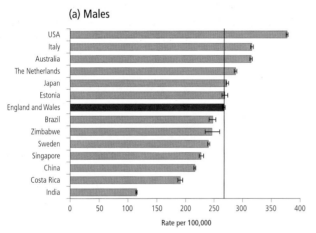
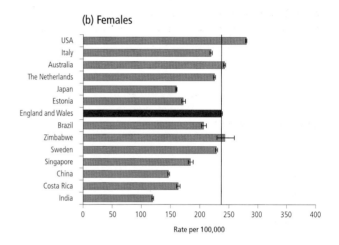

* Directly age-standardised using the World standard population
† All malignant neoplasms excluding non-melanoma skin cancer

Figure 2.33 **Age-standardised international mortality* from all cancers†**

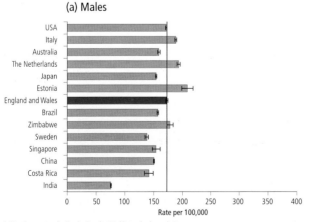
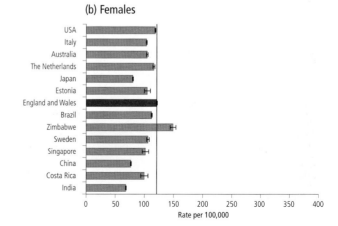

* Directly age-standardised using the World standard population
† All malignant neoplasms excluding non-melanoma skin cancer

mortality, the South West. For both breast and prostate cancers, which have inverse gradients in incidence with deprivation, there was no clear geographical pattern at the regional level, although rates tended to be slightly lower in the south of England than in the north.

International variations in cancer incidence and mortality

The incidence of all cancers combined (excluding non–melanoma skin cancer) in males in England and Wales in 1992 was well below that in the USA, Australia, and several other countries in Europe[19]. Rates in Sweden, and in some countries in the far east (for example Singapore and China) were lower; rates in India were much lower (Figure 2.32). The pattern for females was somewhat different. The overall rates in the USA and Australia were higher than in England and Wales, but not by as much as for males; and incidence in Japan and Estonia was much lower than for their male populations.

The international pattern in mortality for 1990 in males was quite different from that for incidence, with strikingly less variability among the countries (Figure 2.33). In absolute terms, mortality in the USA and Australia was slightly lower than in England and Wales; while rates in Italy, the Netherlands and Estonia were noticeably higher. Mortality rates in all the other countries except India were fairly similar. There was also less variability for

mortality in females than for incidence. The rates broadly reflect those for incidence, except for Estonia where mortality was relatively high (implying poorer survival). Both incidence and mortality were low in Japan, China and India.

There was considerable variability in the international patterns in incidence and mortality among the major cancer sites. For most, rates in England and Wales fell in the middle of the range and were similar to rates in other Western European countries. Both incidence and mortality were, however, relatively high for multiple myeloma and breast cancer and low for cancers of the stomach and of lip, mouth and pharynx.

Prevalence of cancer

Three year prevalence

Just under 320,000 males were diagnosed with cancer during the three year period 1990-92, of whom almost 145,000 (45%) were still alive at the beginning of 1993. Prostate, colorectal and bladder cancers together accounted for almost half of these (around 70,000 cases); lung cancer accounted for 15,000 cases (11%) and the lymphomas and leukaemias together for almost 11,000 (8%) (Table 2.4). Slightly more females than males – just under 330,000 – were diagnosed with cancer during the period and more of these – 190,000 (58%) – were still alive. Almost 75,000 (40%) had been

Table 2.4 Estimated number of cancer patients diagnosed 1990-92 by vital status on 1st January 1993, England and Wales

a) Males

Site	Alive Number	Alive % all alive	Dead	Total cases	% cases alive
Prostate	27,800	19.3	15,500	43,400	64
Colorectal	24,000	16.6	18,900	42,900	56
Bladder	17,400	12.0	7,800	25,100	69
Lung	15,300	10.6	59,400	74,800	20
Non-Hodgkin's lymphoma	6,400	4.4	4,300	10,700	59
Stomach	4,900	3.4	14,000	18,900	26
Leukaemia	4,500	3.1	4,300	8,800	51
Kidney	4,200	2.9	3,700	8,000	53
Lip, mouth and pharynx	3,900	2.7	2,700	6,500	59
Melanoma of skin	3,700	2.6	900	4,700	80
Testis	3,600	2.5	400	4,000	90
Larynx	3,600	2.5	1,300	4,900	73
Oesophagus	2,400	1.6	7,200	9,600	25
Multiple myeloma	2,100	1.5	2,200	4,300	49
Brain	1,900	1.3	3,900	5,700	33
Pancreas	1,300	0.9	7,800	9,100	14
Other	17,500	12.1	19,700	37,200	47
Total cases	144,500	100.0	174,100	318,700	45

b) Females

Site	Alive Number	Alive % all alive	Dead	Total cases	% cases alive
Breast	74,800	39.5	17,200	92,000	81
Colorectal	23,700	12.5	18,900	42,500	56
Uterus	8,900	4.7	2,700	11,600	76
Cervix	8,400	4.4	2,900	11,200	74
Ovary	7,900	4.2	7,900	15,800	50
Lung	7,600	4.0	28,000	35,600	21
Melanoma of skin	6,100	3.2	900	7,000	87
Bladder	6,100	3.2	3,800	9,900	62
Non-Hodgkin's lymphoma	5,400	2.9	3,700	9,100	60
Leukaemia	3,400	1.8	3,600	7,000	49
Stomach	3,000	1.6	8,800	11,800	25
Kidney	2,500	1.3	2,400	4,900	51
Lip, mouth and pharynx	2,300	1.2	1,400	3,700	62
Multiple myeloma	1,900	1.0	2,100	4,000	49
Oesophagus	1,700	0.9	5,100	6,800	25
Brain	1,500	0.8	2,800	4,200	34
Pancreas	1,400	0.7	8,200	9,600	14
Other	22,600	11.9	17,900	40,500	56
Total cases	189,200	100.0	138,200	327,400	58

Table 2.5 Estimated number of cancer patients diagnosed 1983-92 by vital status on 1st January 1993, England and Wales

a) Males

Site	Alive Number	Alive % all alive	Dead	Total cases	% cases alive
Colorectal	47,800	16.5	86,200	134,000	36
Prostate	47,200	16.3	76,300	123,500	38
Bladder	40,700	14.1	38,700	79,400	51
Lung	24,000	8.3	237,900	261,800	9
Non-Hodgkin's lymphoma	13,100	4.5	17,000	30,100	43
Testis	10,500	3.6	1,400	11,900	88
Kidney	9,000	3.1	14,600	23,600	38
Larynx	8,800	3.0	7,200	16,000	55
Leukaemia	8,700	3.0	19,000	27,700	31
Stomach	8,500	2.9	59,500	68,000	13
Melanoma of skin	8,300	2.9	4,600	12,900	64
Lip, mouth and pharynx	8,100	2.8	12,300	20,400	40
Brain	3,500	1.2	14,100	17,600	20
Oesophagus	3,300	1.1	25,200	28,500	12
Multiple myeloma	3,300	1.1	10,200	13,500	24
Pancreas	1,900	0.6	28,400	30,300	6
Other	42,900	14.8	71,200	114,000	38
Total cases	289,400	100.0	723,900	1,013,300	29

b) Females

Site	Alive Number	Alive % all alive	Dead	Total cases	% cases alive
Breast	167,800	38.9	103,200	271,000	62
Colorectal	50,900	11.8	86,800	137,600	37
Cervix	24,500	5.7	16,500	41,000	60
Uterus	24,000	5.6	13,400	37,400	64
Ovary	16,800	3.9	34,300	51,100	33
Melanoma of skin	16,400	3.8	5,300	21,600	76
Bladder	15,200	3.5	16,100	31,300	49
Lung	11,600	2.7	100,400	112,000	10
Non-Hodgkin's lymphoma	11,600	2.7	14,800	26,400	44
Leukaemia	7,100	1.7	15,900	23,000	31
Stomach	5,500	1.3	37,800	43,300	13
Kidney	5,400	1.2	8,900	14,300	37
Lip, mouth and pharynx	5,200	1.2	6,700	12,000	44
Multiple myeloma	3,100	0.7	9,700	12,900	24
Brain	2,900	0.7	10,300	13,200	22
Oesophagus	2,600	0.6	18,200	20,900	13
Pancreas	1,800	0.4	29,500	31,300	6
Other	59,000	13.7	67,800	126,800	47
Total cases	431,400	100.0	595,600	1,027,000	42

diagnosed with breast cancer, 24,000 (13%) with colorectal cancer, and 25,000 (13%) with cancers of the cervix, uterus or ovary.

Ten year prevalence

Of the 1 million males diagnosed with cancer during the ten–year period 1983–92, almost 290,000 (just under 30%) were still alive at the beginning of 1993. Colorectal, prostate and bladder cancers together accounted for almost half of these (around 136,000 cases); lung cancer accounted for 24,000 cases (8%) and the lymphomas and leukaemias together for 22,000 (8%) (Table 2.5). The number of females diagnosed with cancer during the period was almost the same as for males (1 million) but over 430,000 (over 40%) were still alive. Of these, almost 170,000 (39%) had been diagnosed with breast cancer, over 50,000 (12%) with colorectal cancer, and 65,000 (15%) with cancers of the cervix, uterus or ovary. Survival from lung cancer, the most common cancer in men, is poor and only 9% of the 260,000 cases were alive at the end of the ten–year period. In contrast, survival from breast cancer, the most common cancer in women, is good and over 60% of the 270,000 cases were still alive.

Total prevalence

The numbers of patients who were diagnosed with a malignant cancer in each year from 1971 to 1992 and were still alive at the beginning of 1993 are illustrated in Figure 2.34. The cumulative totals of prevalent cases, just over 365,000 in males and 600,000 in females, slightly underestimate the numbers of all such cases because there will be people who were diagnosed before 1971 (the earliest year for which data are available) who were still alive in 1993. The majority of the cancers involved would be those for which higher than average proportions occur early in life and which have good survival, such as breast and cervical cancer. The survivors (almost 1 million) represent just under 2% of the total population, but the proportion of females alive who have been diagnosed with cancer (2.3%) is much higher than for males (1.5%). For males, 39% of the prevalent cases were diagnosed during the previous three years, and 79% during the previous ten; the corresponding proportions for females were lower: 31% and 72%, respectively.

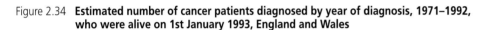

Figure 2.34 **Estimated number of cancer patients diagnosed by year of diagnosis, 1971–1992, who were alive on 1st January 1993, England and Wales**

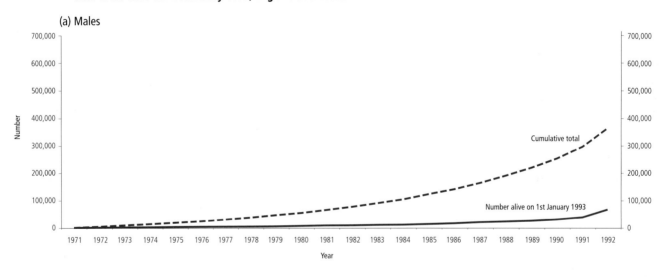

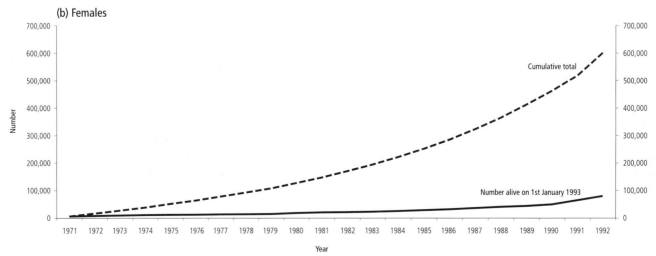

Figure 2.35 **Overall increase in five-year relative survival from 1971-75 to (i) 1986–90 and (ii) 1991–93**

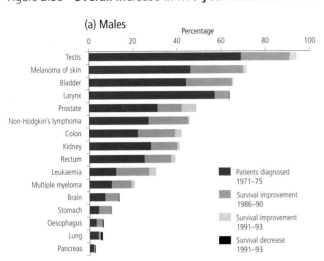

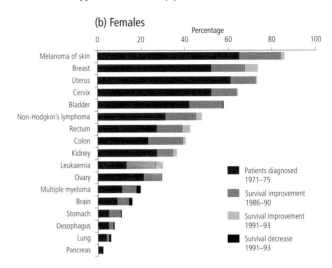

Cancer survival

The remainder of this chapter presents only a small fraction of the extensive and very detailed information on the patterns in cancer survival by cancer site, age group, sex, region and socio–economic deprivation, and over time, which has been published in the *Cancer Survival Trends* volume[6] and in *Health Statistics Quarterly*[20].

Survival for cancer patients in England and Wales diagnosed during 1991–93

For men in England and Wales diagnosed during 1991–93, the highest five–year survival rate, almost 95%, was for testicular cancer (Figure 2.35)[20]. Five–year survival for melanoma of the skin and cancers of the bladder and larynx was in the range 60–70%. Of the remaining major cancers, six had five–year survival in the range 30–50%: prostate, non–Hodgkin's lymphoma, colon, kidney, rectum and all leukaemias combined. Survival was poor (in the range 10–20%) for a further three major cancers: multiple myeloma, brain and stomach. Survival was very poor for cancers of the oesophagus, lung and pancreas.

For women, five–year survival from melanoma of the skin was over 85% (Figure 2.35). Three of the four main female cancers – breast, cervix and uterus (but not ovary) – and cancer of the bladder had five–year survival broadly in the range 60–75%. For the remaining major cancers, survival rates were closely similar to those in men. Six cancers had five–year survival broadly in the range 30–50%: non–Hodgkin's lymphoma, rectum, colon, kidney, all leukaemias combined, and ovary. Survival from multiple myeloma and cancers of the brain, stomach, oesophagus, lung and pancreas, was almost the same as for men.

Women have higher survival than men for many cancers (Figure 2.36). For melanoma of the skin, the difference is almost 15% points overall, and there were differences of at least 10% points in every age-group. There are similar differences for cancers of the tongue and of the salivary glands[6] (data not shown).

Cancer survival by age-group

For most cancers, survival is lower in patients who are older at diagnosis, even after adjustment for higher background mortality in the elderly[6]. For example, five–year survival from bladder cancer for men diagnosed during 1991–93 aged 15–39 years was 90%, but was just 50% for men aged 80–99 at diagnosis[20] (Figure 2.37). Bladder cancer is unusual in that men show a substantial survival advantage over women; for adults diagnosed during 1991–93 the difference was almost 11% points at one year, and 8% points at five years after diagnosis. The advantage for men is most marked in those aged 15–39 or 80–99 at diagnosis; this may be due to the registration of more small transitional cell papillomas in these age-groups. For several cancers, five–year survival falls from 20–50% at the youngest ages to around only 1% for the most elderly. Cancers of the oesophagus in women, pancreas, lung and brain each show this pattern (Figure 2.38 brain tumours). Survival from cancers of the large bowel (colon and rectum) varied less across the age-groups than for most cancers (Figure 2.39 colon cancer). Five–year survival for both cancers is 40–50% in the age range 40–79 years. The patterns of survival by age for cancers of the breast and prostate are unusual: survival is higher in middle age, lower in younger and older patients (Figure 2.40). Five–year survival from breast cancer in women is lower in the 15–39 age-group (70%) than in women aged 40–69 (77–80%). For prostate cancer, five–year survival is lower among the 40–49 age-group (32%) than in men aged 50–69 (50–55%)[20].

The reasons for the generally poorer survival in elderly cancer patients may include less aggressive treatment used for the elderly, and a larger proportion of younger patients being entered into clinical trials (which are associated with higher survival rates).

Survival trends for cancer patients diagnosed during 1971–93

Survival from some cancers, such as melanoma of the skin and testicular cancer, has improved dramatically since the early 1970s, while survival from some of the highly fatal cancers, such as those of the pancreas and lung, has shown little change. This section describes the trends in survival for some of the major cancers in adults and in children.

Figure 2.36 **Differences in survival between women and men, England and Wales, patients diagnosed during 1986–90: five-year relative survival* (%) and number of patients included in analyses**

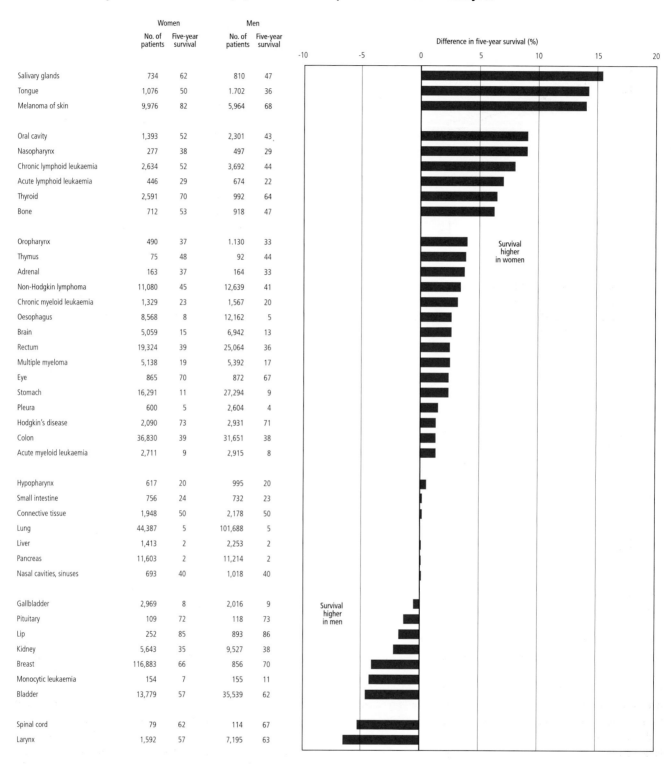

	Women		Men	
	No. of patients	Five-year survival	No. of patients	Five-year survival
Salivary glands	734	62	810	47
Tongue	1,076	50	1.702	36
Melanoma of skin	9,976	82	5,964	68
Oral cavity	1,393	52	2,301	43
Nasopharynx	277	38	497	29
Chronic lymphoid leukaemia	2,634	52	3,692	44
Acute lymphoid leukaemia	446	29	674	22
Thyroid	2,591	70	992	64
Bone	712	53	918	47
Oropharynx	490	37	1.130	33
Thymus	75	48	92	44
Adrenal	163	37	164	33
Non-Hodgkin lymphoma	11,080	45	12,639	41
Chronic myeloid leukaemia	1,329	23	1,567	20
Oesophagus	8,568	8	12,162	5
Brain	5,059	15	6,942	13
Rectum	19,324	39	25,064	36
Multiple myeloma	5,138	19	5,392	17
Eye	865	70	872	67
Stomach	16,291	11	27,294	9
Pleura	600	5	2,604	4
Hodgkin's disease	2,090	73	2,931	71
Colon	36,830	39	31,651	38
Acute myeloid leukaemia	2,711	9	2,915	8
Hypopharynx	617	20	995	20
Small intestine	756	24	732	23
Connective tissue	1,948	50	2,178	50
Lung	44,387	5	101,688	5
Liver	1,413	2	2,253	2
Pancreas	11,603	2	11,214	2
Nasal cavities, sinuses	693	40	1,018	40
Gallbladder	2,969	8	2,016	9
Pituitary	109	72	118	73
Lip	252	85	893	86
Kidney	5,643	35	9,527	38
Breast	116,883	66	856	70
Monocytic leukaemia	154	7	155	11
Bladder	13,779	57	35,539	62
Spinal cord	79	62	114	67
Larynx	1,592	57	7,195	63

*Age-standardised relative survival rates, rounded to nearest whole number: where values are equal for men and women, actual difference may be from -1 to +1.

Adults

Trends in five–year survival are illustrated in Figure 2.35, which shows (ranked for men and women separately) the overall increase between adults diagnosed in 1971–75 and those diagnosed in 1986–90, and any further increases in those diagnosed in 1991–93.

There have been large increases in survival for melanoma of the skin – around 25% points in men and 20% points in women. Long term survival has risen more quickly than short term, strongly suggesting an increase in the proportion of patients effectively cured of their disease[6]. The current five–year survival rate for

22

Figure 2.37 **Five-year survival from bladder cancer, with 95% confidence interval, by age at diagnosis and sex for patients diagnosed during 1991–93**

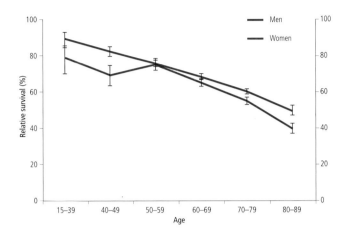

Figure 2.38 **Five-year survival from brain cancer, with 95% confidence interval, by age at diagnosis and sex for patients diagnosed during 1991–93**

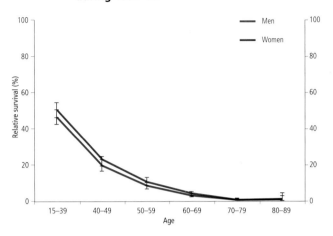

Figure 2.39 **Five-year survival from colon cancer, with 95% confidence interval, by age at diagnosis and sex for patients diagnosed during 1991–93**

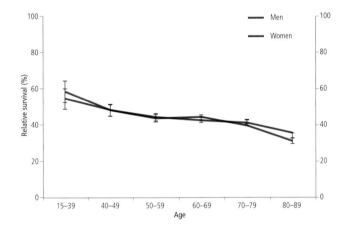

Figure 2.40 **Five-year survival from breast and prostate cancers, with 95% confidence interval, by age at diagnosis and sex for patients diagnosed during 1991–93**

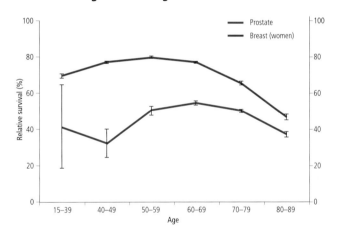

testicular cancer is around 25% points higher than in the early 1970s before the treatment of testicular tumours was transformed with the introduction of cis–platinum drugs; the design of optimal treatment regimens – and for seminoma, radiotherapy – also had an important role. Survival for cancers of the large bowel have shown some of the most consistent improvements among the solid tumours with average gains in both one–year (not shown) and five–year survival of 4–5% points every five years in both men and women (see below). Survival from both acute and chronic leukaemia has also improved by similar amounts to those for the large bowel.

For breast cancer, five–year survival rose from 52% for women diagnosed in 1971–75 to 68% for those diagnosed during 1986–90 (about twice the improvement in one–year survival) (Figure 2.41).

Breast cancer is unusual in that there is excess mortality (the relative survival curve continues to decline) up to 20 years after diagnosis[21,22]; for most cancers, the curve is virtually flat after five years (see, for example, that for colon cancer in men, below: Figure 2.42). For women diagnosed in 1991–93, five–year survival improved further by 6% points overall, with increases of 7–10% in women aged 50–69 at diagnosis[20]. The particularly large increases in this age-group (compared with 1–4% points increases for younger and older women) can be attributed to the impact of the NHS breast screening programme. Earlier diagnosis following mammography will increase survival time even if death is not delayed by treatment (this is called lead time bias), but this is unlikely to be the sole explanation for the increase in survival. For women in the broad age range 15–79 at diagnosis, five–year survival rose by more than one–year survival, suggesting the additional impact of improved treatment.

Compared with survival rates for men diagnosed with prostate cancer in 1986–90, five–year survival for those diagnosed in 1991–93 rose by 7–9% points for those aged 50–79, but by only 1–3% points for younger and older men. These large increases in survival, the first since the early 1980s, may be due in part to the more widespread use of prostate specific antigen (PSA) testing since the early 1990s: the average number of new cases each year rose by over a third between 1986–90 and 1991–93[12,14]. The effect on survival would be analogous to that of mammographic screening for breast cancer. While the PSA test enables invasive prostate cancer to be detected early, it also identifies latent, non–lethal, tumours that show no symptoms; these tumours are very common in elderly men. But five–year survival from prostate cancer rose by nearly 7% points overall, compared with just under 2% points for one–year survival, suggesting (as with breast cancer) more effective treatment.

As noted above, there were clear and substantial improvements in five–year survival from cancers of the colon and rectum, with gains of 5% points and 4% points, respectively, every five years for men and women during the 1970s and 1980s. The improvements were most marked between the late 1970s and early 1980s (Figure 2.42 – colon cancer in men). There were further increases in survival for patients diagnosed in 1991–93: for colon cancer almost 3% for men and 1% for women; and for rectal cancer 2% for men and 4% for women. Gains in one and five–year survival were similar. These trends in survival are consistent with the falls in mortality from bowel cancer despite the gradual increase in incidence, and they suggest the impact of more effective treatment, possibly as a result of earlier diagnosis.

For patients diagnosed in the 1970s and 1980s, survival from bladder cancer improved by 4–6% points every five years, with most of the increase occurring in the earlier part of that period. Survival did not improve for either men or women diagnosed in 1991–93 compared with those diagnosed in 1986–90.

There were on average increases of 2% points every five years in survival from cancer of the stomach for patients diagnosed intake 1970s and 1980s. One–year survival from stomach cancer for men diagnosed in 1991–93 rose by a further 2% points overall compared with men diagnosed in 1986–90, reaching 28%; there were increases in all age-groups. Five–year survival, however, remained unchanged at 10%, and there were no significant

improvements for women (27% and 11% at one and five years, respectively).

Compared with women diagnosed in 1986–90, survival after one year from ovarian cancer for women diagnosed during 1991–93 rose by 2% points to almost 60%, but five–year survival was unchanged at just below 30%. The rise in one–year survival was most marked (4–5% points) in women aged 60–79 years. Most of the improvement in survival from ovarian cancer occurred between women diagnosed in the late 1970s and early 1980s, with rises of 6 and 4% points in one-and five–year survival, respectively. Mortality fell during the 1980s partly as a result of these improvements in survival, but also due to a drop in incidence, with a reduced risk for women born after 1930.

As with ovarian cancer, five–year survival from cancer of the uterus improved on average by almost 3% points every five years for women diagnosed during 1971–90, but there was little further improvement for women diagnosed in 1991–93. The survival trends for cervical cancer were similar, but since 1990 there have been dramatic falls in both incidence and mortality which are directly attributable to the improvements made to the screening programme in the late 1980s (see above).

Improvements in survival for several other major cancers have been very small. In both men and women, five–year survival rates from cancers of the oesophagus, pancreas and lung, which are under 10%, improved by less than 1% point on average every five years between 1971–75 and 1986–90. Five–year survival for lung cancer was 5.0% for both men and women diagnosed in 1991–93, compared with 5.9% for those diagnosed during 1986–90, a small but statistically significant decline of almost 1% point in both sexes. This is likely to be due to the inclusion of more patients with very short survival. Patients whose survival is unknown cannot be included in the analyses; most of these are registered solely from a death certificate, and such cases tend to have had short survival. The proportion of these cases excluded from the analyses for 1991–93 was smaller than for patients diagnosed during 1986–90. This reflects more complete registration of patients with short

Figure 2.41 **Breast cancer survival in women up to ten years: trends by calendar period of diagnosis (1971–93)**

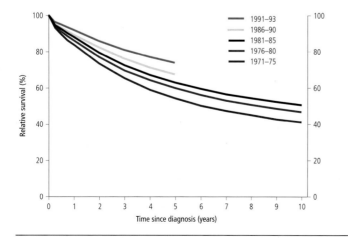

Figure 2.42 **Colon cancer survival in men up to ten years: trends by calendar period of diagnosis (1971–93)**

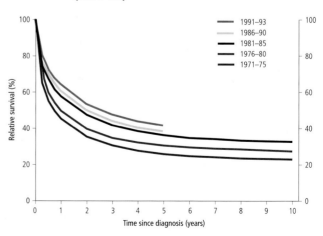

survival who were diagnosed during 1991–93, and hence the overall survival estimate is lower[20].

Children

In contrast to the wide range of changes in survival from the different major cancers in adults, there were substantial improvements during 1971–90 for most of the common malignancies of childhood[6]. For children diagnosed during 1986–90, all but two tumours had five–year survival rates over 50%. Six out of the eleven main cancers had survival rates in excess of 75%. Leukaemias and brain tumours together account for over

half of all cases. Compared with children diagnosed with leukaemia in 1971–75, the five–year survival rate for those diagnosed during 1986–90 had more than doubled – from 33% to 69%. For brain tumours, five–year survival increased from 43% to 58%. These improvements in survival reflect the inclusion of high proportions of children in national or international clinical trials of new treatments, and increased centralisation of care[23].

Region

For some cancers, regional trends in survival since the early 1970s do differ from the overall trend in England and Wales, or from the

Figure 2.43 **Five-year relative survival rate in most affluent group, and difference (gap) between most affluent and most deprived groups, for cancers with more than 1,000 patients diagnosed in England and Wales during 1986–90**

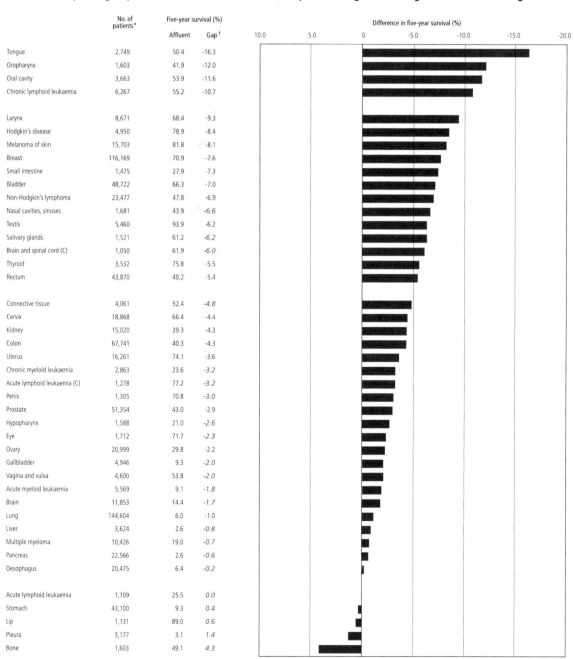

* Patients included in analyses and with a known deprivation category. Adults (15–99 years) unless marked (C) – children (0–14 years).
† Differences in five-year relative survival between affluent and deprived groups – a negative value means survival is lower in the deprived group.
Values in italics are not statistically significant at the 5% level.

25

trends in other regions. But for patients diagnosed in 1986–90 there were, in general, no large differences in survival between regions; and no region showed consistently higher or lower survival for all the common cancers. Detailed figures on survival by region are given in the *Cancer Survival Trends* volume[6]. Survival rates for patients diagnosed with some of the main cancers in 1991–93, however, do vary widely among the health authorities within each region[24].

Variation in cancer survival with socio–economic deprivation

The effect of socio–economic deprivation on survival has been examined for almost 50 different cancers for patients diagnosed in 1971–90 and is described in detail in the *Cancer Survival Trends* volume[6]. For none of the cancers was there a significant survival advantage at either one or five years from diagnosis for the most deprived group. For 44 cancers there was either limited or strong evidence of a negative survival gradient – that is, lower survival in the most deprived group (Figure 2.43). For 36 cancers, there was some evidence of lower five–year survival among patients in the most deprived group, both for those diagnosed during 1981–85 and during 1986–90. Some of the deprivation gaps were small, but for 22 cancers the gap for patients diagnosed during 1986–90 was statistically significant. And for all but two of these cancers both one- and five–year survival were significantly lower among patients in the most deprived group for those diagnosed during both 1981–85 and 1986–90.

Taken together, the results from the analyses of what was a very large data set provide overwhelming evidence that cancer survival among adults in England and Wales is generally lower among patients in deprived groups than among those in more affluent groups, even after allowing for the higher mortality from all causes of death in the more deprived groups[6].

The almost linear deprivation gradients in one- and five–year survival for all cancers in adults, in patients who were diagnosed during the decade 1981–90, are shown in Figure 2.44. Care is required in the interpretation of these patterns because the proportions of the different types of cancer vary among the

deprivation groups. This would affect the overall survival from all cancers combined even if survival rates from each cancer were the same in each deprivation group. Nevertheless, one- and five–year survival rates for all cancers combined are 2–4% points lower among patients in each successive deprivation category from the most affluent to the most deprived, both for those diagnosed during 1981–85 and during 1986–90. The estimated difference in survival between patients in the most affluent and most deprived groups was 11–13% points. There was no evidence that the differences in survival between deprivation categories for all cancers combined became any smaller for patients diagnosed during 1986–90 compared with those diagnosed during 1981–85.

Among children, however, the deprivation gaps in one- and five–year survival were not statistically significant for any of the cancers examined, among patients diagnosed during either 1981–85 or 1986–90.

The overall consistency of cancer survival differences in adults between the deprivation categories raises questions for public health. The reasons for the survival differences are difficult to identify with confidence[25]. They are part of a wide spectrum of socio–economic inequalities in health and disease[26,27]. Wide differences in survival have been found between socio–economic groups defined by housing tenure and other measures for many cancers both in the UK and elsewhere. Possible explanations for lower survival in more deprived groups include more advanced disease at diagnosis, longer delay in diagnosis, worse general health or resistance to malignancy, different histological type or more aggressive disease, poorer access to optimal care, and lower compliance with treatment[28,29]. Differences in stage of disease, however, do not appear to account for the observed survival differences between deprivation categories[30,31].

The public health impact of the differences in survival between deprivation groups can be illustrated by estimating how many fewer deaths would be attributable to cancer if patients in all groups of society had the same survival as those in the most affluent category. For over 750,000 adults diagnosed during

Figure 2.44 **Relative survival by deprivation category, all cancers[†], adults diagnosed in England and Wales, 1981-85 and 1986-90**

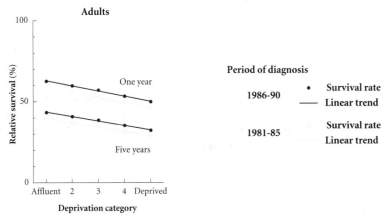

Figure 2.45 **Difference between five-year relative survival in England and Wales (1986–90) and the European average (1985–89)**

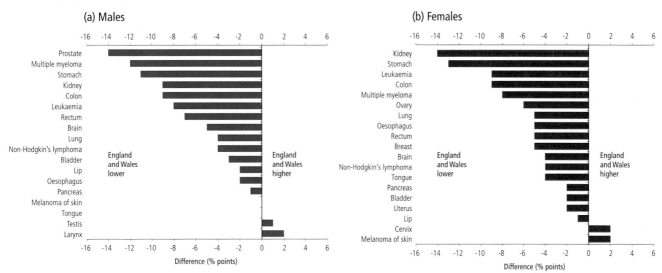

1986–90, there were almost 13,000 more deaths attributable to over 40 cancers (which account for over 90% of the total cases) than there would have been on the above basis. More than 5,000 of these avoidable deaths occurred in patients in the most deprived category, and another 4,200 in the second–most deprived. Of the total, around 2,800 avoidable deaths were from breast cancer in women, and around 1,200 from each of lung, colon, rectum and bladder cancers[6].

International differences in cancer survival

Survival in England and Wales is generally similar to that in Scotland[32]. For most of the common cancers, however, five–year survival for patients diagnosed during 1978–1985 in the UK was lower than in several comparable European countries[33]. Corresponding results for patients diagnosed during 1986–90[34,35] suggest that despite the improvements in cancer survival in England and Wales, the relative patterns had not changed greatly (Figure 2.45). The average five–year survival rate for men in Europe was at least 5% points higher than for men in England and Wales for 10 of 38 cancers, including those of prostate (14% higher, i.e. 56% versus 42% in England and Wales), stomach (11%), kidney (9%) and colon (9%). Survival was up to 5% points below the European average for 19 cancers, including lung (4%) and bladder (3%). For the remaining nine cancers, survival in England and Wales was the same as or higher than the average for Europe. Similarly, for women, the average five–year survival rate for Europe was at least 5% points higher than for women in England and Wales for 17 of 39 cancers, including those of kidney (14%), stomach (13%), colon (9%), ovary (6%), breast (5%) and lung (5%). The deficit was up to 5% points for 15 cancers including those of brain (4%), and non–Hodgkin's lymphoma (4%). For the remaining seven cancers, survival in England and Wales was the same or higher than the European average, including cervical cancer and melanoma of the skin (both 2% higher).

Where the overall England and Wales rates were lower than the European average, the best regional survival rate and/or the

survival rate in the most affluent group was usually fairly close to the European average.

The position compared with the USA is even worse than for the rest of Europe. Five–year survival for men in the USA was more than 5% points higher than in England and Wales for 25 of 39 cancers and up to 5% points higher for a further eight[36]. For women, USA rates were more than 5% points higher for 33 of 40 cancers, and up to 5% points higher for a further five.

Some of the apparently large differences in cancer survival between England and Wales – and Scotland – and other countries, particularly the USA, are likely to be due to the definition of disease and the methods commonly used for diagnosis. This applies especially to prostate cancer, where five–year survival in the USA is 86% compared with 42% (in 1986–90) in England and Wales. The results from the EUROCARE studies[33,34,35] have also been criticised because there may have been differences between registries in the ascertainment of both cases and deaths. And in some countries the registries covered only relatively small proportions of the total population and the results may not have been representative of the country as a whole. It has, however, been shown that the differences in survival between the UK and other European countries for breast and colon cancer arose primarily in the first six months after diagnosis, which suggests that there are international differences in the stage of disease at diagnosis, or in access to optimal treatment[37,38,39]. This has been confirmed for colorectal cancer[40]. At a recent international workshop of experts in the field there was broad consensus that the observed differences in survival rates were at least in part real (as opposed to artefactual); and that the poor survival rates in the UK for some of the common cancers were due at least in part to patients in the UK having more advanced disease at the time of their treatment[41].

27

Chapter 3

BLADDER

Bladder cancer is the fourth most common cancer in males in England and Wales, accounting for 8% of all malignant neoplasms, while in females it is ranked ninth and accounts for 3%. The main risk factors are tobacco smoking, occupational chemical exposure in the rubber, organic dye, metal refining and paint industries, some chemotherapy drugs – including cyclophosphamide – and urinary infection with schistosomiasis.[1] There are well-recognised differences in the classification and registration of bladder papillomas, a non-invasive carcinoma, which is regarded as benign by some registries and pathologists; this contributes to some of the international variation in the incidence and survival patterns of bladder cancer.[2]

About 90-95% of bladder cancers are transitional cell tumours; the remainder are squamous cell carcinomas and adenocarcinomas. Squamous cell carcinomas may account for up to 50% of bladder cancers in areas associated with high levels of schistosomiasis infection, such as parts of Africa and the Middle East. Primary tumours invade locally into the bladder muscle wall and then deeply into neighbouring structures. Metastases arise through lymphatic spread in the pelvic lymph nodes, and blood-borne spread to the lungs and bone. Prognosis is related to grade and stage of the tumour, with five-year survival of around 90% for superficial tumours, but less than 10% if the cancer has metastasised.[1]

Incidence and mortality

There were 12,100 cases diagnosed in England and Wales in 1997 (8,500 in males and 3,600 in females) (Table 3.1) compared with 7,200 cases in 1971, an overall rise of almost 70%. In 1994, virtually all of the cases were of unspecified site within the bladder. Around three-quarters of the cases were transitional cell carcinomas, with papillary and squamous cell carcinomas accounting for 4% of cases and the remainder coded as non-specific epithelial neoplasms.

Bladder cancer is rare in those aged under 40; incidence rates rise from this age, and then more steeply from age 60 to peak in the 85 and over age group in both sexes (at around 320 and 90 per 100,000 in men and women, respectively) (Figures 3.1 and 3.2).

The incidence of bladder cancer increased through the mid-1970s to peak in the late 1980s, reaching an age-standardised rate of around 31 per 100,000 in males and 9 per 100,000 in females (Figure 3.5). The incidence then stabilised to remain at around these levels through the 1990s. The male:female ratio of the age-standardised rate was 3.4:1 in 1997, a gradual decline from 4.1:1 in 1971. The increase in incidence occurred mainly in those aged 75 or over with rises of over 50% compared with the rates in 1971 in both men and women (Figure 3.3). These increases in incidence occurred in cohorts born before the early 1920s for both sexes (Figure 3.6). Incidence then fell for cohorts of men born in later years. After a fall in incidence for cohorts of women born in the 1920s and 1930s, incidence appeared to rise again for the more

recent cohorts (this trend is, however, based on relatively small numbers of cases, with correspondingly wide confidence intervals).

There were over 2,800 deaths in males from bladder cancer in England and Wales in 1999 and almost 1,500 deaths in females in 1999. Mortality from bladder cancer in males rose in the 1950s and 1960s, but has fallen since the early 1970s from just under 13 deaths per 100,000 to 9 per 100,000 in 1999 – a fall of 25% (Figure 3.7). In females mortality has remained almost level at around 3 per 100,000 since 1950. Despite the apparently stable overall rate for females, there has been a large increase in mortality amongst the elderly, particularly in the ten years from 1985 to 1994 (Figure 3.4). There was also a substantial rise in mortality of elderly men from the 1950s, levelling off at around 200 deaths per 100,000 in the 1990s. The cohort trends for mortality are similar to those for incidence, although the peak in mortality for males occurred slightly earlier (Figure 3.8).

In 1992-93, there was only a very slight gradient in the incidence of bladder cancer by deprivation group, with incidence lower in the more affluent groups than in the more deprived groups (Figure 3.9). There was a steeper – but still not large – deprivation gradient in mortality in males, with rates in the more deprived groups around 20% higher than in the more affluent (Figure 3.10).

There was some regional variation in the incidence of bladder cancer but no consistent north-south (or other) pattern (Figure 3.11). There was also no consistent regional variation in mortality from bladder cancer, and the pattern was different from that for incidence (Figure 3.12).

The incidence of bladder cancer in England and Wales is amongst the highest in the world. In the selected countries shown in Figure 3.13, rates in males and females were consistent, except for those in Zimbabwe. The lowest incidence occurred in the Indian sub-continent. Despite the high incidence of bladder cancer in males in England and Wales, mortality was – relatively – much lower, although not as low as in the USA (Figure 3.14); the pattern was similar in females. The comparison may be complicated by differences in the diagnosis of some papillomas of the bladder as malignant disease.

Prevalence

About 69% of males and 62% of females diagnosed in 1990-92 were still alive on 1st January 1993, as were about 50% of both males and females diagnosed in the ten years previously (1983-92) (Figure 3.15).

Survival

Survival from bladder cancer in England and Wales is moderately good, with relative survival at one year after diagnosis of 82% and 71%, and after five years of 66% and 58%, in men and women, respectively (for patients diagnosed in 1991-93). From the early 1970s, five-year relative survival improved by 5-6% every five

Table 3.1: Key statistics for bladder cancer on incidence, mortality and survival by sex, region and country, latest year available

MALES	Incidence Number of cases	CR	ESR*	WSR*	Mortality Number of deaths	CR	ESR*	WSR*	Relative survival (%) patients diagnosed in 1986-90 One-year	Five-year
United Kingdom [a]	9,710	33.7	30.1	19.5	3,300	11.3	9.8	5.9	..	..
Great Britain [a]	9,530	34.0	30.2	19.6	3,250	11.5	9.8	5.9	..	..
England and Wales [b]	8,530	33.2	29.1	18.9	2,850	11.0	9.3	5.6	81	65
England [b]	7,970	32.9	29.0	17.6	2,720	11.1	9.5	5.7	81	65
Northern and Yorkshire [b]	970	31.1	27.3	17.6	340	10.9	9.5	5.8	78	62
Trent [b]	880	34.7	29.6	19.2	310	12.0	9.9	6.0	81	64
Anglia and Oxford [b]	710	26.3	23.9	15.4	300	11.2	9.7	5.8	81	65
North Thames [b]	980	28.4	27.8	17.9	310	9.0	8.6	5.1	85	67
South Thames [b]	1,050	31.3	27.1	17.5	400	12.0	9.9	5.9	84	66
South and West [b]	1,460	44.9	35.3	23.0	420	12.9	9.5	5.7	83	67
West Midlands [b]	920	35.1	31.2	20.4	270	10.3	8.9	5.4	81	64
North West [b]	1,010	31.2	28.3	18.4	360	11.0	9.7	5.9	79	64
Wales [b]	560	38.9	31.5	20.5	130	8.8	6.9	4.2	82	70
Scotland [a]	948	38.1	35.7	23.1	284	11.4	10.5	6.2	82	65
Northern Ireland [a]	187	22.9	24.7	15.8	52	6.3	6.8	4.3	..	..
FEMALES										
United Kingdom [a]	3,890	13.0	8.5	5.6	1,710	5.7	3.2	2.0	..	..
Great Britain [a]	3,800	13.1	8.5	5.6	1,680	5.8	3.2	2.0	..	..
England and Wales [b]	3,550	13.4	8.6	5.7	1,460	5.5	3.0	1.9	72	58
England [b]	3,310	13.2	8.6	5.6	1,370	5.5	3.0	1.9	71	57
Northern and Yorkshire [b]	420	12.9	8.1	5.3	220	6.7	3.7	2.2	70	53
Trent [b]	370	14.1	9.0	6.0	170	6.4	3.5	2.1	72	58
Anglia and Oxford [b]	290	10.7	7.6	5.1	120	4.4	2.5	1.5	69	57
North Thames [b]	400	11.2	7.8	5.0	160	4.4	2.8	1.7	77	65
South Thames [b]	430	12.3	7.5	4.9	170	4.9	2.5	1.5	76	60
South and West [b]	560	16.5	9.8	6.5	180	5.3	2.6	1.6	72	59
West Midlands [b]	410	15.1	10.1	6.7	140	5.2	3.1	1.9	70	58
North West [b]	440	13.2	8.6	5.6	220	6.5	3.8	2.3	69	56
Wales [b]	240	16.1	9.9	6.5	90	5.8	2.8	1.7	77	64
Scotland [a]	461	17.4	11.9	7.9	185	7.0	4.3	2.7	71	57
Northern Ireland [a]	90	10.6	8.6	5.7	25	2.9	1.9	1.1	..	..

* Directly age-standardised rate
[a] 1996 incidence; 1998 mortality.
[b] 1997 incidence; 1999 mortality.
Figures for England and Wales are provisional: incidence - 1995-1997, mortality - 1999

years, although most of the improvement occurred in the 1970s and early 1980s – there was little improvement for patients diagnosed in the late 1980s or early 1990s (Figure 3.16).

Bladder cancer is one of the few cancers in which men have a substantial survival advantage over women. Relative survival is 90% for men diagnosed when aged under 40, and just under 80% for women of the same age. Survival then falls with age and is below 50% for men aged 80 or over at diagnosis and less than 40% for elderly women (Figure 3.17).

There is a deprivation gradient in the survival of bladder cancer: the most affluent groups of patients have around 6% points better survival at both one- and five-years after diagnosis than those patients living in more deprived areas. This is consistent with the patterns of incidence and mortality with deprivation illustrated in Figures 3.9 and 3.10 above. Differences in survival across the regions were not marked for patients diagnosed in 1986-90. Survival from bladder cancer in England and Wales is similar to that in Scotland and only slightly lower than the average for Europe (from the EUROCARE II study). However, survival in the USA is 17-19% higher than in the England and Wales – but, as noted above, this is probably affected by the registration of non-invasive papillomas in the USA.

Figure 3.1 **Age-specific incidence, England and Wales, 1997**

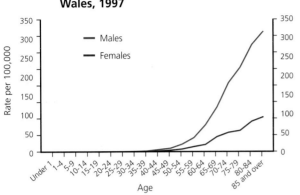

Figure 3.2 **Frequency distribution of new cases by age group, England and Wales, 1997**

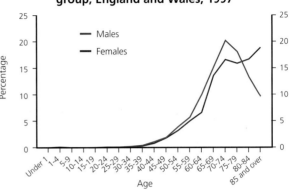

Figure 3.3 **Age-specific incidence, England and Wales, 1971-1997**

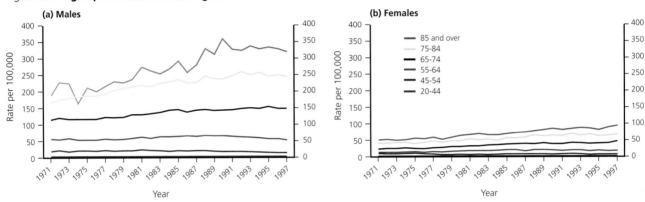

Figure 3.4 **Age-specific mortality, England and Wales, 1950-1999**

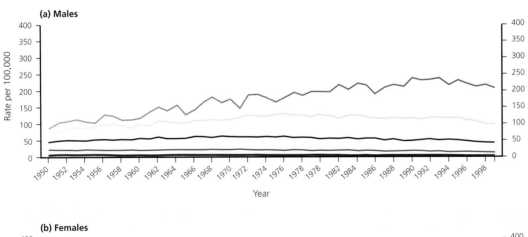

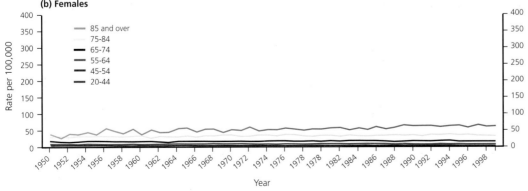

Figure 3.5 **Age-standardised incidence, England and Wales, 1971-1997**

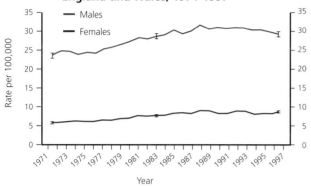

Figure 3.6 **Cohort incidence ratio by sex, England and Wales**

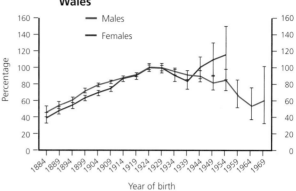

Figure 3.7 **Age-standardised mortality by sex, England and Wales, 1950-1999**

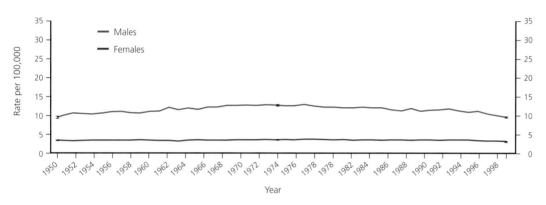

Figure 3.8 **Cohort mortality ratio by sex, England and Wales**

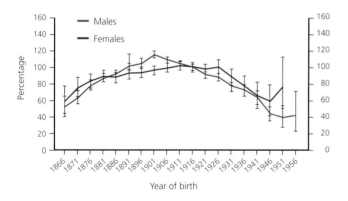

Figure 3.9 **Age-standardised incidence by deprivation category, England and Wales, 1992-93**

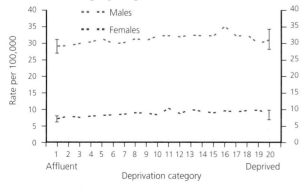

Figure 3.10 **Age-standardised mortality by deprivation category, England and Wales, 1992-93**

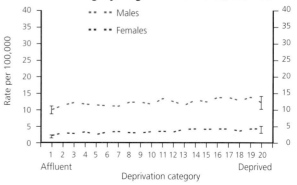

Figure 3.11 **Comparative incidence ratio (CIR) by health region and sex (England and Wales = 100), 1997**

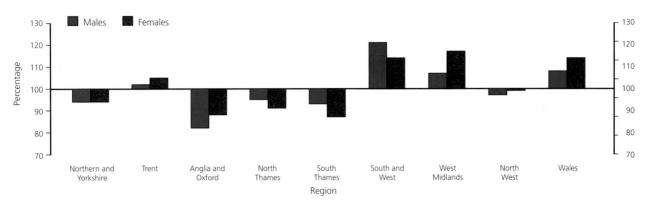

Figure 3.12 **Comparative mortality ratio (CMR) by health region and sex (England and Wales = 100), 1999**

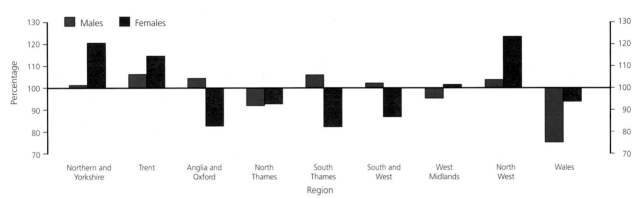

Figure 3.13 **Age-standardised international incidence**

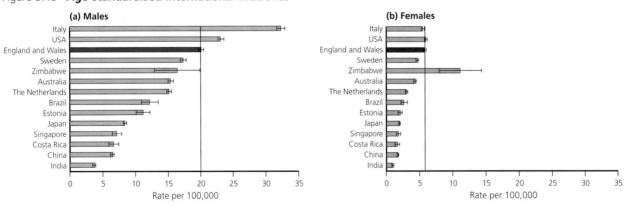

Figure 3.14 **Age-standardised international mortality**

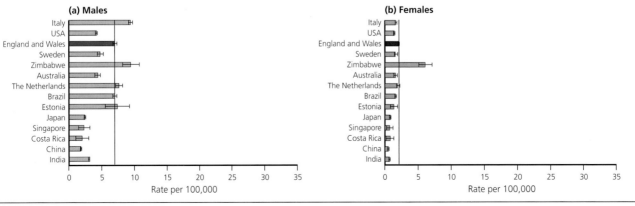

Figure 3.15 **Number and percentage of cases diagnosed in 1990-92 and 1983-92 and alive on 1st January 1993, England and Wales**

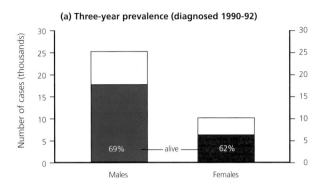

(a) Three-year prevalence (diagnosed 1990-92)

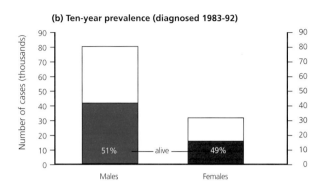

(b) Ten-year prevalence (diagnosed 1983-92)

Figure 3.16 **Relative survival up to ten years, men, England and Wales, 1971-75 to 1991-93**

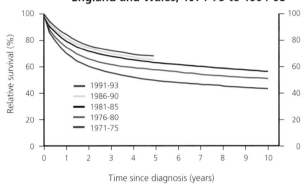

Figure 3.17 **Five-year relative survival by age, England and Wales, patients diagnosed 1991-93**

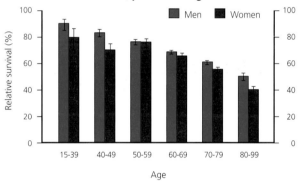

Figure 3.18 **One- and five-year relative survival by deprivation category, England and Wales, adults diagnosed 1986-90**

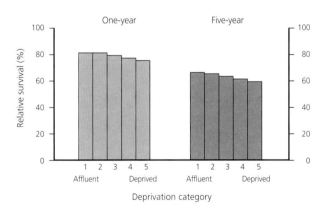

Figure 3.19 **International comparison of five-year relative survival, England and Wales, 1986-90**

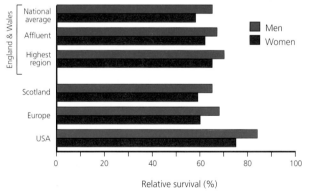

Chapter 4

BRAIN

Tumours of the brain account for 1.6% of all cancers in England and Wales and are the twelfth most common type of cancer in men and the fifteenth most common in women. They comprise several distinct diseases grouped under one general heading. The majority of malignant tumours are gliomas, including astrocytoma, oligodendroglioma, and ependymoma. The other most common histological types, menigiomas and neuromas, are predominantly benign. Astrocytomas occur most frequently in adults, usually in the cerebral hemispheres (especially the frontal and temporal lobes). They are subdivided into low grade (1 and 2) and high grade (3 and 4), with grade 4 astrocytic gliomas termed glioblastomas. Prognosis varies with grade: five-year survival for grade 1 is around 60%, while high grade tumours have one-year survival of only 30%.[1] The brain is a common site for metastases from other parts of the body, particularly lung, but also breast, kidney, gastrointestinal tract, prostate, and malignant melanoma of skin.

The aetiology of the various diseases is generally unknown, although a number of inherited syndromes are known to predispose some individuals to brain tumours (although these may only account for 5% of cases). Ionising radiation is the one known environmental cause of brain tumours, while several occupations have been linked to a raised risk: the petrochemical industry; workers exposed to electromagnetic fields; certain health professions; agricultural workers; the nuclear industry; and workers in the rubber industry, particularly those involved in making tyres.[2]

Accurate diagnosis of brain tumours can be problematic, and it is often difficult to distinguish primary tumours from metastases, as well as to differentiate the various histological types. Improvements in diagnostic practices over the past twenty years have probably led to much of the observed increase in registration of brain cancer[3]. Computed tomography (CT) scans, introduced in the 1970s, and magnetic resonance imaging (MRI) now provide information on the location, size and degree of local invasion of the tumour. Stereotactic biopsy allows the histological diagnosis of tumours, essential when surgical resection is not possible, for example, in the brain stem.

Incidence and Mortality
There were around 3,500 new cases of brain cancer diagnosed in England and Wales in 1997: 2,050 in males and almost 1,500 in females (Table 4.1). The disease is more common in males than females with a male:female ratio in the age-standardised rates of around 1.5:1. There is a bimodal age distribution in tumours of the brain with a small peak in children under 10 and a much larger peak in adults at ages 55 to 80 (Figure 4.2).

Around 40% of cases were coded as located in the frontal, parietal and temporal lobes, and around 4% were in the cerebrum and 3% in the cerebellum, but over 40% of the cases were poorly (or not) specified. Most tumours were gliomas (85%): 30% (of all brain cancer cases) were astrocytomas, 22% glioblastomas, 3%

oligodendroblastomas, and 30% other or unspecified gliomas; the remainder were of poor histological specification.

From 1971 up to the early 1990s, there were overall rises in the age-standardised incidence rates (40% for males and 43% for females), to reach 8 and 5 per 100,000, respectively (Figure 4.5); the rates in both sexes subsequently remained at around these levels. The greatest increases in incidence were in adults aged 65 or over (Figure 4.3). In the early 1970s, the highest incidence occurred in the 55-64 age group for both males and females. The increases since then in incidence in adults aged 65-84 were such that in the 1990s the rates for these ages exceeded those in the middle aged. The substantial rise in incidence in the elderly is likely to be artefactual, partly due to improvements in diagnostic techniques. In turn these advances may have led to a reduction in biopsy rates, and to a rise in secondary malignancies being misclassified as primary brain tumours.[4] A sharp rise in incidence occurred for each successive cohort in both sexes born from the mid-1890s to the mid-1920s (Figure 4.6). For subsequent birth cohorts the incidence continued to rise but by a lesser amount, followed by a sharp fall for those born in the late-1960s.

The trends in mortality resemble those for incidence, with an overall long term but gradual rise in both sexes (Figure 4.7). The age-specific trends also reflect the corresponding incidence rates, with a marked rise in mortality among adults aged 65 or over (Figure 4.4). Mortality remained fairly stable in those aged under 64. As with incidence, mortality increased in successive cohorts up to those born in the 1920s but it then declined gradually (Figure 4.8).

There was a clear difference in incidence by deprivation group: brain tumours were more common in the affluent groups than in more deprived groups with a slightly stronger inverse gradient in males than in females (Figure 4.9). The rates in the most affluent groups in 1991-93 were 25-30% higher than in the most deprived groups for both males and females. There were also distinct inverse deprivation gradients in mortality in 1991-93: the difference of around 40% between males in the most affluent and most deprived groups was larger than that for females (Figure 4.10).

There was no consistent regional variation in either the incidence of or mortality from brain tumours in England and Wales in males and females (Figures 4.11 and 4.12). Both incidence and mortality around the world are highest in developed countries, as shown in Figures 4.13 and 4.14. The rates in England and Wales are similar to the high rates in the USA, Australia, Italy and Sweden. Incidence rates in some countries may be raised through the inclusion of benign and unspecified nervous system tumours.[3]

Prevalence
About one third of both males and females diagnosed in England and Wales in the three years 1990-92 were still alive at the beginning of 1993, as were about one fifth of those diagnosed in the 10 years 1983-92 (Figure 4.15).

Table 4.1: Key statistics for brain cancer on incidence, mortality and survival by sex, region and country, latest year available

MALES	Incidence				Mortality				Relative survival (%) patients diagnosed in 1986-90	
	Number of cases	CR	Rate per 100,000 ESR*	WSR*	Number of deaths	CR	Rate per 100,000 ESR*	WSR*	One-year	Five-year
United Kingdom [a]	2,170	7.5	7.3	5.8	1,790	6.1	5.9	4.4	..	..
Great Britain [a]	2,100	7.5	7.3	5.8	1,740	6.1	5.9	4.4	..	..
England and Wales [b]	2,050	8.0	7.7	6.1	1,620	6.3	6.0	4.6	30	13
England [b]	1,920	7.9	7.7	6.1	1,540	6.3	6.1	4.6	30	14
Northern and Yorkshire [b]	250	7.9	7.8	6.1	210	6.8	6.6	5.0	28	14
Trent [b]	210	8.4	7.9	6.1	200	7.9	7.4	5.6	26	11
Anglia and Oxford [b]	260	9.5	9.2	7.1	170	6.2	6.0	4.6	29	13
North Thames [b]	210	6.1	6.3	5.1	160	4.7	4.8	3.8	39	18
South Thames [b]	220	6.6	6.4	5.2	220	6.6	6.4	4.7	34	16
South and West [b]	320	9.7	9.0	7.4	200	6.1	5.6	4.3	27	11
West Midlands [b]	210	7.8	7.4	5.8	170	6.6	6.3	4.6	31	16
North West [b]	250	7.7	7.6	6.2	200	6.0	5.8	4.5	25	10
Wales [b]	140	9.4	8.4	6.6	80	5.8	5.1	3.7	24	12
Scotland [a]	206	8.3	8.1	6.5	151	6.1	5.9	4.2	28[†]	13[†]
Northern Ireland [a]	71	8.7	9.4	7.8	50	6.0	6.7	5.2	..	..

FEMALES										
United Kingdom [a]	1,780	5.9	5.2	4.3	1,380	4.6	3.9	3.0	..	..
Great Britain [a]	1,730	6.0	5.3	4.3	1,340	4.6	3.9	3.0	..	..
England and Wales [b]	1,480	5.6	5.0	4.1	1,170	4.4	3.7	2.9	31	16
England [b]	1,380	5.5	5.0	4.1	1,090	4.3	3.7	2.8	31	16
Northern and Yorkshire [b]	200	6.0	5.4	4.3	130	4.2	3.6	2.7	27	11
Trent [b]	140	5.4	4.5	3.6	110	4.2	3.4	2.6	29	15
Anglia and Oxford [b]	160	5.9	5.4	4.4	130	4.8	4.2	3.1	24	13
North Thames [b]	160	4.4	4.2	3.6	150	4.2	4.0	3.1	36	20
South Thames [b]	200	5.7	5.3	4.5	160	4.6	4.0	3.1	32	18
South and West [b]	210	6.2	5.3	4.4	170	5.1	4.1	3.1	29	15
West Midlands [b]	150	5.5	4.9	4.0	90	3.3	2.9	2.2	35	17
North West [b]	170	5.1	4.7	3.9	140	4.1	3.4	2.6	34	18
Wales [b]	100	6.5	5.4	4.3	80	5.5	4.5	3.4	29	16
Scotland [a]	156	5.9	5.2	4.3	123	4.7	3.8	2.9	28[†]	14[†]
Northern Ireland [a]	44	5.2	4.9	4.5	34	3.9	3.8	2.9	..	..

* Directly age-standardised rate
[†] ICD9 191 and 192 for Scotland survival data
[a] 1996 incidence; 1998 mortality.
[b] 1997 incidence; 1999 mortality.
Figures for England and Wales are provisional: incidence - 1995-1997, mortality - 1999

Survival

Survival from brain cancer is very poor, with one-year relative survival about 30% in men and women diagnosed in 1991-93, and five-year survival around 13% (Figure 4.16). From the early 1970s up to the late 1980s there was a small but regular increase in survival: 3% points every five years for one-year survival and 2% points for five-year survival. There was, however, no improvement for patients diagnosed during 1991-93 compared with those diagnosed in 1986-90.

Survival from brain cancer falls rapidly with age (Figure 4.17). Five-year relative survival was over 40% for men and women aged 15-39, but fell to 20% in men and 23% in women aged 40-49. Survival was 3% or less in adults aged 60 or over at diagnosis. There is a slight survival advantage for women compared with men (2% points for five-year relative survival).

Regional variation in survival was not marked. One-year survival was 4% points higher in the most affluent group diagnosed in 1986-90 compared with the most deprived group, but there was no difference across the deprivation categories in five-year survival (Figure 4.18). Five-year relative survival in England and Wales was similar to survival in Scotland but around 5% points lower than the average for Europe and 10% points lower than in the USA (Figure 4.19).

Figure 4.1 **Age-specific incidence, England and Wales, 1997**

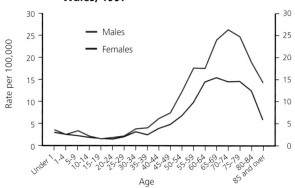

Figure 4.2 **Frequency distribution of new cases by age group, England and Wales, 1997**

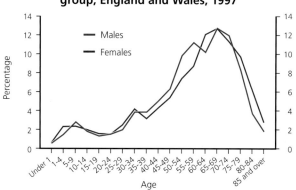

Figure 4.3 **Age-specific incidence, England and Wales, 1971-1997**

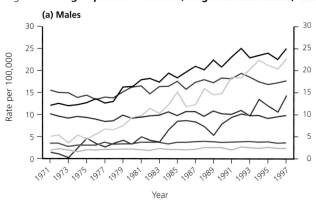

(a) Males

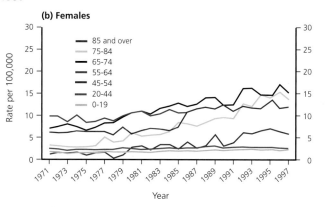

(b) Females

Figure 4.4 **Age-specific mortality, England and Wales, 1950-1999**

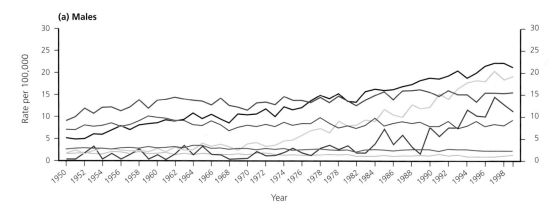

(a) Males

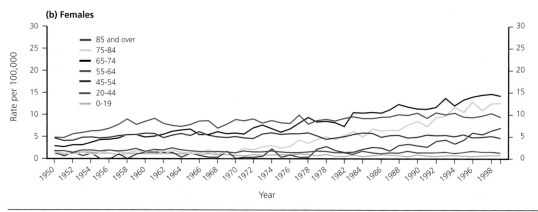

(b) Females

Figure 4.5 **Age-standardised incidence, England and Wales, 1971-1997**

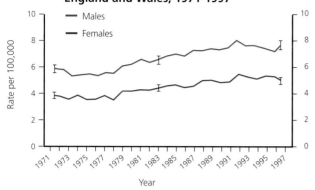

Figure 4.6 **Cohort incidence ratio by sex, England and Wales**

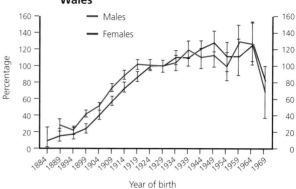

Figure 4.7 **Age-standardised mortality by sex, England and Wales, 1950-1999**

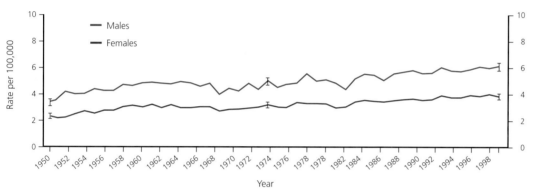

Figure 4.8 **Cohort mortality ratio by sex, England and Wales**

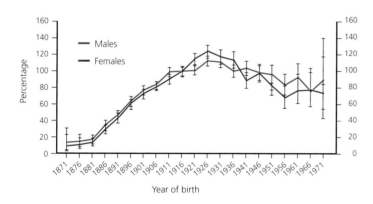

Figure 4.9 **Age-standardised incidence by deprivation category, England and Wales, 1991-93**

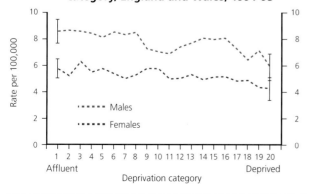

Figure 4.10 **Age-standardised mortality by deprivation category, England and Wales, 1991-93**

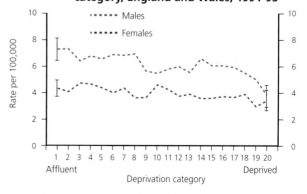

Figure 4.11 **Comparative incidence ratio (CIR) by health region and sex (England and Wales = 100), 1997**

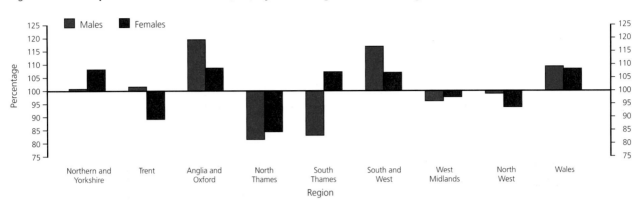

Figure 4.12 **Comparative mortality ratio (CMR) by health region and sex (England and Wales = 100), 1999**

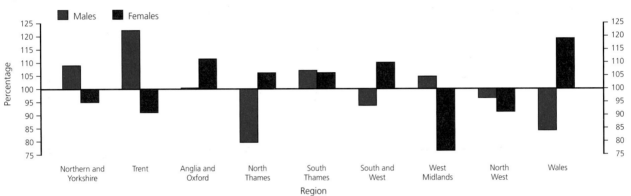

Figure 4.13 **Age-standardised international incidence**

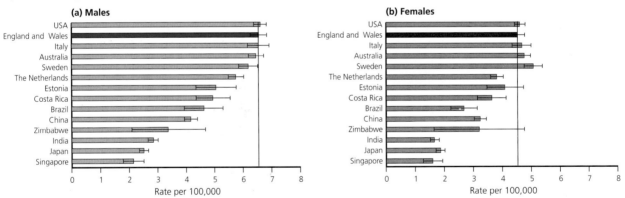

Figure 4.14 **Age-standardised international mortality**

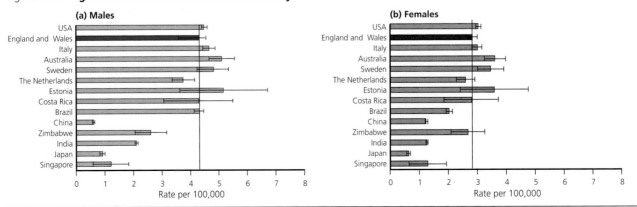

Figure 4.15 **Number and percentage of cases diagnosed in 1990-92 and 1983-92 and alive on 1st January 1993, England and Wales**

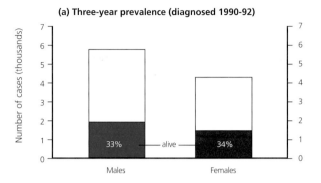

(a) Three-year prevalence (diagnosed 1990-92)

(b) Ten-year prevalence (diagnosed 1983-92)

Figure 4.16 **Relative survival up to ten years, men, England and Wales, 1971-75 to 1991-93**

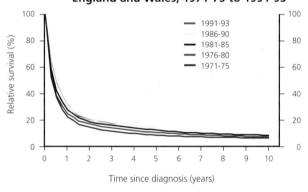

Figure 4.17 **Five-year relative survival by age, England and Wales, patients diagnosed 1991-93**

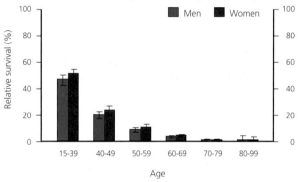

Figure 4.18 **One- and five-year relative survival by deprivation category, England and Wales, adults diagnosed 1986-90**

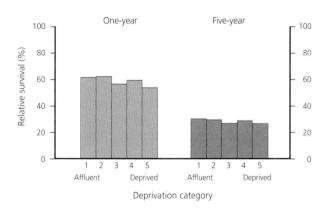

Figure 4.19 **International comparison of five-year relative survival, England and Wales, 1986-90**

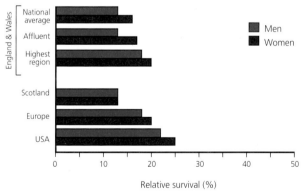

Chapter 5

BREAST

Breast cancer is the most common cancer in women worldwide, although cervical cancer is more frequent in some developing countries. Breast cancer in men is extremely rare. Breast cancer accounts for about 25% of all malignancies in women; the proportion is higher in women in western, developed, countries. Both incidence and mortality vary considerably around the world. Incidence has been rising in many parts of the world, including the USA, Canada, Europe, the Nordic countries, Singapore and Japan.

Most of the known risk factors for breast cancer relate to a woman's reproductive history – early menarche, late first pregnancy, low parity, and late menopause; endogenous hormones, both oestrogens and androgens, probably have an important role. Some types of benign breast disease increase risk. None of these risk factors is currently amenable to primary prevention.[1] Oral contraceptive use and hormonal replacement therapy have been linked to increased risk. There is a positive association between alcohol consumption and breast cancer, but cigarette smoking appears not to increase risk. Studies of migrant populations have suggested that differences in incidence among countries are social and environmental, rather than genetic, in origin; only about 5% of breast cancer is due to highly penetrant dominant genes.[2] Avoidance of obesity may decrease the risk of post-menopausal breast cancer, and switching from a high fat and low vegetable diet to a lower fat, higher vegetable diet may also contribute to a reduced risk.

With this background, and based on the findings from randomised trials that routine screening by mammography could reduce mortality from breast cancer by 25-30% in women aged 50 and over, in 1986 the working group chaired by Forrest recommended that mass population screening for breast cancer of 50-64 year old women should be introduced with single mediolateral oblique view mammography and a three year interval between screens.[3] The NHS breast screening programme began operation in 1988. Screening methods have varied around the country: in some screening units an additional, craniocaudial mammogram was taken; in some units the X ray films were read by more than one person; and in a few areas the screening interval was two years. The prevalence round was not completed everywhere by the target date of March 1993.

Incidence and mortality

In 1997 there were 33,100 new registrations of breast cancer in women in England and Wales (Table 5.1), almost 30% of all cancers in women, and more than twice as many as for the second most common site, colorectal cancer. The lifetime risk of being diagnosed with breast cancer is almost 11% (1 in 9). About 19% of tumours were coded to the upper-outer quadrant of the breast and a further 16% in other quadrants or the nipple, but for 65% there was no anatomic detail. The proportion of tumours described as adenomas or adenocarcinomas had fallen to under 10% (from over 30% in the early 1970s); ductal, lobular and medullary tumours accounted for 60%; and the proportion of cases with a non-specific or no histology was around 30%.

Before the introduction of screening, incidence rates rose with age from the late 20s, but the increases slowed at around 45-54 years, the age of the menopause. The effect of breast screening has been to raise the incidence in women aged 50-54, because many of them were being screened for the first time[4] (Figure 5.3). Rates in women aged 55-64 also rose during the prevalence round, but have since returned to levels expected based on the earlier trends. In 1992, incidence in the screened age groups was about 40% higher than in 1987. Incidence in women aged 65-69 has fallen in recent years: many cancers in these women will have been detected several years earlier (in the previous age group) during the prevalence round; their rates in 1995-97 were lower than those in women aged 50-64.

From the late 1970s up to the introduction of screening in 1988, age-standardised incidence increased by about 2% each year to reach nearly 90 per 100,000 (Figure 5.5). During the prevalence round of screening, the rate increased to a peak of 108 per 100,000 in 1992; it subsequently declined but then rose again to around 107 per 100,000 in 1997. The almost continuous increase in risk is shown in the rise across the successive birth cohorts (Figure 5.6).

During the 1950s and 1960s, mortality from breast cancer increased in women aged 50-69, but not in older women (Figure 5.4). From the early 1980s, mortality started to decline in women aged 45-54. From the cohort of women born at the end of the 19th century, the risk increased to a peak in women born in the mid-1920s to mid-1930s and then declined (Figure 5.8). Thus in the absence of screening, it would have been expected that mortality would fall in the 1990s in women aged under 70, but that rates in older women would increase. Mortality began to fall soon after screening started; there have been falls in all age groups; and the falls were larger in women aged 55-69 than in the other age groups. Compared with the expected mortality based on the cohort models, there has been a fall of just over 20% in women aged 55-69. About a third of this is due directly to screening, and two thirds to improved treatment by tamoxifen and chemotherapy and to indirect effects of screening such as raised awareness leading to earlier presentation and diagnosis outside the screening programme.[5] By 1999, overall mortality was just over 20% lower than the levels in the mid-1980s (Figure 5.7).

In 1993, there was a negative gradient in the incidence of breast cancer by Carstairs deprivation category: the rate was about 30% higher in the most affluent groups than in the most deprived (Figure 5.9). In contrast to the negative relationship of incidence with deprivation, mortality was not related to deprivation (Figure 5.10). This implies that survival is better in the more affluent (see below). There was some variation in incidence by region, with higher than average incidence in the South and West, Anglia and Oxford, and lower than average incidence in Northern and Yorkshire and Trent (Figure 5.11). There was no consistent variation in mortality from breast cancer by region (Figure 5.12).

Table 5.1: Key statistics for female breast cancer on incidence, mortality and survival by region and country, latest year available

| | Incidence | | | | Mortality | | | | Relative survival (%) patients diagnosed in 1986-90 | |
| | Number of cases | Rate per 100,000 | | | Number of deaths | Rate per 100,000 | | | | |
		CR	ESR*	WSR*		CR	ESR*	WSR*	One-year	Five-year
United Kingdom [a]	36,100	120.5	104.3	75.9	13,200	43.8	32.6	22.6	..	..
Great Britain [a]	35,200	121.1	104.3	75.9	12,900	44.1	32.6	22.7	..	..
England and Wales [b]	33,100	124.6	107.0	78.1	11,500	43.3	31.8	22.0	90	68
England [b]	30,800	123.0	105.9	77.3	10,800	43.0	31.8	22.0	90	68
Northern and Yorkshire [b]	3,680	114.1	99.7	73.5	1,320	40.9	30.2	21.1	90	67
Trent [b]	2,660	102.4	86.9	63.5	1,120	43.2	30.4	20.9	88	66
Anglia and Oxford [b]	3,610	132.6	118.4	86.7	1,140	41.5	32.1	22.2	90	69
North Thames [b]	4,030	113.8	104.8	76.2	1,380	38.7	31.7	22.0	92	70
South Thames [b]	4,300	122.1	103.8	75.6	1,550	43.9	31.9	22.1	92	70
South and West [b]	5,150	152.2	122.0	88.9	1,620	47.6	31.8	21.9	89	68
West Midlands [b]	3,310	122.9	104.6	76.2	1,210	45.0	33.2	22.9	91	68
North West [b]	4,060	120.7	104.2	75.9	1,470	43.6	32.7	22.7	89	65
Wales [b]	2,270	152.1	124.6	90.1	730	48.7	32.9	22.6	84	65
Scotland [a]	3,242	122.7	106.9	78.0	1,142	43.3	33.0	22.9	90	67
Northern Ireland [a]	873	102.4	103.3	75.8	297	34.5	31.1	21.6	..	..

* Directly age-standardised rate
[a] 1996 incidence; 1998 mortality.
[b] 1997 incidence; 1999 mortality.
Figures for England and Wales are provisional: incidence - 1995-1997 and mortality - 1999.

Around the world, the highest recorded incidence rates occur in the USA and other western, developed countries; rates in Japan, China and India are only about a quarter of those in the USA (Figure 5.13). The international mortality data indicate that (compared with incidence) mortality was relatively low in the USA and Sweden, and high in India and Costa Rica (Figure 5.14).

Prevalence

As the incidence of breast cancer is high and survival is relatively good compared with many other cancers, there are large numbers of women alive who have been diagnosed with breast cancer. About 81% (75,000) of those diagnosed in 1990-92, and 62% (168,000) of those diagnosed in 1983-92 were still alive at the beginning of 1993 (Figure 5.15).

Survival

Although survival from breast cancer is lower than for cancer of the endometrium, it is better than for cancer of the cervix, and much better than for the other major cancers in women – lung, colorectal, and ovarian. Survival from breast cancer (as for most other cancers) is worse the later the stage of the disease at diagnosis; and the proportion of cases with advanced stage increases with age. Overall results will depend on the stage distribution, which can vary among age groups from one area to another.[6] Five-year relative survival in England in the late 1980s was about 85% for stage I at presentation, compared with only 20% for stage IV.

One-year survival for patients in England and Wales diagnosed in 1991-93 was 92%; five-year survival was 74% (Figure 5.16). Unlike most other cancers, there is still excess mortality (declining curves) for at least 10 years after diagnosis. Survival from breast cancer also differs from that for most cancers in that women aged under 40 at diagnosis had worse survival than those aged 40-49 (Figure 5.17). Five-year survival in 1991-93 fell slightly to 65% in women aged 70-79, and then more sharply to 47% in elderly women. Survival

has improved steadily over time, and in all regions. Five-year survival rose by 14% points between the early 1970s and the late 1980s and by a further 6% for patients diagnosed in 1991-93. Regional differences in survival were not marked. In Scotland, overall relative survival at all periods up to five years from diagnosis, the pattern of survival with age, and improvements in survival since the early 1970s were all closely similar to the results for England and Wales.[7]

There was a clear gradient in survival with deprivation (women in deprived areas having worse survival) (Figure 5.18). The gap in survival between affluent and deprived groups in the 1980s was 6% points at one year after diagnosis, and 9% points five years after. These gaps are unlikely to be due to differences in stage at diagnosis between the deprivation groups.[8]

In the late 1980s, mortality in England and Wales was not only higher than in most western European countries, it was among the highest in the world; incidence, however, was similar to that in other western European countries. The corollary is that survival was worse than elsewhere in Europe. This has been confirmed by the results of the EUROCARE studies.[9] The reasons for this poorer survival are not clear, but there was a high relative risk in England (compared with the rest of Europe) in the first six months after diagnosis, suggesting that later diagnosis and/or adherence to treatment protocols may be important.[10] Mortality in England and Wales has, however, declined substantially since the late 1980s, and with increasing trends in incidence it is likely that (even allowing for the effect of breast screening in reducing the lead time) survival from breast cancer will have improved throughout the 1990s.

Figure 5.1 **Age-specific incidence, England and Wales, 1997**

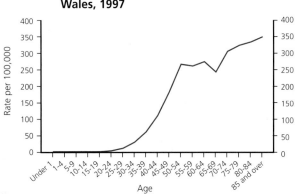

Figure 5.2 **Frequency distribution of new cases by age group, England and Wales, 1997**

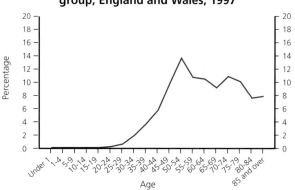

Figure 5.3 **Age-specific incidence, England and Wales, 1971-1997**

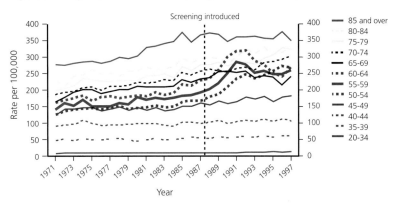

Figure 5.4 **Age-specific mortality, England and Wales, 1950-1999**

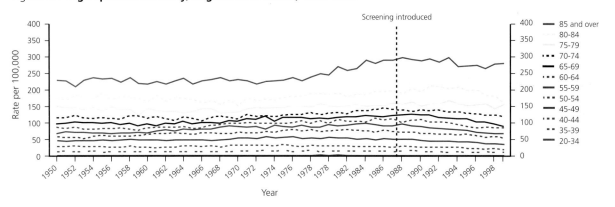

Figure 5.5 **Age-standardised incidence, England and Wales, 1971-1997**

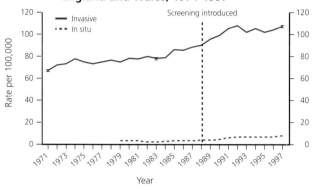

Figure 5.6 **Cohort incidence ratio, England and Wales**

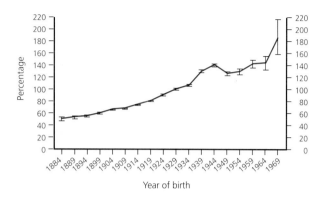

Figure 5.7 **Age-standardised mortality, England and Wales, 1950-1999**

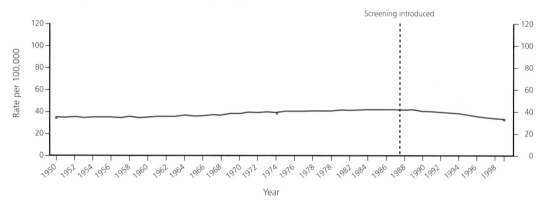

Figure 5.8 **Cohort mortality ratio, England and Wales**

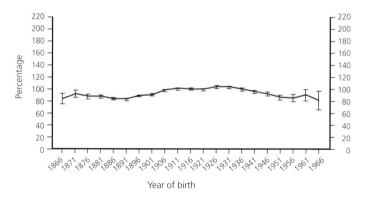

Figure 5.9 **Age-standardised incidence by deprivation category, England and Wales, 1993**

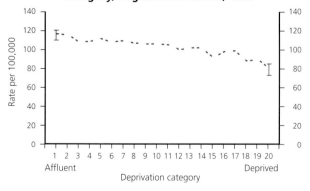

Figure 5.10 **Age-standardised mortality by deprivation category, England and Wales, 1991-93**

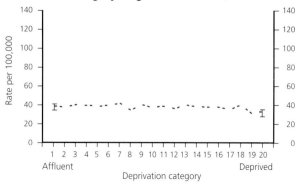

Figure 5.11   **Comparative incidence ratio (CIR) by health region (England and Wales = 100), 1997**

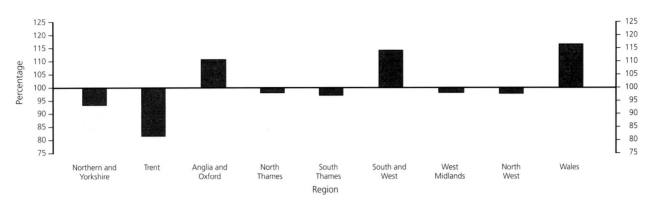

Figure 5.12   **Comparative mortality ratio (CMR) by health region (England and Wales = 100), 1999**

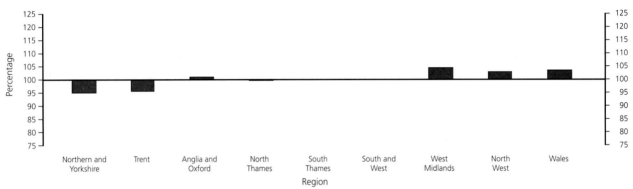

Figure 5.13   **Age-standardised international incidence**

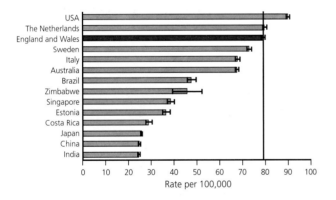

Figure 5.14   **Age-standardised international mortality**

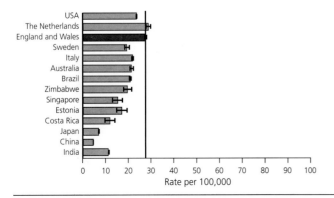

Figure 5.15 **Number and percentage of cases diagnosed in 1990-92 and 1983-92 and alive on 1st January 1993, England and Wales**

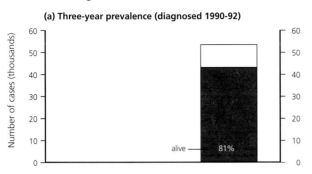

(a) Three-year prevalence (diagnosed 1990-92)

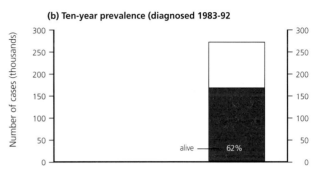

(b) Ten-year prevalence (diagnosed 1983-92

Figure 5.16 **Relative survival up to ten years, England and Wales, 1971-75 to 1991-93**

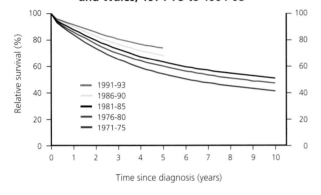

Figure 5.17 **Five-year relative survival by age, England and Wales, 1991-93**

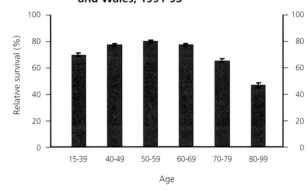

5

Figure 5.18 **One- and five-year relative survival by deprivation category, England and Wales, adults diagnosed 1986-90**

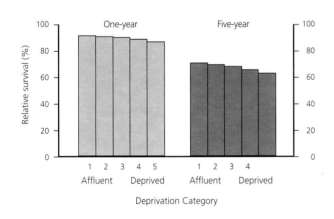

Figure 5.19 **International comparison of five-year relative survival, England and Wales, 1986-90**

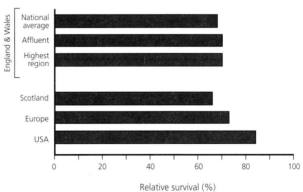

Chapter 6

CERVIX

There were over 370 thousand new cases of cervical cancer in the world in 1990, representing about 10% of all cancers in women, among whom it ranked third after breast and colorectal cancers.[1] About 80% of cases occur in developing countries, where cervical cancer is often the most common cancer. In contrast, cervical cancer accounts for only around 5% of cases in women in North America, Australia and western Europe. Large differences in incidence among different ethnic groups have been reported in some countries. The incidence of the disease had been falling in many Western countries since the second world war – but not, up to the late 1980s, in the UK.

The risk of developing cervical cancer is closely related to sexual behaviour. Very low rates of the disease occur in nuns. A link between cervical cancer and a sexually transmitted infection was first suggested because it was associated with women who had had many sexual partners, or whose husbands or partners had, and an early age at first intercourse.[2] A number of infections have been considered, including herpes simplex virus type 2, genital warts, syphilis and gonorrhoea. Evidence of the aetiological role of human papillomavirus has accumulated from both molecular and epidemiological studies but the mechanism is less clear, since many women with the infection do not go on to develop dysplasia, and of those who do, many do not progress to invasive cervical cancer.[3] Other risk factors include smoking, oral contraceptives and parity. Only the last of these appears to be a risk factor independently of human papillomavirus infection. Folate deficiency also requires further investigation for its possible explanation of the effects of parity – pregnancy is associated with the depletion of maternal folate stores. Other possible explanations are cervical trauma during childbirth and hormonal influences of pregnancy.

Changes in these risk factors over time will have affected the incidence of cervical cancer. Over the past 30 years both sexes have had a tendency to have first sexual relationships at earlier ages, and to have more sexual partners than in the past. In addition, younger cohorts of women have had greater exposure to cigarette smoking and oral contraceptives, although any increase in incidence as a result of this may have been attenuated by the reduction in parity.

Screening
Many tumours appear to arise after the initial development of a precursor premalignant condition – cervical intraepithelial neoplasia – and screening for these lesions has become a cornerstone of attempts to reduce mortality. The Pap smear was developed over 50 years ago, and screening began in Great Britain, in some of the Nordic countries, and in parts of North America, in the 1960s. Although the effectiveness of cervical screening has never been properly demonstrated in randomised, controlled trials, firm evidence comes from the Nordic countries, where the implementation of widely different policies was followed by sharply contrasting trends in incidence and mortality.[4] Although cervical screening in England started in 1964, for over twenty years it failed to achieve sufficient coverage of women or follow-up of all women with positive smears and was largely ineffective.[5,6] During the 1980s, several recommendations were made for improvement of the screening programme. Then in 1987, an Intercollegiate Working Party recommended that screens be repeated at three-year intervals on women aged 20 to 64 years; and in 1988 a national call and recall system was established. In 1996, 60% of district health authorities operated a three-year recall (the remainder five year). Financial incentives were first introduced with general practitioner contracts in 1990.

The target age group for screening is 20 to 64 years; coverage is defined from 1995 onwards as the percentage of women aged 25-64 who had had a smear test in the previous 5 years (for 1988 to 1994, the previous 5½ years). The number of smears taken rose by about 6% each year until the early 1980s, after which the increase was about 8% each year to the end of the decade; since then, about 4.5 million smears have been taken each year (Figure 6A). The coverage of the target age group in the screening programme rose from 42% in 1988 to 85% in 1994, a level subsequently maintained (Figure 6.5). Coverage increased in all age groups, but particularly for older women (55 to 64 years).[7]

Carcinoma in situ
The registrations of in situ cervical cancer are not true incidence because (i) the disease is asymptomatic and cases are only detected by screening, so any changes over time in the number of women screened in different age groups will affect the numbers of registrations; and (ii) as women are not all screened annually, registrations are a mix of cases diagnosed in women screened for the first time, and cumulative incidence since the previous screen for women who have been screened before.

From 1971 to 1983, the registration rate rose broadly in line with the increasing numbers of smears taken, from around 10 to 26 per 100,000 (2,200 to 6,200 cases) (Figure 6A). The apparent large increase in the rates in 1984 and 1985 is an artefact, resulting from the inclusion for the first time of registrations of cervical intraepithelial neoplasia grade III. Subsequently, registrations again increased broadly in line with the number of smears, to reach almost 80 per 100,000 (20,000 cases) in the mid-1990s.

Registration rates increased in all age groups up to the late 1980s. Since then the trends in registrations in women aged 20-24 and 25-29 continued upward until 1996, whereas in women aged from 30 to 49 the rates have fallen (Figure 6B); registrations for older groups were consistently low and declined with age.

Invasive cancer
There were 2,800 cases of cervical cancer in England and Wales in 1997, 2.5% of all cancer cases in women, in whom it ranked seventh. Most cases of cervical cancer (84%) were coded without anatomical detail. Papillary and squamous cell neoplasms accounted for two thirds of the tumours, and nearly 15% were adenocarcinomas; a further 15% were poorly specified carcinomas.

Table 6.1: Key statistics for cervical cancer on incidence, mortality and survival by region and country, latest year available

| | Incidence | | | | Mortality | | | | Relative survival (%) patients diagnosed in 1986-90 | |
	Number of cases	CR	ESR*	WSR*	Number of deaths	CR	ESR*	WSR*	One-year	Five-year
United Kingdom [a]	3,320	11.1	10.0	8.0	1,330	4.4	3.5	2.6	..	..
Great Britain [a]	3,230	11.1	10.0	8.0	1,300	4.4	3.6	2.6	..	..
England and Wales [b]	2,850	10.7	9.6	7.8	1,100	4.1	3.3	2.4	83	64
England [b]	2,670	10.7	9.6	7.7	1,030	4.1	3.3	2.4	83	64
Northern and Yorkshire [b]	460	14.1	12.5	10.2	160	5.1	4.1	3.1	84	67
Trent [b]	250	9.5	8.6	7.0	120	4.4	3.5	2.6	82	63
Anglia and Oxford [b]	220	8.2	7.3	5.8	80	3.1	2.5	1.8	84	67
North Thames [b]	290	8.3	7.5	5.9	100	2.8	2.4	1.8	84	61
South Thames [b]	290	8.4	7.3	5.9	150	4.2	3.5	2.6	85	62
South and West [b]	370	10.8	9.6	7.8	130	3.7	2.6	1.9	84	64
West Midlands [b]	310	11.5	10.6	8.7	110	3.9	3.0	2.2	84	68
North West [b]	480	14.3	13.2	10.8	190	5.6	4.4	3.1	81	61
Wales [b]	180	11.9	10.1	7.9	70	4.8	3.7	2.8	81	63
Scotland [a]	381	14.4	12.9	10.3	145	5.5	4.2	3.0	83	63
Northern Ireland [a]	93	10.9	11.3	9.1	33	3.8	3.5	2.5	..	..

* *Directly age-standardised rate*
[a] *1996 incidence; 1998 mortality.*
[b] *1997 incidence; 1999 mortality.*
Figures for England and Wales are provisional: incidence - 1995-1997, mortality - 1999

The disease was rare below the age of 20, but incidence rates rose rapidly to a first peak in the 35-39 and 40-44 age groups; there was then a slight decline in rates, followed by a rise to a second, slightly higher peak in women in their 70s (Figures 6.1 and 6.2). Cervical cancer was the most common cancer in women aged 25-29, accounting for 21% of all cancers in this age group (breast cancer was the second most common, accounting for 18%) (Appendix B3(b)).

From 1971 to the mid-1980s, age-standardised incidence remained between 14 and 16 per 100,000 (on average 4,000 cases a year) (Figure 6.5). Age-specific incidence rates (Figure 6.1) were influenced by strong cohort effects – see below. After 1990, age-standardised incidence fell in each successive year, reaching 9.6 per 100,000 in 1997, just over 40% lower than in the mid-1980s. Incidence fell in all regions between 1990 and 1997. Age-specific incidence fell, but in different ways in the various age groups (Figure 6.1). In 1997, the overall pattern was similar to that in 1990, but the incidence in every age group from 30-34 to 70-74 was substantially (and statistically significantly) lower – by on average 10 per 100,000 (about 150 cases).

From 1950 to 1987, age-specific mortality changed in different ways in the various age groups – influenced (as was incidence) by strong cohort effects (Figure 6.4). In the youngest women (25-34 years) mortality trebled from around 1 per 100,000 (30 deaths) in the mid-1960s to a plateau of around 3 per 100,000 (100 deaths) in the mid-1980s. Mortality in all the other age groups declined, but at different times. Over the same period, total mortality from cervical cancer fell steadily by just over 1.5% each year from 11.2 per 100,000 (2,700 deaths) in 1950 to 6.2 (1,800 deaths) in 1987 (Figure 6.7). This long term decline in cervical cancer mortality predates the introduction of screening, and may be attributable to improvements in hygiene and nutrition; the shifting of childbearing patterns towards smaller family sizes, delayed childbearing and increased mean age at first birth; and a decline in sexually transmitted diseases.[8] From 1990, the rate of decline in

mortality trebled and by 1999 it had fallen to 3.3 per 100,000 – less than a third of the level in 1950. The 1,100 deaths in 1999 represented 1.7% of cancer deaths in women, and 0.4% of all deaths in women.

The patterns of cervical cancer incidence and mortality by birth cohort lend strong support to the link with a sexually transmitted infection[2] (Figures 6.6 and 6.8). Based on data up to the late 1980s, before national call and recall were introduced in the screening programme, the risk for the cohort of women aged 25 to 34 in the mid-1980s – ie those born in the mid-1950s – mortality was three times that for women born in the mid-1930s. Cervical cancer mortality in each birth cohort historically increased with age up to 60 years. If the raised risk and pattern of mortality are assumed for women born in the mid-1950s, then by 1997 mortality would have increased to around 14 per 100,000 in women aged 35 to 44, and (with a similar projection based on a doubling of mortality for the cohort born in the mid-1940s) to around 19 per 100,000 in women aged 45 to 54. These rates are far higher than those actually observed – around 5 per 100,000 in both age groups (Figure 6.4). Applying the difference between the projected and actual mortality in 1997 to the numbers of women in each age group suggests that screening might have prevented 320 deaths in women aged 35-44 and 430 deaths in women aged 45-54. In addition, mortality in women aged 25-34 in 1997 was one third lower than in the peak in the mid-1980s, so a further 50 deaths may have been prevented in this age group, giving a total of 800 prevented deaths in women aged under 55. The results from formal age-cohort models have broadly confirmed this estimate.[9] Recent trends in both incidence and mortality in Scotland have been similar to those in England and Wales.[10]

In 1990-93 there was a strong positive gradient in the incidence of cervical cancer by Carstairs deprivation category, with the rate in the most deprived group over three times that in the most affluent (Figure 6.9). The relationship, however, was not linear: incidence increased more rapidly with deprivation in categories 12 to 20. The

Figure 6A **Number of smears and age-standardised incidence of carcinoma in situ of the cervix (CIS)**

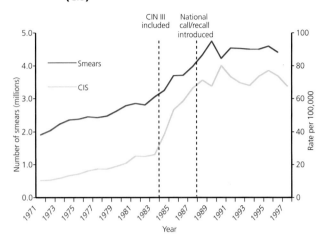

Figure 6B **Age-specific incidence of carcinoma in situ of the cervix, England and Wales, 1971-1997**

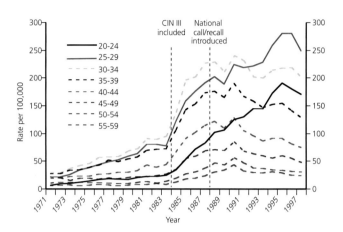

relationship between mortality from cervical cancer and deprivation was closely similar to that for incidence (Figure 6.10).

Incidence was higher than the average for England and Wales in Northern and Yorkshire, the West Midlands and the North West, and below average in the south east of England (Figure 6.11). As with deprivation, the geographical pattern in mortality was broadly similar to that in incidence (Figure 6.12).

Around the world, incidence rates are highest in sub-Saharan Africa, Central and South America (particularly in north eastern Brazil) and South East Asia (Figure 6.13). Incidence in England and Wales was similar to that in other parts of Europe, the USA, Japan and Australia. The international pattern in mortality was similar to that for incidence, except that the rate is relatively higher in China – implying poorer survival than elsewhere – and mortality in England and Wales is now among the lowest in the world (Figure 6.14).

Prevalence
As survival is relatively good compared with many other cancers, the proportion of women who had been diagnosed with cervical cancer in 1990-92 and were still alive at the beginning of 1993 was high – about 75% (8,400); of those diagnosed in 1983-92, 60% (24,500) were still alive (Figure 6.15).

Survival
Survival from cervical cancer for patients diagnosed in 1991-93 was moderately good – one-year survival was 83% and five-year survival 64% – but showed little improvement on survival for women diagnosed in 1986-90 (Figure 6.16). From the early 1970s to the late 1980s there was an improvement in one-year survival of 7%, and in five-year survival of 9%. Survival declined steeply with age. In 1991-93, five-year survival was around 80% in the youngest age group (under 40) but only 21% in the elderly (80 and over) (Figure 6.17).

There was little variation in survival across England and Wales. In England and Wales as a whole, and in most of the regions, there

was a significant gradient in survival with deprivation. In 1986-90, survival was around 4% points higher in the affluent than in the most deprived at both one year and five years after diagnosis (Figure 6.18).

There were no significant improvements in the treatment of cervical cancer in the period covered by the survival analyses. Early invasive cervical cancer is treated equally well by radical hysterectomy or radiotherapy, although the pattern of side effects varies. For more advanced disease, radiotherapy is generally the best treatment; for some patients, surgery is carried out after radiotherapy.

Survival is much poorer for later stage disease: five-year relative survival rates are less than 10% for stage IV, compared with around 80% for stage I. Studies based in the regional cancer registries indicate that since the introduction of national call and recall in the cervical screening programme in 1988, there has been ave shift in the stage distribution towards earlier stage.[11] This would haaffected survival only in part of the period 1986-90.

Survival from cervical cancer varied widely across Europe in the late 1980s: five-year survival ranged from 51% in Poland and 57% in Estonia to almost 70% in Sweden and Switzerland. The lower survival in the former eastern bloc countries may – as in the UK – be related to socio-economic factors. Both one- and five-year survival in England and Wales (and in Scotland) were – in contrast to most other cancers – close to the European average, although below the rates in the USA (Figure 6.19).

49

Figure 6.1 **Age-specific incidence, England and Wales, 1997**

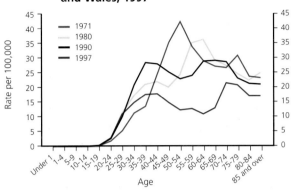

Figure 6.2 **Frequency distribution of new cases by age group, England and Wales, 1997**

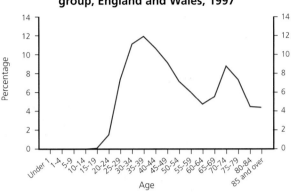

Figure 6.3 **Age-specific incidence, England and Wales, 1971-1997**

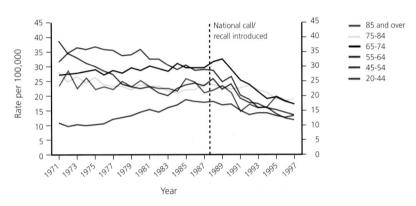

Figure 6.4 **Age-specific mortality, England and Wales, 1950-1999**

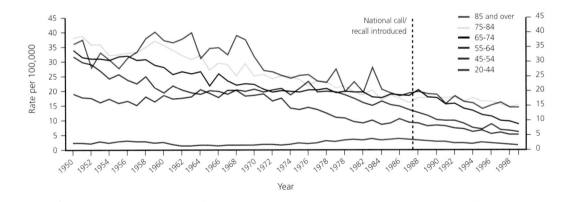

Figure 6.5 **Age-standardised incidence of invasive cervical cancer (total) and of adenocarcinomas of the cervix, England and Wales, 1971-1997**

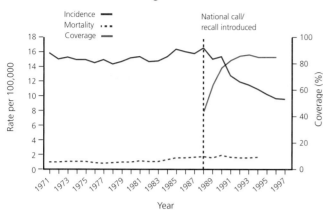

Figure 6.6 **Cohort incidence ratio, England and Wales**

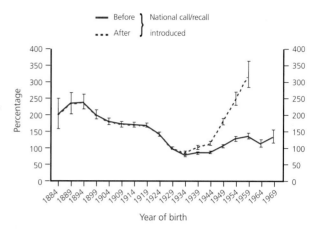

Figure 6.7 **Age-standardised mortality, England and Wales, 1950-1999**

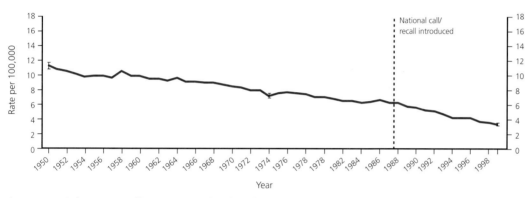

Figure 6.8 **Cohort mortality ratio, England and Wales**

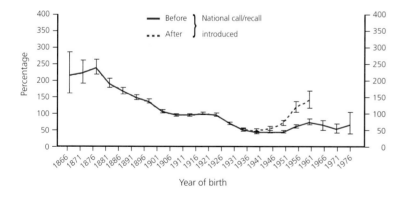

Figure 6.9 **Age-standardised incidence by deprivation category, England and Wales, 1990-93**

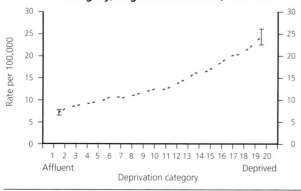

Figure 6.10 **Age-standardised mortality by deprivation category, England and Wales, 1990-93**

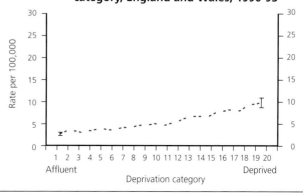

Figure 6.11 **Comparative incidence ratio (CIR) by health region (England and Wales = 100), 1997**

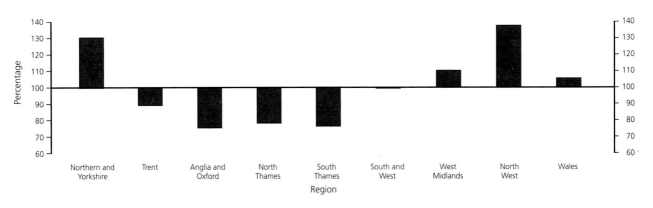

Figure 6.12 **Comparative mortality ratio (CMR) by health region (England and Wales = 100), 1999**

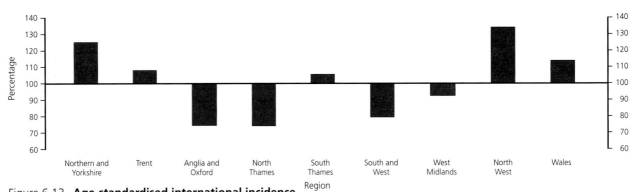

Figure 6.13 **Age-standardised international incidence**

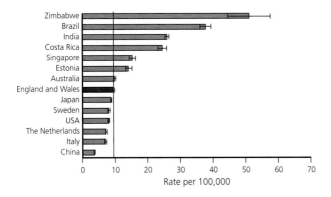

Figure 6.14 **Age-standardised international mortality**

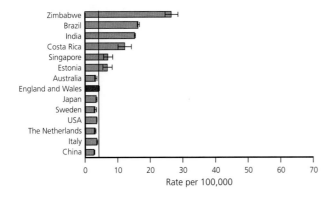

Figure 6.15 **Number and percentage of cases diagnosed in 1990-92 and 1983-92 and alive on 1st January 1993, England and Wales**

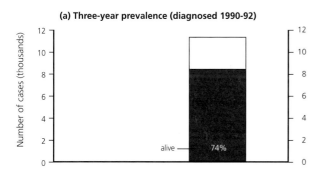

(a) Three-year prevalence (diagnosed 1990-92)

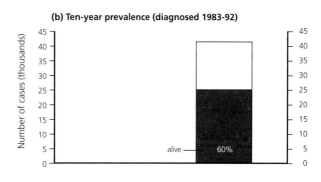

(b) Ten-year prevalence (diagnosed 1983-92)

Figure 6.16 **Relative survival up to ten years, England and Wales, 1971-75 to 1991-93**

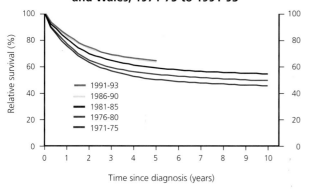

Figure 6.17 **Five-year relative survival by age, England and Wales, patients diagnosed 1991-93**

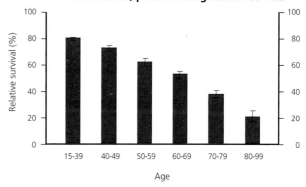

Figure 6.18 **One- and five-year relative survival by deprivation category, England and Wales, adults diagnosed 1986-90**

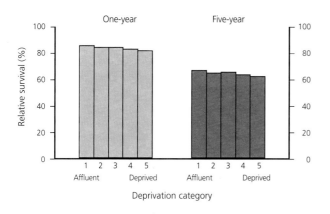

Figure 6.19 **International comparison of five-year relative survival, England and Wales, 1986-90**

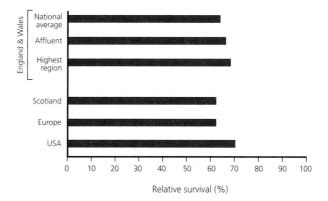

COLORECTAL

Colorectal cancer (cancers of the colon and rectum combined) accounts for 13% of all cancers in England and Wales, and is the second most common cancer in women and the third most common in men. It is a disease mainly associated with developed countries and diets high in fats and animal proteins and low in fruit, vegetables and fibre. Some 15-25% of colorectal cancers may be attributable to high fat intake, while high intake of vegetable or fibre may reduce the risk to between one half and one third. Physical activity has been associated with reduced risk of developing colon cancer, possibly through increasing peristaltic activity and reducing exposure of the colonic mucosa to faecal carcinogens. There is some evidence that colon cancer in women may be related to sex hormones or to reproductive history. This link is consistent with the increased risk of developing colon cancer observed in women previously diagnosed with breast, ovarian or uterine cancers. The risk of developing colorectal cancer is also raised for patients with one or more adenomatous polyps, as occur in familial adenomatous polyposis (FAP) and other hereditary conditions. Screening trials have been conducted for colorectal cancer which suggest that screening may lead to a reduction in mortality from the disease.[1,2]

Incidence and mortality

There were around 28,900 cases of colorectal cancer diagnosed in England and Wales in 1997, of which around two thirds (18,300) were in the colon and around one third (just under 10,600) in the rectum. Colorectal cancers occur more frequently in males than females with a male:female ratio of age-standardised rates of 1.5:1. Incidence increases continuously with age in both sexes for both colon and rectal cancers (Figures 7.1 and 7.2).

The subsite of the colon with the greatest proportion of cases was the sigmoid colon (29%). It is often used interchangeably with the term 'rectosigmoid junction' (part of the rectum), in which there were 19% of rectal cancer cases, and is the reason why these two cancers are often considered together. The other major subsites in the colon were the caecum (18%) and the ascending colon (6%), with almost one-third of cases of unspecified subsite – tumours of the large bowel without further specification are assigned to the colon. Three-quarters of rectal cases were in the rectum, and 3% were in the anus. Two-thirds of colorectal cancer cases are adenocarcinomas, with 5% of cases occurring as mucinous neoplasms and 30% coded as epithelial or unspecified neoplasms.

There was a gradual, long-term rise in the incidence of colorectal cancer throughout the 1970s and 1980s. The overall incidence rates of colon and rectal cancers in males were similar in the early 1970s but diverged over the subsequent decades (Figure 7.5). Overall the age-standardised rates for colon cancer in males rose by 30% between 1971 and 1997, compared with a rise of 6% in rectal cancer; incidence peaked in the early 1990s before falling slightly. In females, the incidence of colon cancer remained at around twice the level of rectal cancer throughout the period and showed a much more gradual rise than in males.

The incidence of colorectal cancer increased during the 1970s and 1980s in men and women aged 45 and over; in the 1990s, however, rates in men aged 75 and over, and in women aged 85 and over, appeared to stabilise and then fall slightly (Figure 7.3).

There was an increase in colorectal cancer incidence for male birth cohorts born from the early 1890s to the late 1930s, followed by a decline (Figure 7.6). In females, there was only a slight rise in the older birth cohorts before rates fell in parallel with those in males.

In contrast to the steady rise in incidence, there was a long-term decline in mortality from colorectal cancer (Figure 7.7). Mortality from colon cancer fell steadily in both sexes between 1950 and the late 1960s. In women, the rates remained level through the 1970s before falling steadily through the 1980s and 1990s, to just under 12 per 100,000 in 1999 (a fall of 53% compared with the rate in 1950). In men, mortality from colon cancer fell from 24 per 100,000 in 1950 to around 18 per 100,000 and then remained at this level until the late 1990s when the rate fell slightly to reach 16 per 100,000 in 1999. Mortality from rectal cancer fell by 56% in both sexes between 1950 and 1999, to reach around 9 and 5 per 100,000 in males and females, respectively.

The overall mortality trend for colorectal cancer in males masks different trends in the age-specific rates, with long term but relatively small rises for the oldest age group (men aged 85 or over), compared with the rate in 1950, and larger falls in mortality for other age groups (Figure 7.4). For both colon and rectal cancers, the falls in mortality for a number of age groups were greatest between 1950 and 1965, after which the decline was more gradual.

In contrast to the trends in incidence by birth cohort, there was a long term decline in mortality in successive cohorts born from the 1880s onwards in both sexes, although the trends may have levelled off for the more recent cohorts (those born in the 1960s onwards) (Figure 7.8).

There was no variation in the incidence of colon cancer by deprivation group for either males or females (Figure 7.9). For rectal cancer, there was a clear (but not large) deprivation gradient for males, with incidence rates around 25% higher in the more deprived groups than in the affluent groups; there was, however, no variation in incidence by deprivation group in females.

The patterns of mortality by deprivation group closely resemble those for incidence (Figure 7.10). There was no variation with deprivation for colon cancer in either sex, or in rectal cancer for females. There was, however, a noticeable deprivation gradient for rectal cancer in males with the rates in the more deprived groups around 50% higher than in the most affluent.

There was no consistent variation in incidence or mortality across the regions in England and Wales (Figures 7.11 and 7.12).

Table 7.1: Key statistics for colorectal cancer on incidence, mortality and survival by sex, region and country, latest year available

MALES	Incidence				Mortality				Relative survival (%) patients diagnosed in 1986-90	
	Number of cases	CR	ESR*	WSR*	Number of deaths	CR	ESR*	WSR*	One-year	Five-year
United Kingdom [a]	17,500	60.6	55.0	36.3	8,910	30.6	26.9	17.2	..	..
Great Britain [a]	17,000	60.6	54.7	36.1	8,610	30.4	26.7	17.1	..	..
England and Wales [b]	14,900	58.1	51.6	34.1	7,450	28.9	25.1	16.0	61	39
England [b]	13,900	57.3	51.2	33.8	6,940	28.5	24.9	15.9	61	39
Northern and Yorkshire [b]	1,920	61.7	54.7	36.0	950	30.5	26.7	17.1	59	37
Trent [b]	1,390	54.7	47.2	31.1	800	31.6	26.7	17.1	61	38
Anglia and Oxford [b]	1,580	58.8	54.3	35.8	690	25.5	23.1	14.7	59	38
North Thames [b]	1,480	43.0	42.7	28.3	770	22.0	21.8	13.9	66	41
South Thames [b]	1,650	49.2	43.4	28.5	900	26.7	23.0	14.7	65	41
South and West [b]	2,200	67.5	54.1	35.8	1,050	32.2	24.7	15.8	65	43
West Midlands [b]	1,680	63.9	57.4	37.9	830	31.4	27.5	17.5	60	39
North West [b]	1,990	61.6	56.3	37.2	940	29.1	26.2	16.5	56	35
Wales [b]	1,020	71.4	58.4	38.3	510	35.7	28.8	18.4	53	37
Scotland [a]	1,846	74.3	70.3	46.2	857	34.5	31.9	20.5	61	40
Northern Ireland [a]	498	61.0	67.0	45.0	302	36.5	38.8	24.3	..	..

FEMALES										
United Kingdom [a]	16,800	56.0	37.5	24.8	8,200	27.2	16.5	10.5	..	..
Great Britain [a]	16,300	56.1	37.3	24.6	8,000	27.4	16.5	10.5	..	..
England and Wales [b]	14,000	52.6	34.8	23.0	7,090	26.6	16.1	10.3	59	39
England [b]	13,000	52.1	34.6	22.8	6,630	26.4	16.1	10.3	60	39
Northern and Yorkshire [b]	1,730	53.5	35.4	23.4	780	24.2	14.4	9.2	56	37
Trent [b]	1,200	46.3	30.2	20.0	690	26.7	16.3	10.5	60	39
Anglia and Oxford [b]	1,420	52.0	37.0	24.5	700	25.5	16.3	10.4	57	38
North Thames [b]	1,510	42.7	31.5	20.7	740	20.6	14.0	8.9	66	42
South Thames [b]	1,810	51.4	32.9	21.6	1,020	28.8	17.0	10.9	65	42
South and West [b]	2,140	63.3	37.9	25.1	1,040	30.7	16.0	10.2	64	45
West Midlands [b]	1,380	51.2	34.3	22.5	720	26.6	17.1	11.1	59	40
North West [b]	1,870	55.6	37.1	24.5	940	28.1	17.4	11.2	54	35
Wales [b]	900	60.5	37.3	24.7	460	30.7	16.9	10.7	54	37
Scotland [a]	1,765	66.8	45.9	30.4	803	30.5	19.2	12.3	59	40
Northern Ireland [a]	467	54.8	45.2	30.3	199	23.1	17.2	11.1	..	..

* Directly age-standardised rate   [a] 1996 incidence; 1998 mortality.   [b] 1997 incidence; 1999 mortality.

Figures for England and Wales are provisional: incidence - 1995-1997, mortality - 1999

The incidence of colorectal cancer is three to four times greater in developed countries than in developing countries. This is illustrated in Figure 7.13 with higher rates occurring in Australia, USA, Japan, Singapore and European countries, and lower rates in India, Costa Rica, Zimbabwe and Brazil. Comparison of the international variation in colorectal cancer mortality with that for incidence suggests that survival for both males and females is better in Australia and the USA than in England and Wales (Figure 7.14).

Prevalence
Over 55% of males and of females diagnosed with colon or rectal cancers in England and Wales in the three years 1990-92 were still alive at the beginning of 1993, as were 36-37% of those diagnosed in the 10 years 1983-92 (Figure 7.15).

Survival
Survival from both colon and rectal cancer is only moderate with relative survival five years after diagnosis of around 40% for patients diagnosed in 1991-93. One-year survival is 5-9% lower for colon cancer at 61-64%, compared with 69% for rectal cancer. There is only a small decline in survival beyond five years after diagnosis, with a relative survival of 33% after ten years for colon cancer and 31% for rectal cancer for adults diagnosed in the early 1980s (five-year relative survival for these patients was 36% for both cancers). Survival from colon and rectal cancers improved

substantially over the 1970s and 1980s, by 10-15% for men and women diagnosed in 1986-90 compared with those diagnosed in 1971-75 (Figure 7.16 – colon cancer). The greatest increase occurred between the late 1970s and early 1980s.

Survival from colon and rectal cancers has a smaller variation with age than many other cancers. Five-year relative survival from colon cancer was between 40 and 50% for all adults aged from 40 to 79 years and around 30-35% for those aged 80-99 (Figure 7.17). There was little difference in survival between men and women in all age groups.

There was a significant deprivation gradient in survival from colon and rectal cancers for patients diagnosed in 1986-90: survival was 6-7% points lower for the most deprived at one year after diagnosis and 4-5% points lower at five years after diagnosis (Figure 7.18 – colon cancer). Among the regions, survival was highest in the South and West for both sexes and both cancers, and lowest in the North West. Five-year relative survival from colon and rectal cancer in England and Wales for patients diagnosed in the late 1980s was similar to that for Scotland but lower than the average for Europe in the Eurocare II study (8% points lower for colon cancer and 4-6% points lower for rectal cancer). Five-year survival from colon and rectal cancers in the USA was over 60% for both sexes, compared with less than 40% in England and Wales (Figure 7.19 – colon cancer).

Figure 7.1 **Age-specific incidence, England and Wales, 1997**

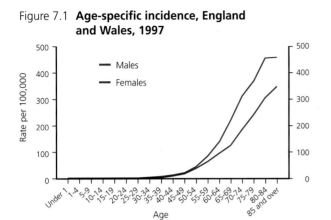

Figure 7.2 **Frequency distribution of new cases by age group, England and Wales, 1997**

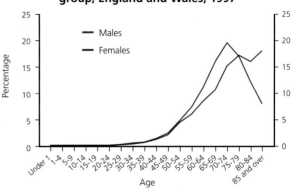

Figure 7.3 **Age-specific incidence, England and Wales, 1971-1997**

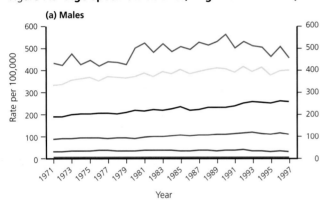

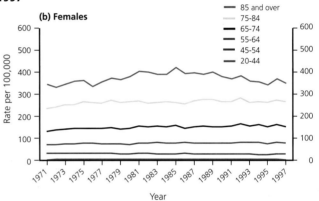

Figure 7.4 **Age-specific mortality, England and Wales, 1950-1999**

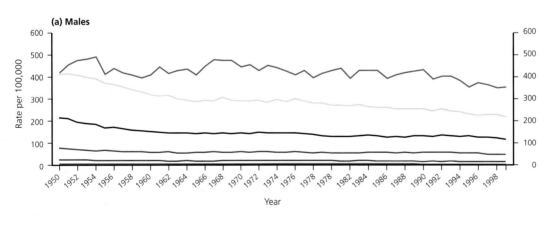

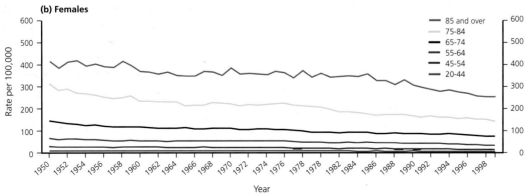

Note: charts are for colorectal cancer unless otherwise indicated.

Figure 7.5 **Age-standardised incidence by site, England and Wales, 1971-1997**

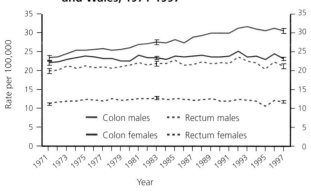

Figure 7.6 **Cohort incidence ratio by sex, England and Wales**

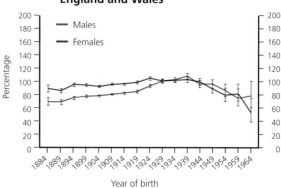

Figure 7.7 **Age-standardised mortality by site, England and Wales, 1950-1999**

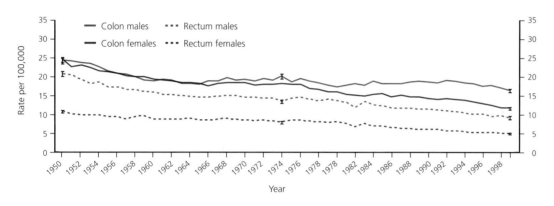

Figure 7.8 **Cohort mortality ratio by sex, England and Wales**

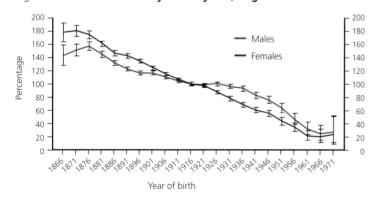

Figure 7.9 **Age-standardised incidence by deprivation category and site, England and Wales, 1992-93**

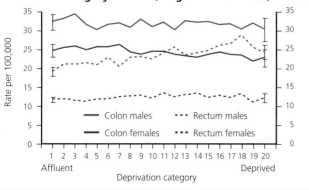

Figure 7.10 **Age-standardised mortality by deprivation category and site, England and Wales, 1992-93**

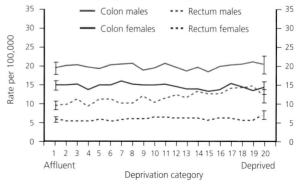

Figure 7.11 **Comparative incidence ratio (CIR) by health region and sex (England and Wales = 100), 1997**

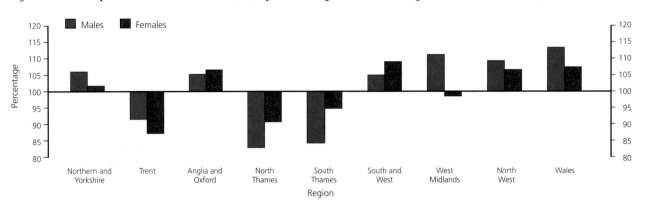

Figure 7.12 **Comparative mortality ratio (CMR) by health region and sex (England and Wales = 100), 1999**

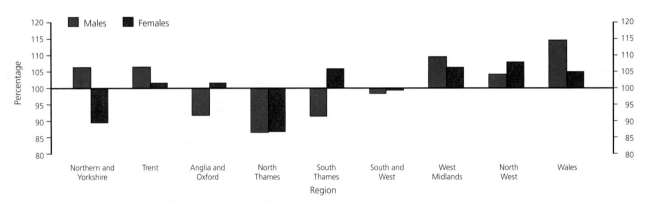

Figure 7.13 **Age-standardised international incidence**

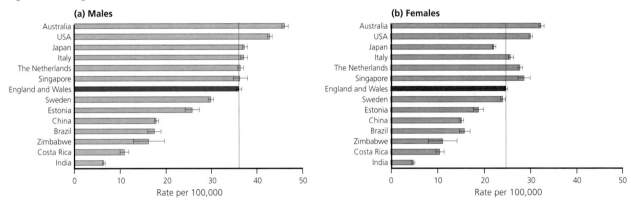

Figure 7.14 **Age-standardised international mortality**

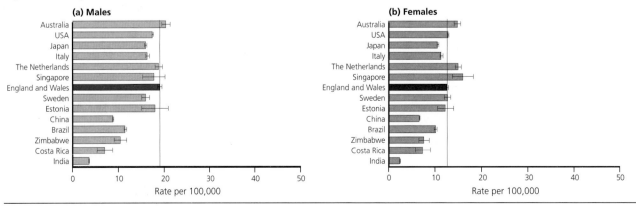

Figure 7.15 **Number and percentage of cases diagnosed in 1990-92 and 1983-92 and alive on 1st January 1993, England and Wales**

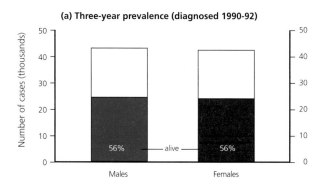

Figure 7.16 **Relative survival from colon cancer up to ten years, men, England and Wales, 1971-75 to 1991-93**

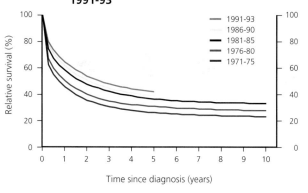

Figure 7.17 **Five-year relative survival from colon cancer by age, patients diagnosed 1991-93**

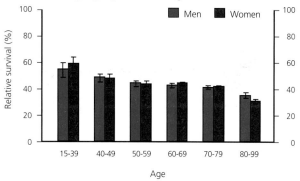

Figure 7.18 **One- and five-year relative survival from colon cancer by deprivation category, England and Wales, adults diagnosed 1986-90**

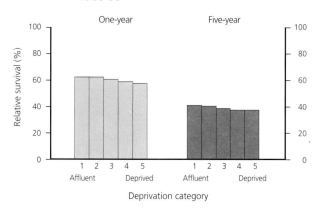

Figure 7.19 **International comparison of five-year relative survival from colon cancer, England and Wales, 1986-90**

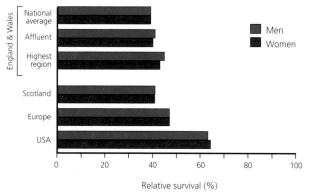

Chapter 8

KIDNEY

Cancers of the kidney are the eighth most common cancer in males and fourteenth most common in females in England and Wales. In adults, over three-quarters of malignant tumours of the kidney are renal cell carcinomas, the remainder being mainly transitional cell carcinomas of the renal pelvis. Other renal tract cancers (mainly of the ureter) are rare. Wilms' tumour (or nephroblastoma) is the most common childhood cancer of the kidney.

It is estimated that there are 100,000 new cases each year of kidney cancer worldwide, accounting for 1.5% of all cancers. Generally the incidence in North America and northern Europe is about two to three times the rate in Africa, Asia and South America. Rural areas seem to have lower rates of kidney cancer than urban areas. There are consistently more cases in males than in females, with sex ratios generally between 1.5 and 2.5 to 1. The incidence of kidney cancer has increased in both sexes around the world. This is largely due to improved detection and registration.[1]

Renal cell carcinomas have a high tendency to spread to other parts of the body, generally via the bloodstream rather than the lymph nodes, and give rise to metastases especially in the lungs. Treatment is generally surgery – complete removal of the kidney and some surrounding tissue – with elective pre- or post-operative radiotherapy. Renal cell carcinoma is among the few types of cancer that have been shown to respond to biological response-modifying agents, the most important of these being interleukin-2 and interferon for this particular cancer.

Kidney cancer has been associated with an affluent lifestyle (perhaps related to diet), cigarette smoking, phenacetin, paracetamol, obesity, hypertension and renal injury. In women (but not men) a high level of coffee drinking has been identified as a risk factor. Kidney cancer has been associated with cadmium exposure and with coke oven workers and insulation workers where asbestos is implicated.[2]

Incidence and mortality

In 1997, there were 4,800 new cases of kidney cancer diagnosed in England and Wales – just over 3,000 in males and 1,800 in females (Table 8.1). Almost 90% of tumours in both sexes occurred in the kidney (excluding the pelvis), with around a further 5% in the renal pelvis and 5% in the ureter. Over half the malignancies were coded specifically as adenocarcinomas – most were clear cell adenocarcinomas; transitional cell carcinomas accounted for 14% of the cases. About 30% of tumours had non-specific morphology. Around 2% of cases were nephroblastomas in children.

The disease is rare before the age of 40 (Figure 8.1). Over the age of 35 the rate for men is consistently higher than that for women in all age groups. In both sexes the incidence rate generally increases with age. Over 45% of cases occurred in those aged 65-79 (Figure 8.2).

From 1971 to 1997, the age-standardised rate in both males and females almost doubled, reaching 11 per 100,000 in males and 5

per 100,000 in females (Figure 8.5). Incidence in all age groups increased during 1971 to 1997, with a more than two-fold rise in the rates for men and women aged over 60 (Figure 8.3). The male:female ratio of the age-standardised rates was 2.1:1. The lifetime risk of being diagnosed with cancer of the kidney was 1.1% for males (1 in 90) compared with 0.6% in females (1 in 160). Incidence rose for each successive cohort born from the mid-1860s to the mid-1950s (Figure 8.6).

Mortality from kidney cancer has increased steadily since 1950 in both sexes, in line with the rise in incidence (seen since 1971) (Figure 8.7). By the early 1990s age-standardised mortality had risen by over 85% since 1950 in both males and females; rates then levelled off at around 6 per 100,000 in males, and around 3 per 100,000 in females. The increase in mortality was greatest for the elderly, particularly in males (Figure 8.4). The age-specific mortality for men aged 85 and over showed a three-fold rise over the period, while mortality in elderly women doubled. Mortality by birth cohort for both males and females showed similar rises to those in incidence for people born up to the mid-1920s; mortality then levelled off for those born from the 1930s to the early 1950s and appears to have fallen in subsequent cohorts (Figure 8.8).

For cases diagnosed in 1990-93, there was – despite the apparent association of kidney cancer with smoking – no gradient in incidence by Carstairs deprivation index for males; incidence in females in the more deprived groups was slightly higher than in the more affluent (Figure 8.9). The patterns in the variation in mortality by deprivation category closely resemble those for incidence (Figure 8.10).

There was a suggestion of a north-south gradient in both the incidence of and mortality from kidney cancers in England and Wales (Figures 8.11 and 8.12).

There is wide variation in the incidence of cancer of the kidney around the world. Higher incidence occurs in Europe and North America with age-standardised rates of 11-12 per 100,000 (world standard) in males and 5-6 per 100,000 in females. Low incidence occurs in India, China, Africa and South America with rates of 1-3 and 1-2 per 100,000 in males and females, respectively (Figure 8.13). The incidence in England and Wales was in the middle range: 7 and 3 per 100,000, in males and females, respectively. The patterns in international mortality were similar to those in incidence, with higher rates in Europe, Australia and the USA (Figure 8.14). Comparison of the incidence and mortality suggests that survival is better in Italy and the USA than in England and Wales.

Prevalence

Over 50% of both males and females diagnosed in England and Wales in the three years 1990-92 were still alive at the beginning of 1993, as were 38% of those diagnosed in the ten years 1983-92 (Figure 8.15).

Table 8.1: Key statistics for kidney cancer on incidence, mortality and survival by sex, region and country, latest year available

MALES	Incidence				Mortality				Relative survival (%) patients diagnosed in 1986-90	
	Number of cases	Rate per 100,000			Number of deaths	Rate per 100,000			One-year	Five-year
		CR	ESR*	WSR*		CR	ESR*	WSR*		
United Kingdom [a]	3,450	11.9	11.2	7.9	1,880	6.5	5.9	4.0	..	..
Great Britain [a]	3,350	11.9	11.2	7.8	1,840	6.5	5.9	4.0	..	..
England and Wales [b]	3,040	11.8	11.0	7.7	1,680	6.5	5.9	4.0	58	40
England [b]	2,830	11.7	10.9	8.1	1,570	6.4	5.9	4.0	58	40
Northern and Yorkshire [b]	380	12.3	11.4	8.1	210	6.8	6.2	4.2	55	38
Trent [b]	290	11.5	10.2	7.1	170	6.8	6.0	4.0	57	41
Anglia and Oxford [b]	290	10.7	10.3	7.2	170	6.4	6.0	4.0	54	37
North Thames [b]	320	9.4	9.6	6.7	180	5.1	5.3	3.6	64	45
South Thames [b]	400	11.8	11.0	7.9	210	6.3	5.7	3.8	63	40
South and West [b]	410	12.6	10.8	7.7	210	6.3	5.1	3.5	59	41
West Midlands [b]	320	12.0	11.1	7.8	180	6.9	6.2	4.2	55	38
North West [b]	420	13.0	12.5	8.8	230	7.1	6.5	4.4	56	38
Wales [b]	210	14.8	12.7	8.8	120	8.1	6.9	4.8	53	38
Scotland [a]	373	15.0	14.4	9.9	190	7.6	7.1	4.9	54	38
Northern Ireland [a]	95	11.6	13.0	9.2	42	5.1	5.6	3.6	..	..

FEMALES	Number of cases	CR	ESR*	WSR*	Number of deaths	CR	ESR*	WSR*	One-year	Five-year
United Kingdom [a]	2,160	7.2	5.5	4.0	1,230	4.1	2.8	1.9	..	..
Great Britain [a]	2,100	7.2	5.5	4.0	1,190	4.1	2.8	1.9	..	..
England and Wales [b]	1,800	6.8	5.2	3.8	1,030	3.9	2.6	1.7	53	36
England [b]	1,670	6.7	5.2	3.8	960	3.8	2.6	1.7	54	36
Northern and Yorkshire [b]	250	7.8	5.8	4.1	130	4.1	2.6	1.7	52	36
Trent [b]	170	6.6	5.1	3.9	110	4.4	2.9	2.0	54	36
Anglia and Oxford [b]	170	6.3	5.3	3.8	90	3.4	2.4	1.6	53	33
North Thames [b]	180	5.0	4.3	3.2	120	3.5	2.6	1.7	60	39
South Thames [b]	200	5.7	4.4	3.3	120	3.5	2.3	1.5	57	36
South and West [b]	270	7.9	5.9	4.4	130	3.9	2.5	1.7	53	38
West Midlands [b]	170	6.3	4.8	3.5	100	3.8	2.6	1.8	53	39
North West [b]	270	7.9	6.1	4.4	140	4.2	2.9	2.0	50	32
Wales [b]	130	8.4	5.7	3.9	70	4.5	2.7	1.8	45	31
Scotland [a]	261	9.9	7.7	5.6	119	4.5	3.0	2.1	50	34
Northern Ireland [a]	60	7.0	6.2	4.4	37	4.3	3.3	2.1	..	..

* Directly age-standardised rate

[a] 1996 incidence; 1998 mortality.

[b] 1997 incidence; 1999 mortality.

Figures for England and Wales are provisional: incidence - 1995-1997, mortality - 1999

Survival

Survival from cancer of the kidney in England and Wales is moderate, with five-year relative survival of 41% in men and 37% in women (for patients diagnosed during 1991-93). Ten-year relative survival for patients diagnosed in the early 1980s was around 30% in both sexes. Survival improved from the 1970s with a rise every five years in both one- and five-year relative survival of around 3% points (Figure 8.16). There was very little improvement in survival for patients diagnosed in 1991-93 compared with those diagnosed in 1986-90.

Survival decreases with age: relative survival decreased from 50% or higher in patients aged under 50 at diagnosis to around 35% or lower in those aged 70 or over (Figure 8.17). Cancer of the kidney is unusual as it is one of the few cancers to show a survival advantage for males. Survival is higher in the more affluent groups than in the more deprived groups, with an advantage of almost 7% points at one year after diagnosis and of 4% points at five years after diagnosis for patients diagnosed in 1986-90 (Figure 8.18).

Survival from kidney cancer in England and Wales was similar to that for Scotland but 9-16% points below the average for Europe and 20% points lower than in the USA (Figure 8.19).

Figure 8.1 **Age-specific incidence, England and Wales, 1997**

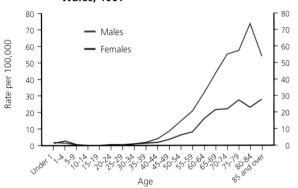

Figure 8.2 **Frequency distribution of new cases by age group, England and Wales, 1997**

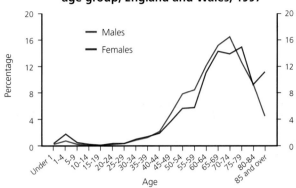

Figure 8.3 **Age-specific incidence, England and Wales, 1971-1997**

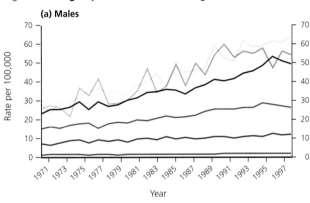
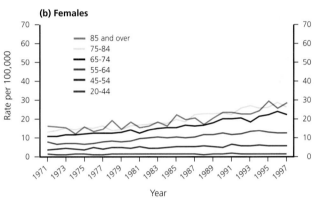

Figure 8.4 **Age-specific mortality, England and Wales, 1950-1999**

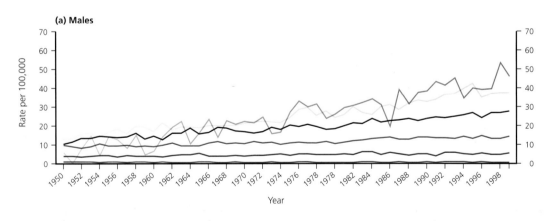
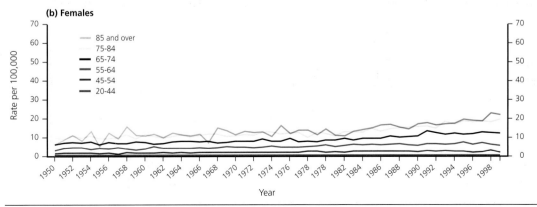

Figure 8.5 **Age-standardised incidence, England and Wales, 1971-1997**

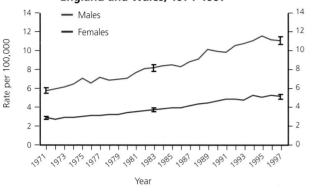

Figure 8.6 **Cohort incidence ratio by sex, England and Wales**

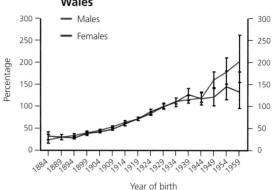

Figure 8.7 **Age-standardised mortality by sex, England and Wales, 1950-1999**

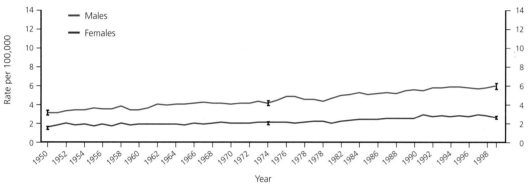

Figure 8.8 **Cohort mortality ratio by sex, England and Wales**

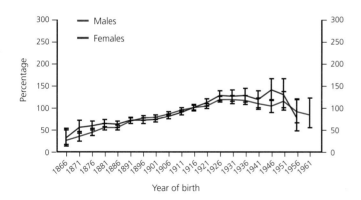

Figure 8.9 **Age-standardised incidence by deprivation category, England and Wales, 1990-93**

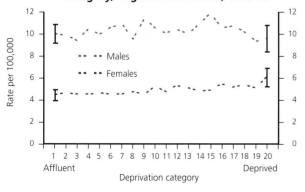

Figure 8.10 **Age-standardised mortality by deprivation category, England and Wales, 1990-93**

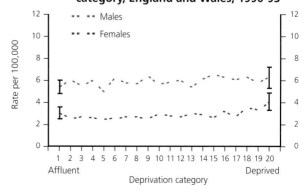

Figure 8.11 **Comparative incidence ratio (CIR) by health region and sex (England and Wales = 100), 1997**

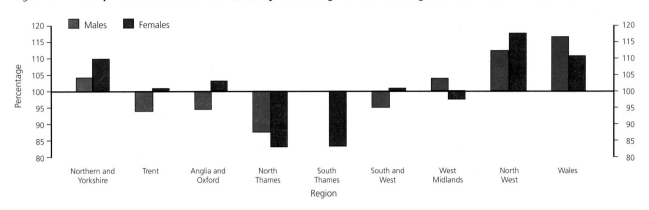

Figure 8.12 **Comparative mortality ratio (CMR) by health region and sex (England and Wales = 100), 1999**

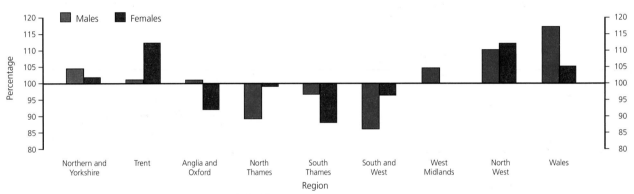

Figure 8.13 **Age-standardised international incidence**

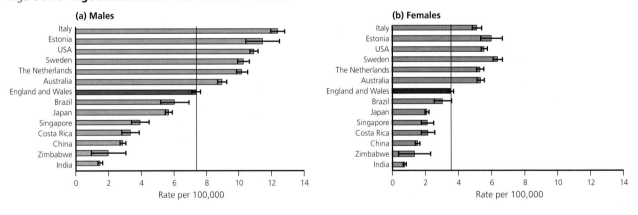

Figure 8.14 **Age-standardised international mortality**

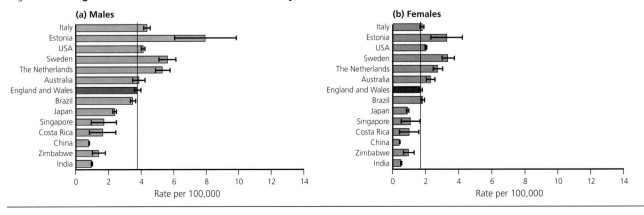

Figure 8.15 **Number and percentage of cases diagnosed in 1990-92 and 1983-92 and alive on 1st January 1993, England and Wales**

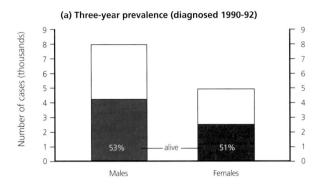

(a) Three-year prevalence (diagnosed 1990-92)

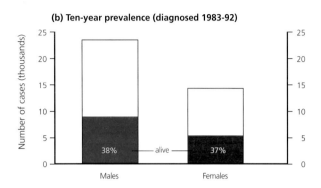

(b) Ten-year prevalence (diagnosed 1983-92)

Figure 8.16 **Relative survival up to ten years, men, England and Wales, 1971-75 to 1991-93**

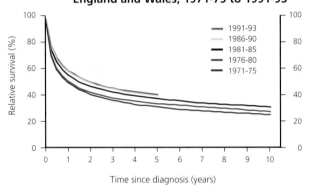

Figure 8.17 **Five-year relative survival by age, England and Wales, patients diagnosed 1991-93**

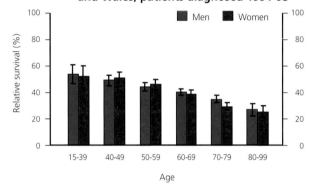

Figure 8.18 **One- and five-year relative survival by deprivation category, England and Wales, adults diagnosed 1986-90**

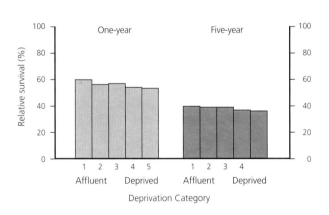

Figure 8.19 **International comparison of five-year relative survival, England and Wales, 1986-90**

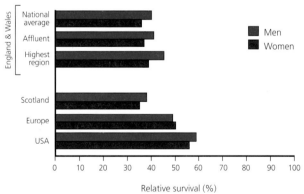

Chapter 9

LARYNX

Laryngeal cancer is the fourteenth most common cancer in males in England and Wales accounting for 1.4% of all cancers. In females the numbers of newly diagnosed cases and deaths each year are small (under 400 cases and 160 deaths) and are not discussed in detail in this chapter.

The larynx is classified into four main anatomical regions: glottis, supraglottis, subglottis and laryngeal cartilages. In males, nearly half of cases arise in the glottis. These have the most favourable prognosis as early symptoms, such as hoarseness, occur and medical advice is sought earlier. In addition, the probability of metastatic disease is low until there is a spread of the disease into more lymphatic regions. Around 15% of cases occur in the supraglottic region where the prognosis is not so favourable. There are usually no early symptoms and the area is richly supplied with lymphatics, so metastases often occur.

The main risk factors for cancer of the larynx are tobacco and alcohol, which have a multiplicative effect. Some recent studies have suggested that risk may be affected by various components of diet.[1] Smoking is the dominant risk factor for glottal cancers, while alcohol is the major risk factor for cancers of the supraglottis. Survival is poor for cancers of the supraglottis and hypopharynx but fair for cancers of the glottis. Differences in the diagnosis and registration practices of laryngeal cancers combined with variations in the proportion of cancers of the glottis make international comparisons difficult to interpret.[2]

Incidence and mortality

In males there were over 1,500 new cases diagnosed in England and Wales in 1997 compared with 1,300 in 1971, an overall rise of 18% (Table 9.1). In 1994, over 47% of cases occurred in the glottis, 16% in the supraglottis, and just under 2% in the subglottis and laryngeal cartilages; the sub-site was, however, unspecified in almost 40% of cases. Just under 80% of cases were papillary and squamous cell neoplasms, with non-specific epithelial neoplasms accounting for 11% and non-specific neoplasms 9%.

The incidence of laryngeal cancer is rare in males aged under 40 but rates rise quickly after this age, reaching a peak in the 75-79 age group (27 per 100,000 in 1997) (Figures 9.1 and 9.2). Although the total number of cases has risen by almost 20%, the incidence rates of laryngeal cancer in all age groups in the mid-1990s were largely unchanged from those in the early 1970s (Figure 9.3). Overall, age-standardised incidence has fluctuated around 6 per 100,000 over the same period, rising slightly in the late 1980s and subsequently declining (Figure 9.5). While there was little apparent variation in the age-specific incidence, the risk of developing the disease increased for men born in the 1880s and 1890s (Figure 9.6). This rise, which also occurs for the incidence of lung cancer, is related to smoking patterns. Unlike lung cancer, however, the risk then remains fairly stable across subsequent cohorts, with possibly a small decline in those born in the late 1940s onwards.

There were just under 600 deaths in males from laryngeal cancer in England and Wales in 1999 (Table 9.1). As with incidence, mortality from laryngeal cancer is rare in the under 40s but rises steeply after this (Figure 9.4). Age-standardised mortality fell markedly in the 1950s and 1960s, from around 4.5 per 100,000 to 2.5 per 100,000 in the early 1970s (Figure 9.7). Subsequently the rate of decline slowed, but the age-standardised rate in 1999 was only just over 2 per 100,000, well under half the rate in the early 1950s. Mortality by birth cohort decreased up to those born at the beginning of the 20th century (Figure 9.8); the subsequent pattern of risk was similar to that for incidence.

There is a steep – but non-linear – gradient in the incidence of laryngeal cancer by deprivation group, with incidence in the more affluent groups (1-5) of around 4 per 100,000 (Figure 9.9). The subsequent increase in incidence across the deprivation groups steepens from groups 13 and 14 to reach around 12 per 100,000 in the most deprived groups. The variation in mortality by deprivation category closely resembles the pattern for incidence (Figure 9.10). The most affluent groups have the lowest rates: those in groups 1 to 6 are around 1.5 per 100,000; mortality in the most deprived groups is around 4 times that in the most affluent groups. The steeper gradient with deprivation in mortality than in incidence suggests that survival is worse in the more deprived groups (see below).

As for other smoking-related cancers, there was a north-south divide in the incidence of laryngeal cancer (Figure 9.11). Incidence was substantially higher in the Northern and Yorkshire and North West regions with a rate around 30% above the average for England and Wales. Incidence in Anglia and Oxford, South Thames, Trent, South West and West Midlands was below average. The regional variation in mortality was generally similar to that for incidence (Figure 9.12).

The incidence of and mortality from laryngeal cancer in England and Wales was low compared with several other countries in Europe, Australia and the USA: in the selected countries shown in Figure 9.13 only Costa Rica, Japan, China and Sweden had lower age-standardised rates. The incidence and mortality rates for Sweden were around half those for England and Wales, while the rates for Italy, with the highest levels of laryngeal cancer, were around three times greater.

Prevalence

Over 70% of males diagnosed in 1990-92 were still alive on 1st January 1993, as were 55% of males diagnosed in the ten years previously (1983-92) (Figure 9.15).

Survival

Survival from cancer of the larynx in England and Wales is moderately good, with one-year relative survival of 83% and after five years of 64% for patients diagnosed in 1991-93. Five-year relative survival from laryngeal cancer increased by on average 2-3% every five years from the early 1970s to the late 1980s, but

Table 9.1: Key statistics for laryngeal cancer on incidence, mortality and survival by region and country, latest year available

MALES	Incidence				Mortality				Relative survival (%) patients diagnosed in 1986-90	
	Number of cases	Rate per 100,000			Number of deaths	Rate per 100,000			One-year	Five-year
		CR	ESR*	WSR*		CR	ESR*	WSR*		
United Kingdom [a]	1,870	6.5	6.2	4.3	700	2.4	2.3	1.5	..	..
Great Britain [a]	1,830	6.5	6.2	4.2	680	2.4	2.3	1.5	..	..
England and Wales [b]	1,530	5.9	5.6	3.9	580	2.2	2.0	1.3	84	64
England [b]	1,430	5.9	5.6	5.1	540	2.2	2.0	1.3	84	64
Northern and Yorkshire [b]	240	7.7	7.3	5.1	80	2.5	2.4	1.6	80	59
Trent [b]	140	5.4	5.0	3.5	60	2.3	1.9	1.2	87	68
Anglia and Oxford [b]	120	4.5	4.3	2.9	40	1.6	1.5	0.9	86	70
North Thames [b]	190	5.4	5.7	4.0	80	2.2	2.2	1.5	87	65
South Thames [b]	170	5.1	4.8	3.3	60	1.9	1.7	1.1	88	67
South and West [b]	190	5.8	5.0	3.5	70	2.1	1.6	1.0	82	61
West Midlands [b]	140	5.2	4.9	3.4	70	2.6	2.3	1.5	84	63
North West [b]	250	7.6	7.2	5.0	90	2.7	2.5	1.7	81	63
Wales [b]	100	6.8	6.0	4.2	40	2.4	2.0	1.3	83	63
Scotland [a]	259	10.4	10.3	7.2	102	4.1	4.0	2.8	86	61
Northern Ireland [a]	47	5.8	6.7	4.7	24	2.9	3.0	1.9	..	..

* Directly age-standardised rate
[a] 1996 incidence; 1998 mortality.
[b] 1997 incidence; 1999 mortality.
Figures for England and Wales are provisional: incidence - 1995-1997, mortality - 1999

there was no improvement compared with survival for men diagnosed in 1986-90 (Figure 9.16).

Five-year relative survival decreased only slowly with increasing age at diagnosis, from 75% in the youngest age group (15-39) to just over 50% in the oldest (80-99) (Figure 9.17).

Five-year relative survival for patients diagnosed in 1986-90 was 10% points higher in the most affluent groups than in the most deprived (Figure 9.18).

Laryngeal cancer is one of the few cancers for which five-year survival for men in England and Wales was slightly higher than the average for Europe (Figure 9.19). While the national average was 4% points lower than the USA, the survival for the most affluent was 3% higher.

Figure 9.1 **Age-specific incidence, England and Wales, 1997**

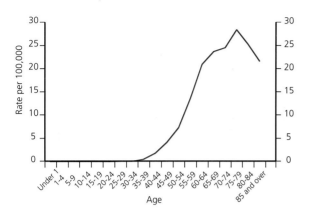

Figure 9.2 **Frequency distribution of new cases by age group, England and Wales, 1997**

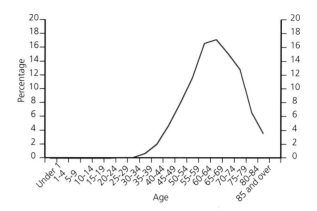

Figure 9.3 **Age-specific incidence, England and Wales, 1971-1997**

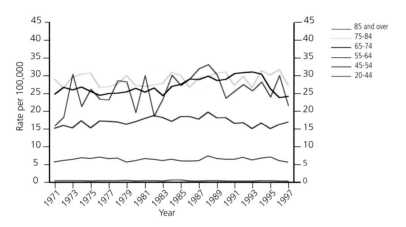

Figure 9.4 **Age-specific mortality, England and Wales, 1950-1999**

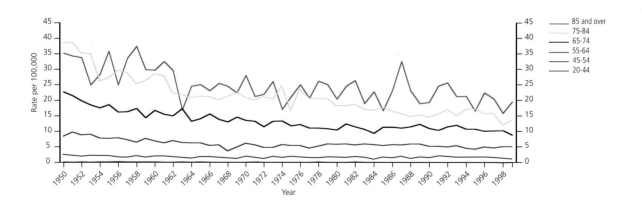

Figure 9.5 **Age-standardised incidence, England and Wales, 1971-1997**

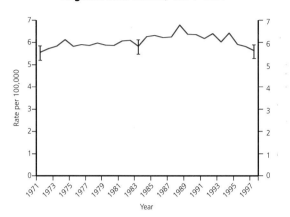

Figure 9.6 **Cohort incidence ratio, England and Wales**

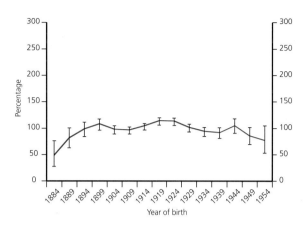

Figure 9.7 **Age-standardised mortality, England and Wales, 1950-1999**

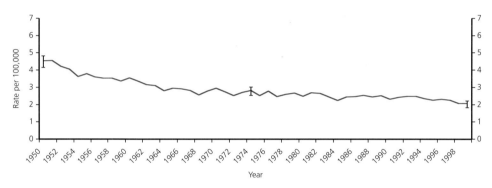

Figure 9.8 **Cohort mortality ratio, England and Wales**

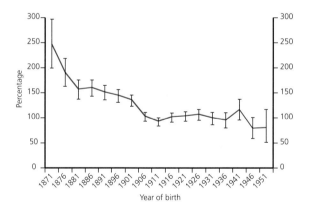

Figure 9.9 **Age-standardised incidence by deprivation category, England and Wales, 1988-93**

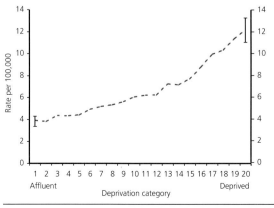

Figure 9.10 **Age-standardised mortality by deprivation category, England and Wales, 1988-93**

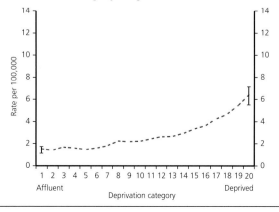

Figure 9.11 **Comparative incidence ratio (CIR) by health region (England and Wales = 100), 1997**

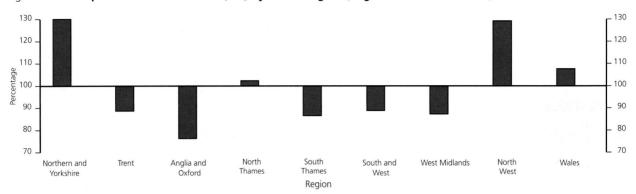

Figure 9.12 **Comparative mortality ratio (CMR) by health region for males (England and Wales = 100), 1999**

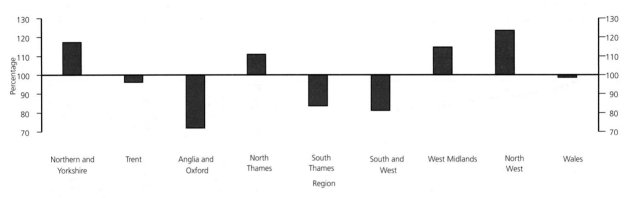

Figure 9.13 **Age-standardised international incidence**

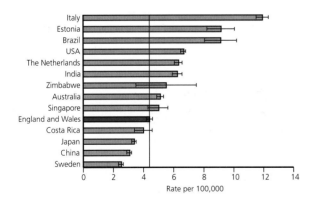

Figure 9.14 **Age-standardised international mortality**

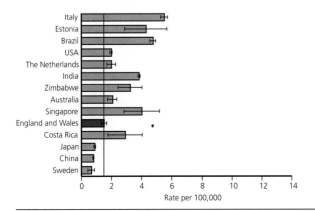

Figure 9.15 **Number and percentage of cases diagnosed in 1990-92 and 1983-92 and alive on 1st January 1993, England and Wales**

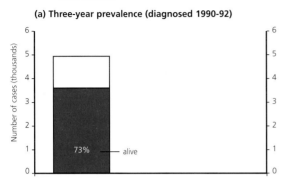

(a) Three-year prevalence (diagnosed 1990-92)

(b) Ten-year prevalence (diagnosed 1983-92)

Figure 9.16 **Relative survival up to ten years, men, England and Wales, 1971-75 to 1991-93**

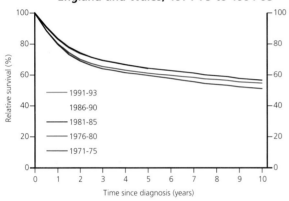

Figure 9.17 **Five-year relative survival by age, England and Wales, men diagnosed 1991-93**

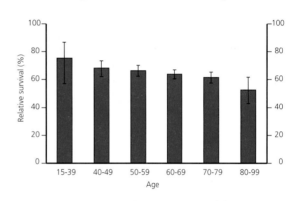

Figure 9.18 **One- and five-year relative survival by deprivation category, England and Wales, men diagnosed 1986-90**

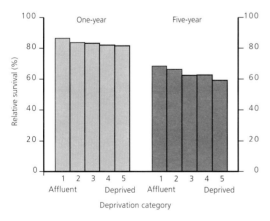

Figure 9.19 **International comparison of five-year relative survival, England and Wales, men diagnosed 1986-90**

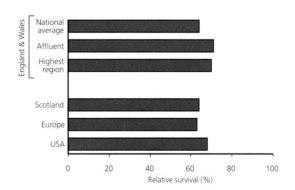

Chapter 10

LEUKAEMIA

The leukaemias are a diverse group of malignancies deriving from the precursor cells of blood and tissue leukocytes. They are conventionally distinguished both by the cell type of origin (lymphocytes, myelocytes, monocytes) and by their clinical and pathological behaviour (acute, subacute, chronic). As a group, the leukaemias represent about 3% of cancer incidence world-wide, with relatively little geographical variation. In general, males are more affected than females, although the sex ratio is greatest for chronic lymphoid leukaemia (CLL) (about 2 to 1) and much closer to 1 to 1 for other types. The age distribution of incidence shows a bimodal curve with a first peak in childhood, a trough at ages 15-29, and then a slow rise. The patterns are, however, quite different for the different subtypes: CLL is a disease of the elderly; the risk for myeloid leukaemia increases with age; and acute lymphoid leukaemia (ALL) is infrequent (relative to the other types) in adults.

Not a great deal is known about the aetiology of leukaemia. Although radiation is well recognised as a cause of leukaemia, it accounts for only a small proportion of cases and there appears to be no effect of radiation in increasing the risk of CLL. Other agents that are accepted as causal risk factors are occupational exposure to benzene and certain drugs used for cancer chemotherapy. It has long been thought that leukaemia may be induced by viruses (especially given the associations in animals).[1] It seems likely that control of the known causes would prevent only a small proportion of cases. More recently, an association has been observed with smoking for both myeloid and lymphoid types of acute leukaemia.[2] There is no evidence that screening for leukaemia using routine blood tests is effective.

Incidence and mortality
Leukaemias account for 2.5% of all cancers in England and Wales and are the ninth most common form of cancer in males and the twelfth in females. There were around 5,300 new cases diagnosed in England and Wales in 1997, just over 2,900 in males and 2,300 in females (Table 10.1). In 1994, almost one-third of cases were acute myeloid leukaemia (AML) and around a further third were chronic lymphoid leukaemia (CLL). Chronic myeloid (CML) and acute lymphoid leukaemias each accounted for just over one-tenth of cases. Monocytic leukaemia and other specified leukaemia each accounted for only 1% of cases; and 9% were other or unspecified forms of leukaemia.

The age-standardised incidence rates in 1997 were 10.5 and 6.5 per 100,000 for males and females, respectively (Figure 10.5). The rates have fluctuated around these levels since the mid 1970s. While there were only small changes in incidence amongst the younger age-groups, the rates in those aged 75 or over increased by 40% in males and 65% in females (Figure10.3). Incidence rose for cohorts born from the mid 1880s to around the early 1900s in both sexes and subsequently remained fairly stable (Figure 10.6).

There were over 3,600 deaths from leukaemia in England and Wales in 1999, 1,900 in males and 1,700 in females (Table 10.1).

Mortality from leukaemia increased throughout the 1950s, but the rates in both sexes levelled off between the late 1960s and mid 1970s and subsequently declined (Figure 10.7). Mortality has increased sharply in the elderly: over 10-fold in women aged 85 or over (from around 5 to 53 deaths per 100,000) and around seven-fold in men of the same age (from 13 to 87 per 100,000 – the rate peaked at almost 100 per 100,000 in 1991) (Figure 10.4). The rates in the younger age-groups showed little change over the period. As with incidence, the cohort mortality showed increases in both sexes for those born during the 19th century – but then fell gradually in successive cohorts (Figure 10.8).

There was only a slight variation in incidence by deprivation category, with marginally higher rates in the more affluent groups compared with the most deprived groups (Figure 10.9). There was no significant difference in mortality across the deprivation categories for either males or females (Figure 10.10). There were no consistent geographical patterns in either the incidence of or mortality from leukaemia in England and Wales (Figures 10.11 and 10.12).

Internationally, incidence is greater in the developed countries: the USA, Australia, European countries including Italy, Sweden and England and Wales, and particularly low in India. The patterns are similar for males and females (Figure 10.13). There was less variation in mortality from leukaemia than for incidence: mortality in England and Wales was amongst the lowest (Figure 10.14).

Prevalence
About 50% of both males and females diagnosed in the three years 1990-92 were still alive at the beginning of 1993, as were just over 30% of those diagnosed in the 10 years 1983-92 (Figure 10.15).

Survival
One-year relative survival from all leukaemia was around 55% and five-year survival around 30% for patients diagnosed in 1991-93 (Figure 10.16). These rates represent a substantial rise on the survival of patients diagnosed in the early 1970s. There was an increase of almost 5% points every five years in one-year survival and of 4% in five-year survival for patients diagnosed between 1971 and 1993. The largest improvement in five-year survival occurred during the late 1970s and early 1980s.

There is wide variation in the levels of survival for the different forms of leukaemia. For the two most common types, AML had a five-year survival for patients diagnosed in 1986-90 of less than 10%, compared with nearer to 50% for CLL. Five-year survival was 20-30% for ALL and CML.

Survival from the leukaemias as a group does not vary much by age except for the very elderly. Five-year survival was around 40% for adults aged 15-59 but only around 15% at ages 80-99 (Figure 10.17).

72

Table 10.1: Key statistics for leukaemia on incidence, mortality and survival by sex, region and country, latest year available

MALES	Incidence				Mortality				Relative survival (%) patients diagnosed in 1986-90	
	Number of cases	CR	ESR*	WSR*	Number of deaths	CR	ESR*	WSR*	One-year	Five-year
United Kingdom [a]	3,380	11.7	10.8	8.3	2,140	7.3	6.5	4.6	..	..
Great Britain [a]	3,310	11.8	10.9	8.4	2,080	7.3	6.5	4.6	..	..
England and Wales [b]	2,920	11.4	10.5	8.1	1,930	7.5	6.6	4.6	53	27
England [b]	2,690	11.1	10.3	7.9	1,810	7.4	6.6	4.6	53	27
Northern and Yorkshire [b]	370	11.8	11.0	8.6	200	6.4	5.8	4.1	51	27
Trent [b]	280	11.0	10.1	7.5	190	7.6	6.7	4.7	51	25
Anglia and Oxford [b]	270	10.1	9.6	7.3	190	7.2	6.5	4.6	48	20
North Thames [b]	370	10.8	10.7	8.2	230	6.6	6.4	4.5	61	29
South Thames [b]	380	11.2	10.2	7.9	240	7.1	6.4	4.5	62	33
South and West [b]	440	13.4	11.5	9.0	280	8.6	6.8	4.6	57	31
West Midlands [b]	290	10.8	10.0	7.7	220	8.1	7.3	5.0	54	26
North West [b]	310	9.6	9.0	6.9	260	7.9	7.2	5.0	45	21
Wales [b]	230	15.8	13.8	10.4	120	8.0	6.6	4.5	48	27
Scotland [a]	317	12.8	12.4	9.4	176	7.1	6.6	4.7	52	31
Northern Ireland [a]	65	8.0	8.3	6.2	58	7.0	7.6	5.5	..	..
FEMALES										
United Kingdom [a]	2,850	9.5	7.0	5.6	1,800	6.0	4.0	2.8	..	..
Great Britain [a]	2,790	9.6	7.0	5.6	1,770	6.0	4.0	2.8	..	..
England and Wales [b]	2,340	8.8	6.5	5.2	1,720	6.4	4.4	3.2	49	27
England [b]	2,170	8.7	6.4	5.1	1,610	6.4	4.4	3.2	49	27
Northern and Yorkshire [b]	280	8.6	6.4	5.0	240	7.6	5.1	3.8	48	27
Trent [b]	200	7.7	5.5	4.3	150	5.8	3.9	2.9	43	24
Anglia and Oxford [b]	230	8.4	6.2	4.5	180	6.5	4.7	3.4	42	20
North Thames [b]	280	7.9	6.3	5.1	180	5.0	3.7	2.6	55	32
South Thames [b]	350	10.1	7.2	5.6	240	6.7	4.2	3.0	58	33
South and West [b]	320	9.5	6.7	5.6	250	7.2	4.4	3.2	58	34
West Midlands [b]	240	9.0	6.9	5.8	160	5.8	4.3	3.1	47	20
North West [b]	270	8.1	6.1	4.9	220	6.6	4.7	3.5	42	20
Wales [b]	160	11.0	8.1	6.4	100	6.8	4.4	3.2	48	29
Scotland [a]	260	9.8	7.2	5.5	153	5.8	3.9	2.7	47	24
Northern Ireland [a]	60	7.0	6.3	5.5	35	4.1	3.3	2.5	..	..

* Directly age-standardised rate
[a] 1996 incidence; 1998 mortality.
[b] 1997 incidence; 1999 mortality.
Figures for England and Wales are provisional: incidence - 1995-1997, mortality - 1999

There is little variation in survival across the regions of England and Wales, although five-year survival is below the average for England and Wales in the North West region for both males and females and above average in the South and West and South Thames regions. There was a survival disadvantage for the most deprived group compared with the most affluent diagnosed in 1986-90, with deprivation gaps of 8% and 5% for one- and five-year survival, respectively (Figure 10.18). Survival in England and Wales was similar to that in Scotland but 7% points lower than the average for Europe, and 11-14% points lower than in the USA (Figure 10.19).

Figure 10.1 **Age-specific incidence, England and Wales, 1997**

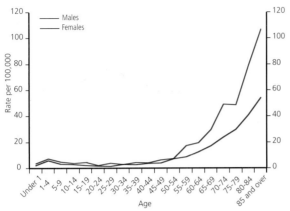

Figure 10.2 **Frequency distribution of new cases by age group, England and Wales, 1997**

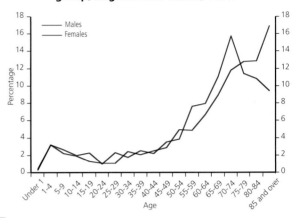

Figure 10.3 **Age-specific incidence, England and Wales, 1971-1997**

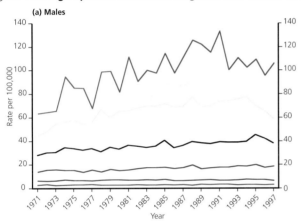

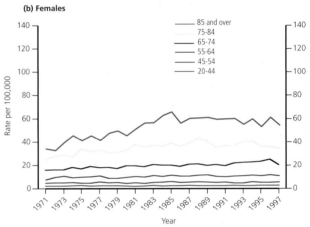

Fig 10.4 **Age-specific mortality, England and Wales, 1950-1999**

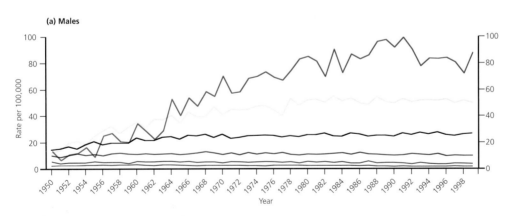

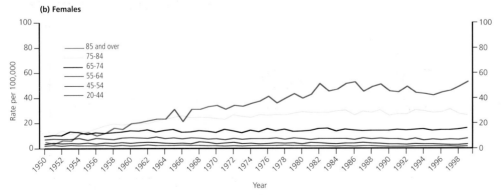

Figure 10.5 **Age-standardised incidence, England and Wales, 1971-1997**

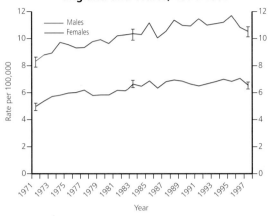

Figure 10.6 **Cohort incidence ratio by sex, England and Wales**

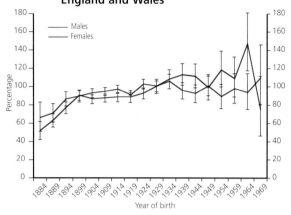

Figure 10.7 **Age-standardised mortality by sex, England and Wales, 1950-1999**

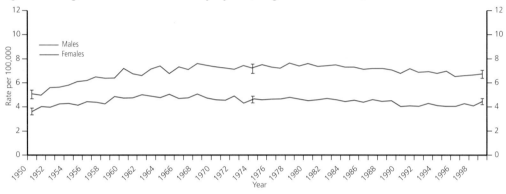

Figure 10.8 **Cohort mortality ratio by sex, England and Wales**

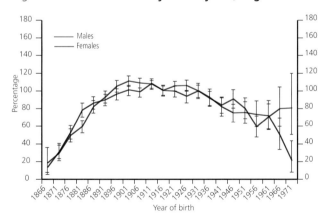

Figure 10.9 **Age-standardised incidence by deprivation category, England and Wales, 1992-93**

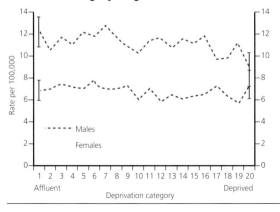

Figure 10.10 **Age-standardised mortality by deprivation category, England and Wales, 1992-93**

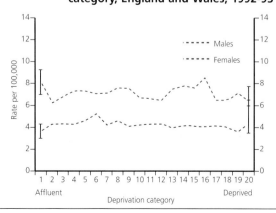

Figure 10.11 **Comparative incidence ratio (CIR) by health region and sex (England and Wales = 100), 1997**

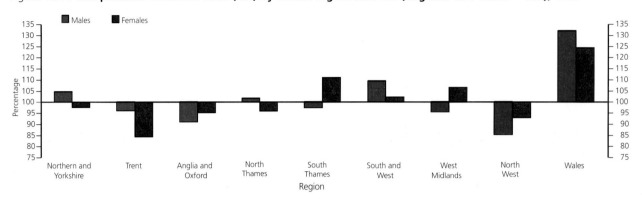

Figure 10.12 **Comparative mortality ratio (CMR) by health region and sex (England and Wales = 100), 1999**

Figure 10.13 **Age-standardised international incidence**

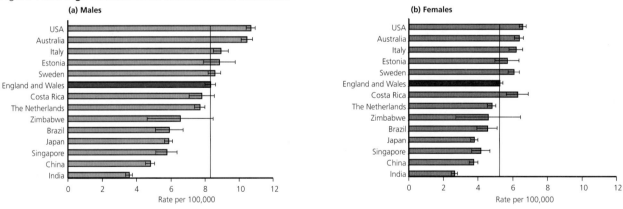

Figure 10.14 **Age-standardised international mortality**

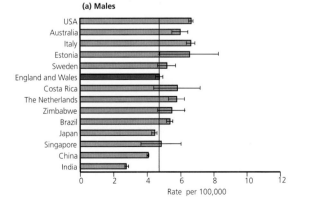
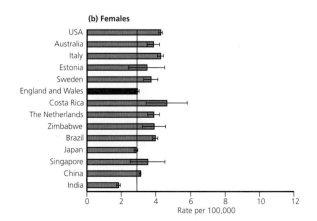

Figure 10.15 **Number and percentage of cases diagnosed in 1990-92 and 1983-92 and alive on 1st January 1993, England and Wales**

(a) Three-year prevalence (diagnosed 1990-92)

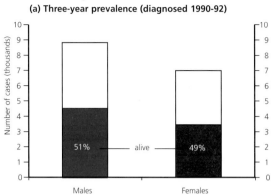

(b) Ten-year prevalence (diagnosed 1983-92)

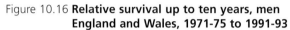

Figure 10.16 **Relative survival up to ten years, men England and Wales, 1971-75 to 1991-93**

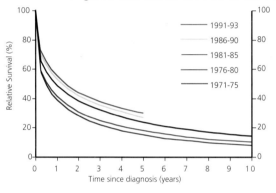

Figure 10.17 **Five-year relative survival by age, England and Wales, patients diagnosed 1991-93**

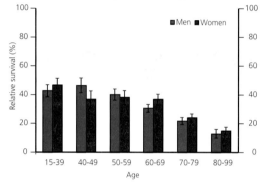

Figure 10.18 **One- and five-year relative survival by deprivation category, England and Wales, adults diagnosed 1986-90**

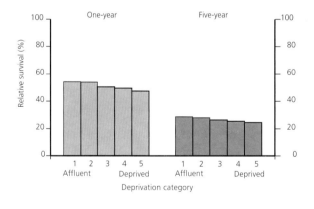

Figure 10.19 **International comparison of five-year relative survival, England and Wales, 1986-90**

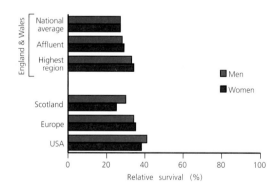

Chapter 11

LIP, MOUTH AND PHARYNX

The separate sites covered in this chapter are shown in Table 11A below, together with the numbers of cases and age-standardised rates in 1994 in England and Wales. Trends with age, deprivation and time are described for these cancers as a single group except where stated otherwise.

Cancers of the lip, mouth and pharynx combined are the eleventh most common malignancy in males and the sixteenth most common in females in England and Wales. There are around 3,800 new cases diagnosed each year. A major risk factor for oral cancer is tobacco smoking, affecting cancers of the tongue, mouth, and pharynx. Pipe smoking has also been linked to lip cancer, and chewing tobacco to gum and cheek tumours. Alcohol ingestion also features as a risk factor; high alcohol mouthwashes have been implicated as well. Long exposure to sunlight has been linked with lip cancer. The consumption of fresh fruits and vegetables appears to be protective. A very large proportion of this group of cancers could be prevented.

Incidence and mortality

Pharyngeal cancers comprised 34% of all lip, mouth and pharyngeal tumours, with mouth the second most common site (26%), followed by cancers of the tongue (22%). A further 10% occurred in the salivary glands, mainly the parotid – the largest of the glands. The remainder occurred in the lip (9%). In 1994, nearly 80% of male, and over 70% of female, lip, mouth and pharyngeal tumours were papillary and squamous cell neoplasms; epithelial neoplasms accounted for 12% of cases for both males and females, and 12% were adenocarcinomas.

There were almost 2,400 new cases of lip, mouth and pharyngeal cancer diagnosed in males in England and Wales in 1997 (Table 11.1) compared with 1,900 in 1971, an increase of 24%. Over the same period the number of cases in females rose by 21% to over 1,400. Cancers of the lip, mouth and pharynx are rare in the under 40s (Figures 11.1 and 11.2). In both sexes the incidence rates increased with age. The incidence of this group of cancers in males is around twice that in females in all age groups. Incidence in elderly men has fallen dramatically, from 100 per 100,000 to 38 (a drop of over 60%) (Figure 11.3). There have, however, been increases in rates in the 55-64 age group of over 40% in men and 25% in women. For both males and females, the overall age-standardised incidence rate of lip, mouth and pharyngeal cancers declined very slightly during the 1970s to reach around 8 and 4 per 100,000, respectively (Figure 11.5). The rates remained stable during the 1980s before rising gradually in the 1990s. The incidence of lip, mouth and pharyngeal cancer decreased rapidly in cohorts born in the late 19th century, before stabilising in those born from the turn of the century up to the early 1930s; for subsequent cohorts the risk appears to be higher in males (Figure 11.6).

Mortality from lip, mouth and pharyngeal cancer has decreased since 1950 in both sexes (Figure 11.7). The fall was particularly marked in males during the 1950s and 1960s; mortality in the 1990s was under half that in the early 1950s. In females, the rate fell more gradually, and by about 40% over the whole period. Although rates decreased in men aged 65 and over, there were increases in mortality from the early 1970s in those aged 45-64. In females, however, mortality declined in all age groups (Figure 11.4). From the early 1970s, mortality rose in younger females (under 65) while the long term decline continued in those aged over 65. The patterns in mortality by birth cohort (Figure 11.8) were similar to those in incidence.

In 1989-93, there were significant patterns in the incidence of lip, mouth and pharyngeal cancers by Carstairs deprivation index for both males and females (Figure 11.9). The increases in incidence with deprivation were, however, not linear. For males, incidence in those groups in the middle of the deprivation range was around 40% higher than in the most affluent; but the rate in the most deprived group was over 3 times higher than in the most affluent. There was a similar, but less pronounced, pattern in females. The variations in mortality by deprivation category very closely resemble those for incidence (Figure 11.10).

The incidence of lip, mouth and pharyngeal cancer in males also showed regional variation with above average rates in the north of England and in Wales (Figure 11.11). The regional pattern in mortality (Figure 11.12) was similar to that for incidence, although in the Thames regions incidence was below average and mortality slightly above. Both the incidence of and mortality from lip, mouth and pharyngeal cancer in England and Wales were amongst the lowest internationally, for both sexes (Figure 11.13). Mortality in Australia, compared with its incidence, was low, suggesting overall better survival than in India and Singapore, the countries with the highest mortality and incidence (Figure 11.14).

Prevalence

About 60% of males and of females diagnosed in the three years 1990-92 were still alive at the beginning of 1993. The corresponding proportions of those diagnosed in the ten years 1983-92 were 40 and 44%, respectively (Figure 11.15).

Survival

Survival for cancers of the lip and mouth is generally better than for those of the pharynx. There are, however, several common features in the patterns in five-year relative survival for cancers of the tongue, salivary glands, oral cavity (mouth), oropharynx and nasopharynx (which together constitute nearly 90% of all lip, mouth and pharygeal cancers) diagnosed in 1986-90. First, five-year survival was noticeably higher in women than in men: tongue, 50% and 36%, respectively; salivary glands, 62% and 47%; oral cavity, 52% and 43%; oropharynx, 37% and 33%; and nasopharynx, 38% and 29%. Second, there was little or no improvement in survival during the 1970s and 1980s. And third, there were strong gradients in survival with deprivation: the differences between the most affluent and most deprived groups for the five cancers (same order) were 16%, 6%, 12%, 12% and 17%. As with most cancers, survival in both men and women declined with age at diagnosis.

Table 11A: Individual site breakdown of incidence and mortality for latest year by sex, England and Wales

MALES

ICD10	Site	Number of cases	Incidence[a] CR	ESR*	WSR*	% of all lip, mouth and pharynx	Rank	Number of deaths	Mortality[b] CR	ESR*	WSR*	% of all lip, mouth and pharynx	Rank
C00-C14	Lip, mouth and pharynx	2,364	9.3	9.1	6.4	100	..	1,075	4.2	3.9	2.7	100	..
C00	Lip	202	0.8	0.7	0.5	9	7	11	0.0	0.0	0.0	1	10
C01-C02	Tongue	510	2.0	2.0	1.4	22	1	234	0.9	0.9	0.6	22	1
C03	Gum	82	0.3	0.3	0.2	3	10	53	0.2	0.2	0.1	5	8
C04	Floor of mouth	240	0.9	0.9	0.7	10	5	51	0.2	0.2	0.1	5	9
C05-C06	Other and unspecified parts of mouth	292	1.2	1.1	0.8	12	3	134	0.5	0.5	0.3	12	3
C07-C08	Salivary glands	236	0.9	0.9	0.6	10	6	89	0.3	0.3	0.2	8	5
C09-C10	Oropharynx	311	1.2	1.2	0.9	13	2	206	0.8	0.8	0.5	19	2
C11	Nasopharynx	122	0.5	0.5	0.4	5	9	87	0.3	0.3	0.2	8	7
C12-C13	Hypopharynx	244	1.0	0.9	0.6	10	4	88	0.3	0.3	0.2	8	6
C14	Pharynx unspecified	125	0.5	0.5	0.3	5	8	122	0.5	0.4	0.3	11	4

FEMALES

ICD10	Site	Number of cases	Incidence[a] CR	ESR*	WSR*	% of all lip, mouth and pharynx	Rank	Number of deaths	Mortality[b] CR	ESR*	WSR*	% of all lip, mouth and pharynx	Rank
C00-C14	Lip, mouth and pharynx	1,306	5.0	3.8	2.7	100	..	591	2.2	1.5	1.1	100	..
C00	Lip	78	0.3	0.2	0.1	6	7	11	0.0	0.0	0.0	2	10
C01-C02	Tongue	277	1.1	0.8	0.6	21	1	135	0.5	0.3	0.2	23	1
C03	Gum	67	0.3	0.2	0.1	5	8	49	0.2	0.1	0.1	8	6
C04	Floor of mouth	86	0.3	0.3	0.2	7	6	24	0.1	0.1	0.0	4	9
C05-C06	Other and unspecified parts of mouth	225	0.9	0.6	0.5	17	2	92	0.3	0.2	0.1	16	2
C07-C08	Salivary glands	189	0.7	0.6	0.4	14	3	71	0.3	0.2	0.1	12	4
C09-C10	Oropharynx	142	0.5	0.4	0.3	11	4	72	0.3	0.2	0.2	12	3
C11	Nasopharynx	61	0.2	0.2	0.2	5	9	36	0.1	0.1	0.1	6	8
C12-C13	Hypopharynx	123	0.5	0.4	0.3	9	5	53	0.2	0.1	0.1	9	5
C14	Pharynx unspecified	58	0.2	0.2	0.1	4	10	48	0.2	0.1	0.1	8	7

* Directly age-standardised rate   [a] 1994   [b] 1999

Trends in survival from cancer of the oral cavity for up to 10 years from diagnosis are shown in Figure 11.16. The decline in five-year relative survival with age for cancer of the tongue is shown in Figure 11.17. Patterns of one- and five-year survival with deprivation for cancer of the oral cavity are shown in Figure 11.18. Full details of survival from each of the cancers of the lip, mouth and pharynx are given in the *Cancer Survival Trends* volume.[1]

Five-year survival from this group of cancers in England and Wales was generally comparable to rates in Scotland and to the average for Europe. For cancers of the oral cavity, survival in women was, however, considerably lower than in the USA (Figure 11.19).

Table 11.1: Key statistics for lip, mouth and pharyngeal cancers on incidence, mortality and survival by sex, region and country, latest year available

MALES

	Incidence Number of cases	CR	ESR*	WSR*	Mortality Number of deaths	CR	ESR*	WSR*	Relative survival (%) patients diagnosed in 1986-90 One-year	Five-year
United Kingdom [a]	2,940	10.2	9.9	7.0	1,290	4.4	4.2	2.9	..	..
Great Britain [a]	2,840	10.1	9.8	7.0	1,250	4.4	4.2	2.9	..	..
England and Wales [b]	2,380	9.3	8.8	6.3	1,080	4.2	3.9	2.7	74	44
England [b]	2,200	9.1	8.7	6.2	1,000	4.1	3.9	2.7	..	..
Northern and Yorkshire [b]	350	11.2	10.6	7.4	150	4.9	4.6	3.2	73	41
Trent [b]	200	8.0	7.4	5.3	110	4.2	3.8	2.6	76	47
Anglia and Oxford [b]	230	8.5	8.2	5.8	70	2.7	2.6	1.8	78	50
North Thames [b]	240	6.8	7.2	5.2	140	4.0	4.2	2.9	77	48
South Thames [b]	280	8.3	8.0	5.6	150	4.4	4.2	2.8	76	49
South and West [b]	280	8.7	7.6	5.5	110	3.5	2.9	2.0	73	46
West Midlands [b]	230	8.6	8.1	5.8	100	3.9	3.7	2.5	70	42
North West [b]	400	12.4	11.9	8.5	160	4.8	4.7	3.3	75	43
Wales [b]	180	12.3	11.1	7.8	80	5.5	5.0	3.7	67	40
Scotland [a]	394	15.8	15.8	11.1	167	6.7	6.6	4.6	70	42
Northern Ireland [a]	101	12.4	13.9	10.1	37	4.5	4.9	3.4	..	..

FEMALES

	Incidence Number of cases	CR	ESR*	WSR*	Mortality Number of deaths	CR	ESR*	WSR*	Relative survival (%) patients diagnosed in 1986-90 One-year	Five-year
United Kingdom [a]	1,620	5.4	4.3	3.1	710	2.4	1.7	1.1	..	..
Great Britain [a]	1,570	5.4	4.3	3.1	690	2.4	1.7	1.1	..	..
England and Wales [b]	1,440	5.4	4.3	3.1	590	2.2	1.5	1.1	73	52
England [b]	1,340	5.3	4.3	3.1	550	2.2	1.5	1.0	..	..
Northern and Yorkshire [b]	200	6.1	4.7	3.4	70	2.2	1.4	1.0	73	54
Trent [b]	130	5.2	4.0	3.0	60	2.2	1.5	1.0	73	55
Anglia and Oxford [b]	140	5.0	4.3	3.0	60	2.0	1.3	0.8	78	54
North Thames [b]	150	4.3	3.7	2.6	70	2.1	1.6	1.1	80	54
South Thames [b]	170	4.7	3.8	2.7	80	2.3	1.6	1.1	73	49
South and West [b]	230	6.7	4.7	3.3	70	2.0	1.2	0.8	74	60
West Midlands [b]	130	4.6	3.8	2.7	60	2.2	1.6	1.1	73	47
North West [b]	200	6.0	5.0	3.6	80	2.5	1.9	1.4	71	51
Wales [b]	100	7.0	5.0	3.6	40	2.9	1.9	1.3	61	41
Scotland [a]	247	9.3	7.4	5.3	88	3.3	2.4	1.6	72	47
Northern Ireland [a]	50	5.9	5.0	3.7	18	2.1	1.7	1.1	..	..

* Directly age-standardised rate   [a] 1996 incidence; 1998 mortality.   [b] 1997 incidence; 1999 mortality.   Figures for England and Wales are provisional: incidence - 1995-1997, mortality - 1999

Figure 11.1 **Age-specific incidence, England and Wales, 1997**

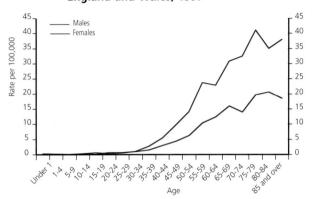

Figure 11.2 **Frequency distribution of new cases by age group, England and Wales, 1997**

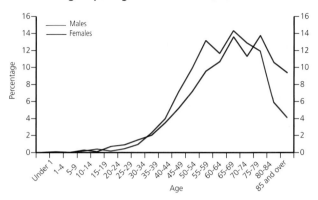

Figure 11.3 **Age-specific incidence rates, England and Wales, 1971-1997**

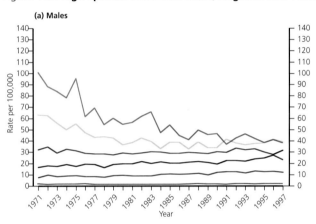

(a) Males

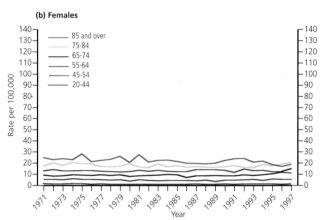

(b) Females

FIgure 11.4 **Age-specific mortality, England and Wales, 1950-1999**

(a) Males

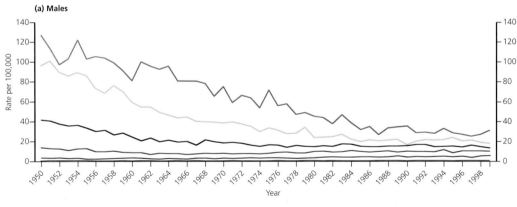

(b) Females

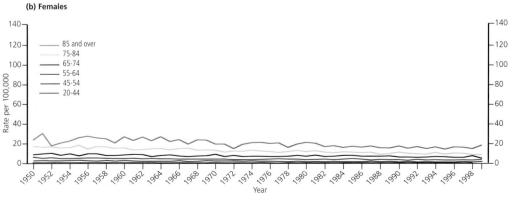

Figure 11.5 **Age-standardised incidence rates by sex, England and Wales, 1971-1997**

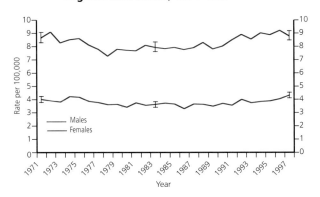

Figure 11.6 **Cohort incidence ratio by sex, England and Wales**

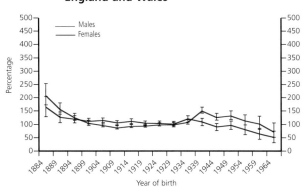

Figure 11.7 **Age-standardised mortality, England and Wales, 1950-1999**

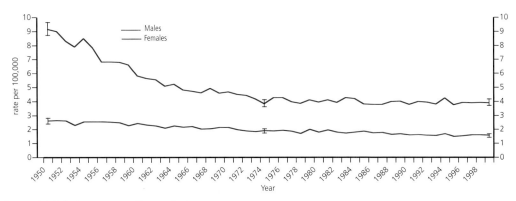

Figure 11.8 **Cohort mortality ratio by sex, England and Wales**

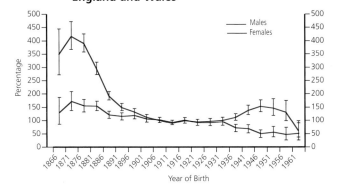

Figure 11.9 **Age-standardised incidence by deprivation category, England and Wales, 1989-93**

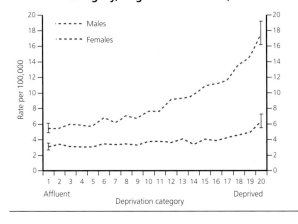

Figure 11.10 **Age-standardised mortality by deprivation category, England and Wales, 1989-93**

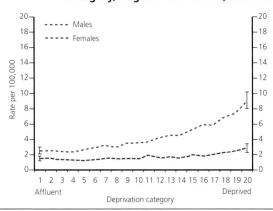

Figure 11.11 **Comparative incidence ratio (CIR) by health region and sex (England and Wales = 100), 1997**

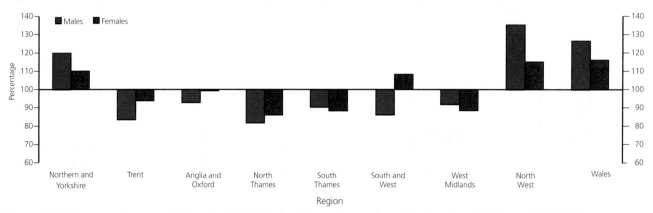

Figure 11.12 **Comparative mortality ratio (CMR) by health region and sex (England and Wales = 100), 1999**

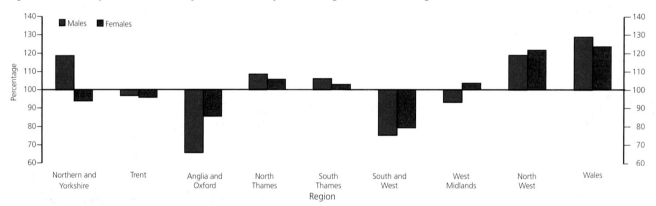

Figure 11.13 **Age-standardised international incidence**

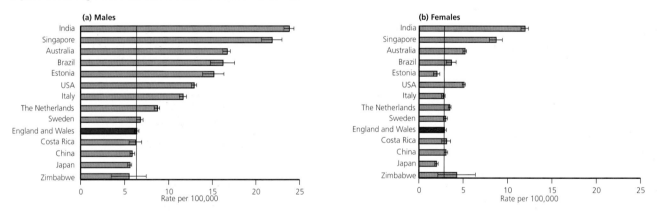

Figure 11.14 **Age-standardised international mortality**

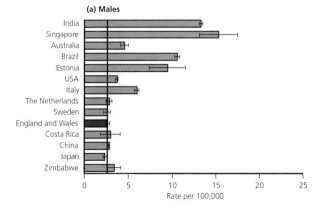
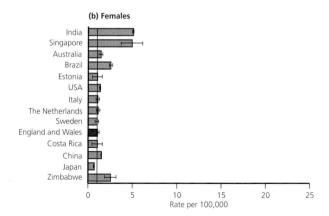

Figure 11.15 **Number and percentage of cases diagnosed in 1990-92 and 1983-92 and alive on 1st January 1993, England and Wales**

(a) Three-year prevalence (diagnosed 1990-92)

(b) Ten-year prevalence (diagnosed 1983-92)

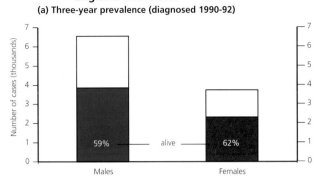
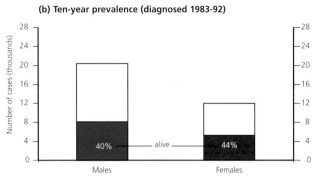

Figure 11.16 **Relative survival from cancer of oral cavity up to ten years, England and Wales, men diagnosed 1971-75 to 1986-90**

Figure 11.17 **Five-year relative survival for cancer of the tongue, England and Wales, adults diagnosed in 1991-93**

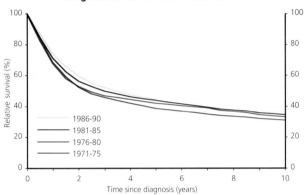
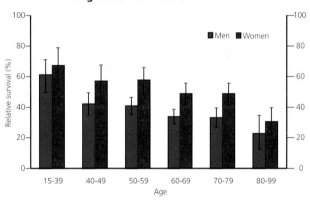

Figure 11.18 **One- and five-year relative survival from oral cavity cancer by deprivation category, England and Wales, adults diagnosed 1986-90**

Figure 11.19 **International comparison of five year relative survival from oral cavity cancer, England and Wales, adults diagnosed 1986-90**

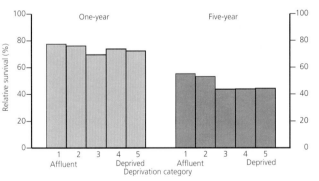
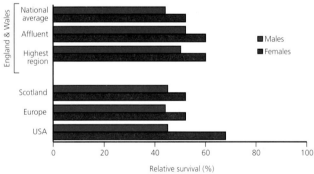

Chapter 12

LUNG

Lung cancer was a relatively rare disease at the beginning of the 20th century, but is currently the most common cancer in the world; 90% of cases are caused by tobacco smoking. Incidence is generally four to six times higher in men than in women; differences in the observed rates are largely explained by smoking habits – in almost all countries, fewer women smoke, or (where this is no longer the case) started to smoke more recently than men, started later in life, smoke less and use brands of cigarette containing less tar.

The latency period for lung cancers attributable to smoking is at least 20 years. Many specific occupations and occupational exposures have been associated with raised risks for lung cancer, and in some cases the link has been established to be causal. Other risk factors include air pollution and indoor radon and its decay products; but without smoking, lung cancer would be a rare disease. Proportions of other cancers are also attributable to smoking, including cancers of the mouth, pharynx, oesophagus, pancreas, larynx, cervix, bladder and kidney. Overall, about a third of all cancer deaths are estimated to be the result of smoking.[1] Exposure to other people's smoke puts non-smokers at risk. About a quarter of all lung cancers in non-smokers may be attributable to passive smoking and the risk is higher for non-smokers living with smokers.

Strategies for preventing tobacco-related cancer show a great discrepancy between knowledge and action. A gap of 20 years divided the clear demonstration of tobacco's causative role from the first effective campaigns for smoking control. Major economic, social and cultural factors were the main determinants of this delay. Large scale intervention programmes in the USA have, however, already achieved noticeable results in reducing tobacco consumption and lung cancer rates – at least in men; success will take longer for women because of the rapid increase over the past several decades in the proportions who smoke, but there is evidence of success in younger and better educated women. Education, particularly of children and adolescents, to dissuade people from taking up smoking is effective, but multiple interventions are needed. Legislative and administrative actions are the most rapid and probably most effective ways to control tobacco manufacture, sale and consumption. In 1989 a former Surgeon General of the USA remarked that "It is a curious public policy that we, as a society, allow the most important preventable cause of death to be one of the most heavily advertised consumer products".[2] Despite "voluntary agreements" and increasing restrictions on advertising and sponsorship of sporting events, in the mid-1990s the tobacco industry spent over £100m each year on promotion of smoking in the UK[3].

Most patients when diagnosed have advanced disease for which no curative treatment is available. Small cell carcinoma accounts for about a quarter of cases. It is aggressive; more than 60% of patients have regional lymph node involvement and metastases in liver, bone or brain when diagnosed, and long term survival remains poor. About 75% of lung tumours are non-small cell cancers (squamous carcinoma, adenocarcinoma and large cell carcinoma), for which surgery is potentially curative, but less than 20% of patients are eligible for surgery.

Incidence and mortality

In 1997 there were 21,000 newly diagnosed cases of lung cancer in males in England and Wales, in whom it was the most common cancer; and 12,300 cases in females, in whom it was the third most common (Table 12.1). The male:female ratio of the number of cases was around 1.7:1; the ratio of the directly age-standardised rates – 72 and 34 per 100,000 in males and females, respectively – was slightly higher at 2.1:1. The lifetime risk of being diagnosed with lung cancer was 8.0% (1 in 13) for males and 4.3% (1 in 23) for females; the risks for non-smokers are only a small fraction of these figures, which are averaged across the whole population. Almost 60% of cases of lung cancer were coded as "unspecified" (in terms of sub-site). About a quarter of the tumours were papillary or squamous cell neoplasms, and 10% were adenomas or adenocarcinomas; about two thirds were poorly specified carcinomas.

Incidence was low in those aged under 50, but rates then increased steeply with age in both males and females to peaks of around 620 per 100,000 in men aged 80-84 and 220 per 100,000 in women aged 70-84 (Figure 12.1). The peaks in the age distributions occurred in males and females aged 70-74 (Figure 12.2). Since the early 1970s, incidence has fallen in males aged up to 74. In men aged 75-84, incidence rose to a peak in the early 1980s and then started to decline in parallel with rates in younger men (Figure 12.3). In the elderly (85 and over) incidence continued to rise until the late 1980s, since when the rates have been similar to those in men aged 75-84. These trends reflect the smoking habits of the cohort of men born around the beginning of the last century (Figure 12.6). The age-specific trends in incidence in females show a completely different pattern from those in males, but again reflect smoking habits in the various cohorts. For women aged 55-64 there was a peak in incidence in the late 1980s, resulting from increases in smoking during and after the second world war in women born in the mid 1920s (Figure 12.3). Incidence in women aged 65-84 is still rising. In the elderly, in whom rates have been lower than those in the immediately younger women, the upward trend in incidence levelled off in the late 1980s. The age-standardised rates show an overall long term decline in males, with rates in the mid-1990s some 70% of those in the late 1970s and early 1980s; in contrast, the rate in females in 1997 was almost 80% higher than in 1971 (Figure 12.5).

In 1999, lung cancer accounted for 26% of cancer deaths in males, and 17% in females in England and Wales; overall, lung cancer accounted for 5% of all deaths. Time trends in lung cancer mortality are closely similar to those in incidence, owing to the very low survival rates (Figure 12.7). The age-standardised rate in males approximately doubled from 1950 to the early 1970s. The long upward trend in mortality in females levelled off in the early 1990s. Lung cancer mortality peaked in males born at the turn of

Table 12.1: Key statistics for lung cancer on incidence, mortality and survival by sex, region and country, latest year available

MALES	Incidence Number of cases	Rate per 100,000 CR	ESR*	WSR*	Mortality Number of deaths	Rate per 100,000 CR	ESR*	WSR*	Relative survival (%) patients diagnosed in 1986-90 One-year	Five-year
United Kingdom [a]	25,700	88.9	79.6	52.1	21,800	74.9	65.8	42.6	..	..
Great Britain [a]	25,100	89.5	79.7	52.2	21,300	75.4	65.9	42.7	..	..
England and Wales [b]	21,000	81.6	72.0	47.2	18,300	70.7	61.3	39.5	21	6
England [b]	19,600	80.8	71.6	57.6	17,200	70.4	61.3	39.5	21	6
Northern and Yorkshire [b]	3,050	97.9	87.1	57.6	2,710	86.9	75.6	48.8	19	5
Trent [b]	2,160	85.1	73.2	48.0	1,920	75.6	63.3	40.7	18	5
Anglia and Oxford [b]	1,790	66.5	60.4	38.9	1,600	59.0	52.6	33.4	20	5
North Thames [b]	2,440	70.7	69.5	45.1	2,080	59.7	58.1	37.1	28	7
South Thames [b]	2,530	75.7	65.5	42.2	2,150	63.8	55.4	35.8	25	6
South and West [b]	2,340	71.7	56.7	37.0	2,140	65.3	50.0	32.0	20	6
West Midlands [b]	2,190	83.1	73.6	48.6	1,870	70.9	61.9	40.0	20	5
North West [b]	3,100	95.8	87.5	57.9	2,690	83.0	74.4	48.5	19	6
Wales [b]	1,370	95.7	78.0	51.1	1,100	76.5	60.7	39.2	20	8
Scotland [a]	2,835	114.0	107.0	70.7	2,332	93.9	85.8	56.0	19	6
Northern Ireland [a]	563	69.0	74.7	49.6	478	57.8	60.7	39.5	..	..
FEMALES										
United Kingdom [a]	15,200	50.9	37.4	25.4	13,100	43.5	30.3	20.2	..	..
Great Britain [a]	14,900	51.3	37.5	25.5	12,800	43.7	30.4	20.3	..	..
England and Wales [b]	12,300	46.4	33.7	22.8	11,100	41.7	28.8	19.2	20	6
England [b]	11,500	46.1	33.6	22.8	10,400	41.6	28.9	19.2	20	6
Northern and Yorkshire [b]	2,030	62.8	46.1	31.4	1,830	56.8	39.3	26.3	18	5
Trent [b]	1,080	41.8	29.6	20.0	1,020	39.4	27.0	18.0	17	5
Anglia and Oxford [b]	1,010	37.0	28.0	18.8	920	33.4	24.2	15.9	19	6
North Thames [b]	1,550	43.8	35.0	23.5	1,320	36.9	28.7	19.1	28	8
South Thames [b]	1,510	42.8	29.5	19.6	1,350	38.2	25.6	16.9	24	6
South and West [b]	1,280	37.7	24.4	16.5	1,210	35.5	21.7	14.3	19	6
West Midlands [b]	1,090	40.5	29.4	19.9	1,000	37.2	26.0	17.3	19	6
North West [b]	1,990	59.2	45.0	30.9	1,790	53.3	37.6	25.1	19	6
Wales [b]	780	52.2	35.2	24.0	650	43.8	28.4	18.9	23	9
Scotland [a]	2,023	76.6	56.1	38.1	1,652	62.7	43.7	29.1	19	6
Northern Ireland [a]	310	36.4	32.5	22.4	297	34.5	28.5	19.4	..	..

* Directly age-standardised rate   [a] 1996 incidence; 1998 mortality.   [b] 1997 incidence; 1999 mortality.   Figures for England and Wales are provisional: incidence - 1995-1997, mortality - 1999

the 20th century and in females born in the late 1920s (Figure 12.8).

In 1993, there were very strong positive gradients in the incidence of lung cancer by Carstairs deprivation category (Figure 12.9). For males, incidence in the most deprived groups was almost $2\frac{1}{2}$ times that in the most affluent; for females, the ratio was around 3. The relationship between incidence and deprivation was not linear: incidence increased more rapidly with deprivation, in both males and females, in categories 12 to 20. There was also a marked north-south gradient in incidence, with rates above the average for England and Wales in Northern and Yorkshire and in the North West, and below average in the whole of southern England (Figure 12.11). The patterns in lung cancer mortality with deprivation (Figure 12.10) and by region (Figure 12.12) were closely similar to those in incidence.

Lung cancer is the leading form of cancer in most western European and North American countries, but rates are comparatively low in India and several African and Latin American countries (Figure 12.13). Trends by age group and sex in different countries can be accounted for almost entirely by national tobacco smoking habits, as reflected in the prevalence of smoking in different birth cohorts and the tar content of the cigarettes they smoked. The very high rates of lung cancer in industrialised countries reflect the substantial increases in the consumption of cigarettes by men since the First World War and in women since the Second. The international patterns in lung cancer mortality were similar to those in incidence (Figure 12.14).

Prevalence

Only about 20% of both males and females diagnosed in the three years 1990-92 were still alive at the beginning of 1993; and only 10% of those diagnosed in the 10 years 1983-92 (Figure 12.15).

Survival

Survival from lung cancer in England and Wales in 1986-90 was very poor – only around 20% after one year and 5% after five years in both men and women. One-year survival had improved by around 5% points since the early 1970s, but five-year survival had hardly improved at all and was similar for patients diagnosed in 1991-93 (Figure 12.16). There was little regional variation in survival.

For patients diagnosed in 1991-93, five-year survival declined continuously with age from 13% in men, and 29% in women, aged under 40 to only 3% or less for adults aged 70 or over (Figure 12.17). In England and Wales as a whole (and in most regions) rates for both one- and five-year survival were lower in the deprived group than in the affluent (Figure 12.18), the gaps being 3% points and 1% point, respectively; this is unlikely to be due to differences in the stage of disease at diagnosis between the deprivation groups.[4]

The results from the second EUROCARE study showed that in the late 1980s, five-year survival for both males and females in England and Wales was similar to that in Scotland, and to the rates in Denmark, Slovenia, Poland and Estonia, but about 5% points lower than the European average, and 7 to 10% points lower than in the USA. Proportionally, these are very large differences. Even the best regional survival rates, or survival in England and Wales for the most affluent group, were significantly below the average rates in Europe and the USA (Figure 12.19).

Figure 12.1 **Age-specific incidence, England and Wales, 1997**

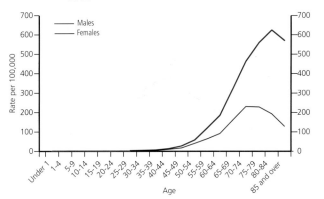

Figure 12.2 **Frequency distribution of new cases by age group, England and Wales, 1997**

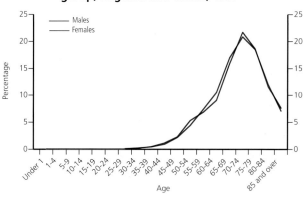

Figure 12.3 **Age-specific incidence, England and Wales, 1971-1997**

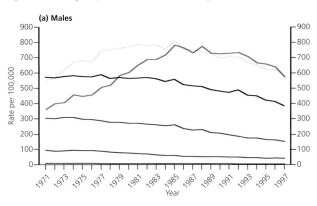

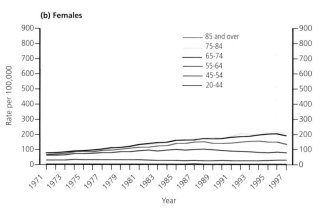

Figure 12.4 **Age-specific mortality, England and Wales, 1950-1999**

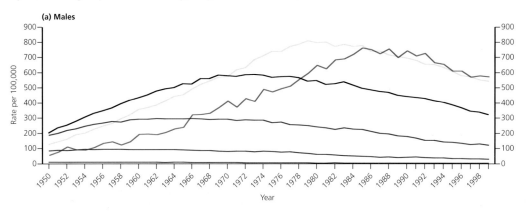

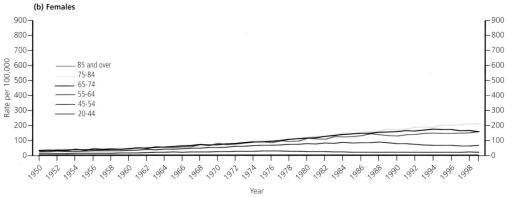

Figure 12.5 **Age-standardised incidence, England and Wales, 1971-1997**

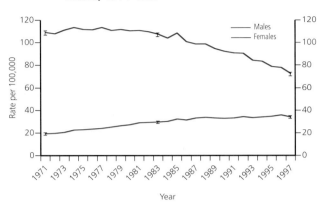

Figure 12.6 **Cohort incidence ratio by sex, England and Wales**

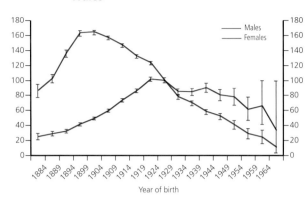

Figure 12.7 **Age-standardised mortality, England and Wales, 1950-1999**

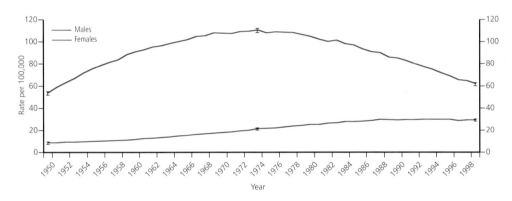

Figure 12.8 **Cohort mortality ratio by sex, England and Wales**

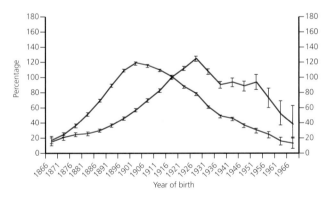

Figure 12.9 **Age-standardised incidence by deprivation category, England and Wales, 1993**

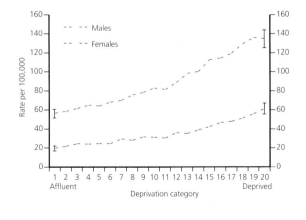

Figure 12.10 **Age-standardised mortality by deprivation category, England and Wales, 1993**

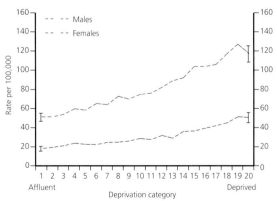

Figure 12.11 **Comparative incidence ratio (CIR) by health region and sex (England and Wales = 100), 1997**

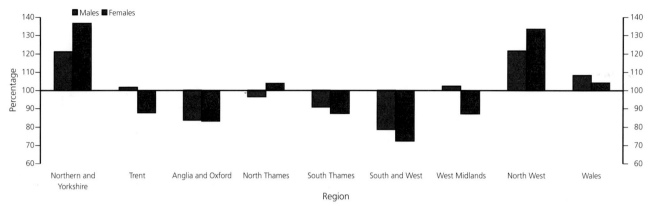

Figure 12.12 **Comparative mortality ratio (CMR) by health region and sex (England and Wales = 100), 1999**

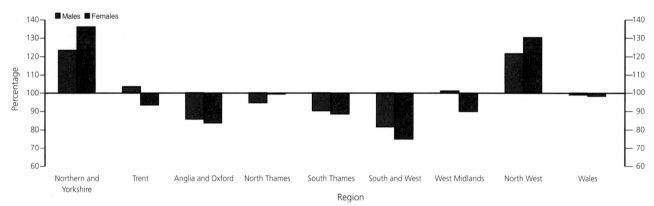

Figure 12.13 **Age-standardised international incidence**

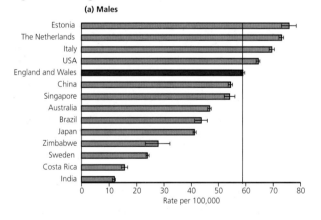

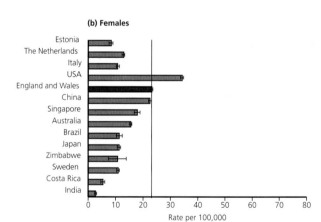

Figure 12.14 **Age-standardised international mortality**

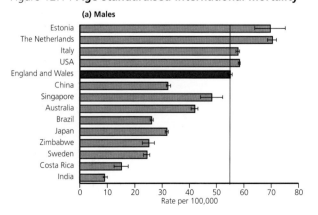

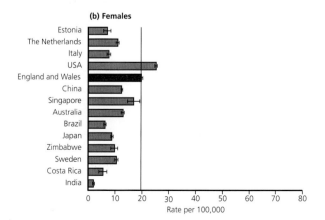

Figure 12.15 **Number and percentage of cases diagnosed in 1990-92 and 1983-92 and alive on 1st January 1993, England and Wales**

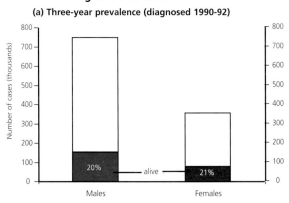

(a) Three-year prevalence (diagnosed 1990-92)

(b) Ten-year prevalence (diagnosed 1983-92)

Figure 12.16 **Relative survival up to ten years, men, England and Wales, 1971-75 to 1991-93**

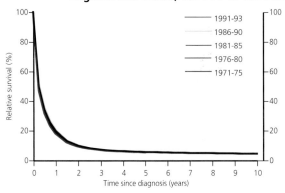

Figure 12.17 **Five-year relative survival by age, England and Wales, patients diagnosed 1991-93**

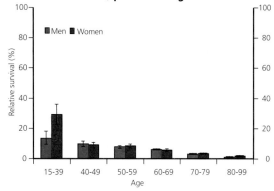

Figure 12.18 **One- and five-year relative survival by deprivation category, England and Wales, adults diagnosed 1986-90**

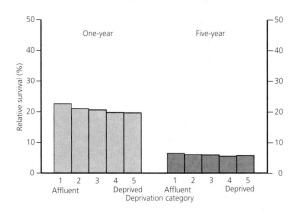

Figure 12.19 **International comparison of five-year relative survival, England and Wales, adults diagnosed 1986-90**

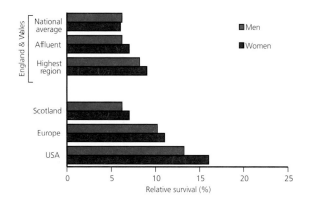

Chapter 13

MELANOMA OF SKIN

Malignant melanoma is a comparatively rare but serious form of skin cancer, which develops in the pigment-producing cells of the skin. There are four major reasons for concern over what may largely be a preventable cancer. First, incidence rates (and to a lesser extent mortality) have increased significantly in recent years. Second, although the disease is extremely rare in children, it is one of the few cancers that have an impact on young adults as well as on older people. Third, survival rates could be markedly improved if all patients were treated at an early stage of disease. And fourth, depletion of the ozone layer will increase exposure to solar ultraviolet radiation, the main aetiological factor.[1]

There are four types of melanoma. Superficial spreading and nodular melanomas occur in younger people and most commonly on body sites that are not continuously exposed to the sun. A third type, lentigo maligna, occurs in older people with sun damaged skin; its epidemiological characteristics are similar to those of non-melanoma skin cancer. The fourth type, acral, occurs on the palms and soles and is proportionally infrequent in Europeans.

There are two major groups of risk factors for the disease. The constitutional factors include fair skin, blue eyes, fair hair and a tendency to burn but not tan after exposure to the sun; a person may also be at risk if their skin has a large number of naevi and a tendency to freckle, and if there is a family history of melanoma. The importance of exposure to sunlight is reflected in an association of melanoma with latitude, although the situation in Europe is less clear than in Australia, Canada and the USA. It is almost certain that in addition to differential susceptibility, intermittent (recreational) exposure to sunlight (and sun beds) is important, and severe sunburn is a risk factor. There is an unusually strong inverse social class gradient in risk, with higher rates among professional and non-manual workers.

Incidence and mortality

There were 4,700 cases of malignant melanoma in England and Wales in 1997, around 2% of all cancers (Table 13.1). This is one of the few tumours for which there was an excess in women – the female:male ratio of cases was about 1.3:1; that of the age-standardised rates was about 1.2:1. The age-specific incidence rates began to rise in young adults, but subsequently did not increase with age as rapidly as for most other cancers (Figure 13.1). About 20% of melanoma cases occurred in young adults (15-39 years) (Figure 13.2) compared with 4% of all cancers. Proportionally, nearly three times as many cases occurred on the trunk in men as in women (35% and 14%, respectively); a similar ratio, but in reverse, occurred on the lower limbs (18% in men and 50% in women). Virtually all cases were histologically coded to naevi and melanomas.

Incidence rates in most age groups in both males and females increased gradually during the 1970s and early 1980s (Figure 13.3). Rates then increased more steeply to a peak in 1988, dropped slightly, and then rose to another peak in 1993. This pattern may be due to the bringing forward of the diagnosis of some cases as a

result of publicity about skin cancer in the late 1980s which attempted to reduce people's exposure to the sun.[2] A steep rise in incidence occurred for cohorts born from around the 1930s onwards, particularly for men (Figure 13.6).

Although the incidence of malignant melanoma of the skin is higher in males than in females at all ages, age-specific mortality is closely similar (Figure 13.4). During the 1950s and 1960s, age-standardised mortality in both males and females more or less doubled (Figure 13.7). The upward trend then increased slightly to reach around 2 per 100,000 at the end of the 1980s. In the 1990s, mortality in men continued to increase, while in women it levelled off. Mortality rose in successive cohorts (in parallel with the increase in incidence) until around the mid-1940s when it levelled off (Figure 13.8).

There is a strong inverse pattern in the incidence of malignant melanoma of the skin with deprivation in England and Wales, with higher rates in the more affluent groups (Figure 13.9). The trend with deprivation is, however, not linear. Incidence declines slightly with increasing deprivation as far as the middle of the range (categories 9 and 10); it then declines much more rapidly, to the extent that incidence in the most deprived groups is only around 40% of that in the more affluent for men, and 30% for women. There is also an inverse pattern in the mortality from malignant melanoma (Figure 13.10) but it is far less pronounced than that in incidence. It is, however, also not linear, with little difference in mortality in deprivation categories 1 to 10 and declining mortality with increasing deprivation thereafter. Mortality in the most deprived groups is around 60% of that in the more affluent for men and 70% for women.

There is some, but not consistent, evidence of a north-south gradient in malignant melanoma of the skin in England and Wales, with generally higher incidence and mortality in the south, particularly the south west, and lower than average rates in the north (Figures 13.11 and 13.12).

The incidence of malignant melanoma is around three times higher in Australia, in both men and women, than in the USA and Sweden, where rates are in turn about twice those in England and Wales (Figure 13.13). Incidence in India, China and Japan is very low. The international patterns in mortality broadly reflect those in incidence (Figure 13.14).

Prevalence

About 80% of males and 87% of females diagnosed in the three years 1990-92 were still alive at the beginning of 1993, as were 64% of males and 76% of females diagnosed in the 10 years 1983-92 (Figure 13.15).

Survival

Survival from malignant melanoma of the skin in England and Wales was among the highest of all cancers. However, as would be expected from the higher incidence in women but similar

Table 13.1: Key statistics for melanoma on incidence, mortality and survival by sex, region and country, latest year available

	Incidence				Mortality				Relative survival (%) patients diagnosed in 1986-90	
MALES	Number of cases	Rate per 100,000			Number of deaths	Rate per 100,000			One-year	Five-year
		CR	ESR*	WSR*		CR	ESR*	WSR*		
United Kingdom [a]	2,240	7.7	7.5	5.6	800	2.7	2.5	1.8	..	..
Great Britain [a]	2,170	7.7	7.4	5.5	790	2.8	2.6	1.8	..	..
England and Wales [b]	2,020	7.8	7.5	5.6	760	2.9	2.7	1.9	91	70
England [b]	1,900	7.8	7.5	4.5	700	2.9	2.7	1.9	91	71
Northern and Yorkshire [b]	200	6.5	6.2	4.5	90	2.8	2.6	1.9	91	71
Trent [b]	150	5.8	5.4	3.9	60	2.2	2.0	1.4	89	69
Anglia and Oxford [b]	260	9.8	9.5	7.2	80	2.9	2.7	1.9	93	74
North Thames [b]	180	5.2	5.3	3.8	100	2.8	2.9	2.0	93	70
South Thames [b]	220	6.6	6.3	4.7	110	3.3	3.1	2.2	89	68
South and West [b]	440	13.5	12.3	9.3	130	3.8	3.2	2.3	90	72
West Midlands [b]	180	6.9	6.5	4.9	70	2.5	2.3	1.6	94	74
North West [b]	270	8.3	8.1	6.1	80	2.5	2.3	1.6	91	67
Wales [b]	120	8.0	7.3	5.4	60	4.1	3.6	2.6	78	54
Scotland [a]	276	11.1	10.8	8.1	68	2.7	2.7	1.9	94	77
Northern Ireland [a]	71	8.7	9.7	7.5	12	1.5	1.5	1.0	..	..
FEMALES										
United Kingdom [a]	3,110	10.4	9.1	7.1	840	2.8	2.1	1.5	..	..
Great Britain [a]	3,000	10.3	9.0	7.0	820	2.8	2.1	1.5	..	..
England and Wales [b]	2,670	10.0	8.8	6.9	710	2.7	2.0	1.4	95	84
England [b]	2,520	10.1	8.9	6.9	670	2.7	2.0	1.4	96	85
Northern and Yorkshire [b]	270	8.5	7.5	5.9	70	2.3	1.5	1.0	96	87
Trent [b]	230	9.0	7.7	5.9	60	2.4	1.7	1.3	93	85
Anglia and Oxford [b]	340	12.4	11.7	9.3	70	2.5	2.0	1.4	96	85
North Thames [b]	250	6.9	6.1	4.8	80	2.4	1.9	1.3	96	85
South Thames [b]	290	8.2	6.9	5.2	110	3.0	2.2	1.6	96	84
South and West [b]	550	16.3	13.9	10.8	120	3.4	2.2	1.6	95	83
West Midlands [b]	260	9.6	8.6	6.6	80	2.9	2.2	1.6	98	87
North West [b]	330	9.9	8.8	7.0	80	2.5	2.0	1.4	95	84
Wales [b]	140	9.6	8.1	6.5	40	2.5	1.7	1.2	88	78
Scotland [a]	403	15.3	13.3	10.4	76	2.9	2.2	1.6	98	91
Northern Ireland [a]	110	12.9	12.7	10.1	16	1.9	1.4	1.0	..	..

* Directly age-standardised rate
[a] 1996 incidence; 1998 mortality.
[b] 1997 incidence; 1999 mortality.
Figures for England and Wales are provisional: incidence - 1995-1997, mortality - 1999

mortality to that in men, women have a marked survival advantage. For melanomas diagnosed in 1986-90, one-year relative survival was 91% in men and 95% in women; five-year survival was 70% and 84%, respectively. Consistent with the larger increases in incidence than in mortality during the 1970s and 1980s, survival improved markedly during this period (Figure 13.16). One-year relative survival rose by 12% points in men and 5% in women; and five-year survival by 22% and 17%, respectively. Broadly similar improvements in survival occurred in each successive five-year period. Survival showed little improvement for patients diagnosed with melanoma of skin in 1991-93.

Five-year survival declined only slightly with age up to 70-79 years in both men and women, but then fell more sharply in the very elderly (80-99 years) (Figure 13.17). Survival at one, five and ten years varied little among the regions of England, but both one- and five-year survival in Wales (based on fairly small numbers) was significantly below levels in England.

The patterns in incidence and mortality with deprivation (Figures 13.9 and 13.10) are consistent with the gaps in survival between the affluent and deprived groups (Figure 13.18). It has been shown that melanomas with good prognosis are more common in the affluent, possibly because of higher awareness, leading to earlier diagnosis. But after allowing for this, survival was still inversely related to socio-economic status – possibly as a result of poor nutrition or immunological defects.[3]

Survival from malignant melanoma of the skin varied widely across Europe in the late 1980s: for men, five-year survival ranged from under 50% in Slovenia, Poland and Estonia to over 80% in Sweden and Switzerland; the pattern was similar for women, with corresponding rates of just under 70% and around 90%. The lower survival in the former eastern bloc countries may – as in the UK – be related to socio-economic factors.[4] Survival in England and Wales was slightly lower than in Scotland, but – unlike for most cancers – was around the European average for both men and women (Figure 13.19). Survival in the UK was, however, below that in the USA, particularly for men.

There were no significant improvements in the treatment of malignant melanoma of the skin in the period covered by the survival analyses. Small tumours can be effectively treated by surgical removal; the prospects of cure for patients with disseminated disease have improved, but remain poor. The smaller increases in mortality than in incidence are likely to be due to earlier diagnosis and/or changes in sunbathing behaviour and use of sunscreens.

Figure 13.1 **Age-specific incidence, England and Wales, 1997**

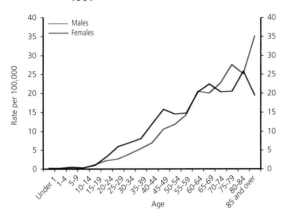

Figure 13.2 **Frequency distribution of new cases by age group, England and Wales, 1997**

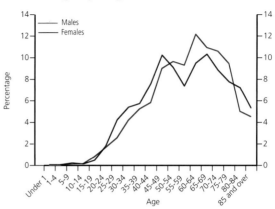

Figure 13.3 **Age-specific incidence, England and Wales, 1971-1997**

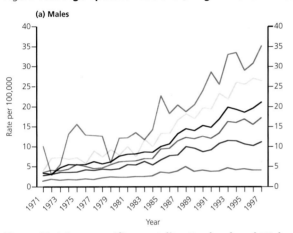

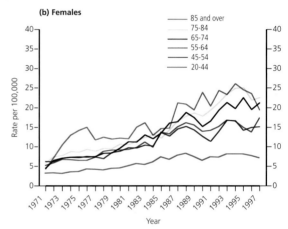

Figure 13.4 **Age-specific mortality, England and Wales, 1950-1999**

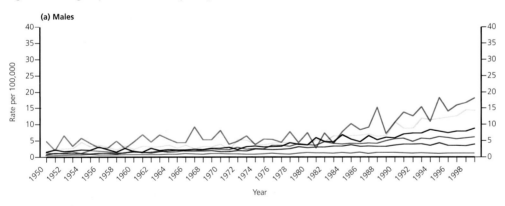

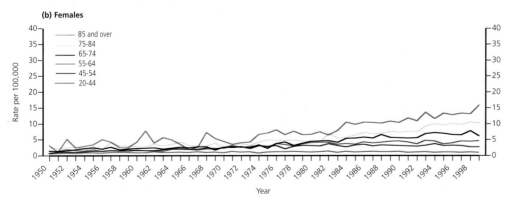

Figure 13.5 **Age-standardised incidence, England and Wales, 1971-1997**

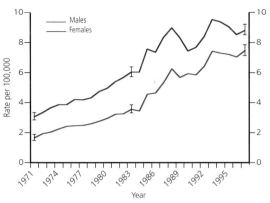

Figure 13.6 **Cohort incidence ratio by sex, England and Wales**

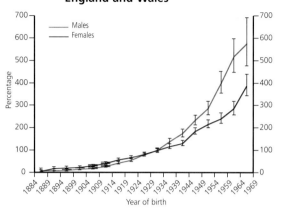

Figure 13.7 **Age-standardised mortality, England and Wales, 1950-1999**

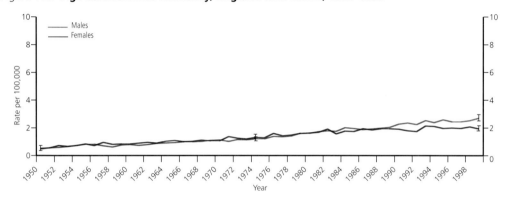

Figure 13.8 **Cohort mortality ratio by sex, England and Wales**

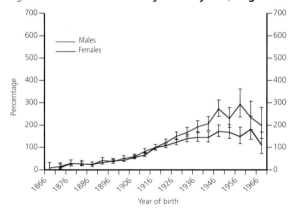

Figure 13.9 **Age-standardised incidence by deprivation category, England and Wales, 1988-93**

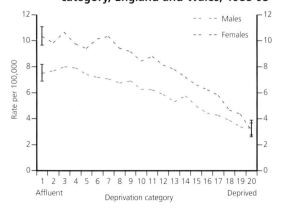

Figure 13.10 **Age-standardised mortality by deprivation category, England and Wales, 1988-93**

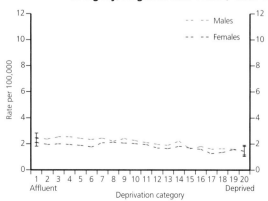

Figure 13.11 **Comparative incidence ratio (CIR) by health region and sex (England and Wales = 100), 1997**

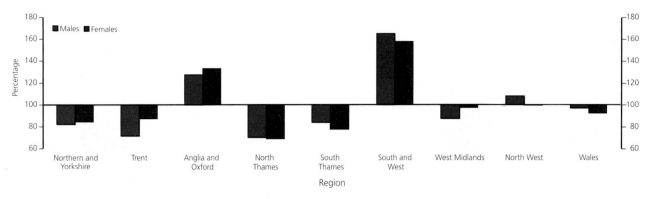

Figure 13.12 **Comparative mortality ratio (CMR) by health region and sex (England and Wales = 100), 1999**

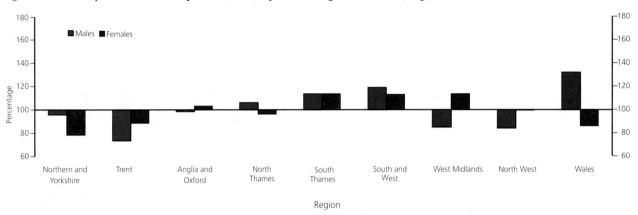

Figure 13.13 **Age-standardised international incidence**

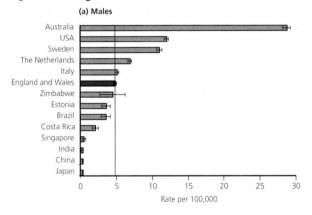
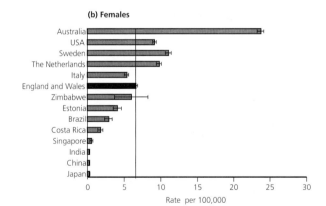

Figure 13.14 **Age-standardised international mortality**

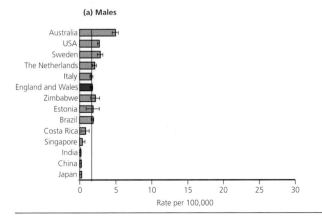
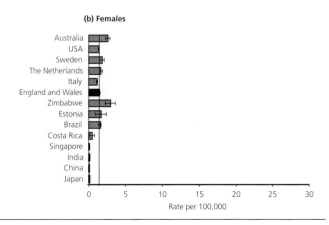

Figure 13.15 **Number and percentage of cases diagnosed in 1990-92 and 1983-92 and alive on 1st January 1993, England and Wales**

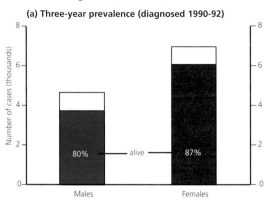

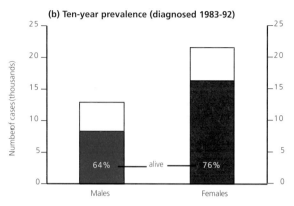

Figure 13.16 **Relative survival up to ten years, men, England and Wales, 1971-75 to 1991-93**

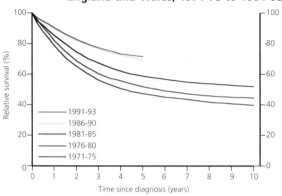

Figure 13.17 **Five-year relative survival by age, England and Wales, patients diagnosed in 1991-93**

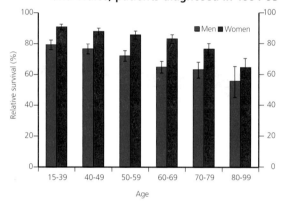

Figure 13.18 **One- and five-year relative survival by deprivation category, England and Wales, adults diagnosed 1986-90**

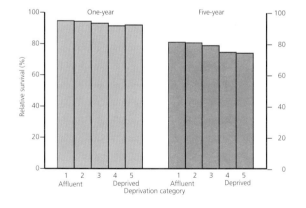

Figure 13.19 **International comparison of five-year relative survival, England and Wales, adults diagnosed 1986-90**

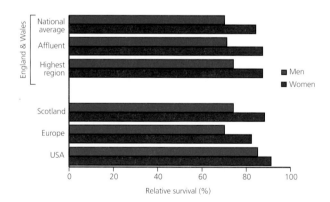

Chapter 14

MULTIPLE MYELOMA

Multiple myeloma is a fairly rare type of cancer, accounting for under 1% of all cancers in most countries. It is a malignant clonal proliferation of the plasma cells in the bone marrow, and it has some similarities with chronic lymphocytic leukaemia; in leukaemia, however, malignant white cells move into the bloodstream, while malignant myeloma cells only rarely move from the bone marrow. A related condition, Waldenström's macroglobilinaemia, is a type of lymphoma that occurs in cells which are similar to plasma cells.[1]

Multiple myeloma interferes with the normal functioning of the bone marrow, often causing anaemia, and compromising the body's ability to fight infection. Myeloma cells attack and weaken the surrounding bone, thus affecting the spine, pelvis, ribs, skull and the upper arms and legs, and sometimes leading to fractures. The cells can also disturb the balance of salts, particularly calcium, in the blood and can also interfere with the normal functioning of the kidney and nerves.

Incidence and mortality have increased around the world, particularly in countries where the rates initially were low; these increases were greater in the older age groups. A substantial proportion of the observed increases is probably due to changes in diagnostic methods and in the accuracy of death certification.[2]

The causes of multiple myeloma are unknown. It does not appear to be related to smoking or any other lifestyle factors; or to occupational or environmental exposures except farming and employment in the food industry. Risk is increased by ionising radiation, whether from nuclear weapons, low dose exposure in the nuclear industry, or radiotherapy (eg for cervical cancer or ankylosing spondylitis). The disease, which affects men slightly more than women, is very rare in people aged under 40; rates subsequently increase rapidly with age, suggesting the involvement of an age dependent loss in regulatory cell function.[1]

Diagnosis of multiple myeloma has, in addition to a characteristic radiological appearance and microscopic examination of bone marrow aspirate, been made on the basis of detection of particular proteins in urine and blood. Earlier detection of the disease owing to the use of more sensitive diagnostic procedures may have influenced survival as well as incidence. The three main factors which influence survival are kidney function, anaemia and general mobility of the patient. If a person has normal kidney function, is not anaemic and can walk and carry out normal daily tasks, survival is good. The disease is usually treated with chemotherapy, but radiotherapy is also sometimes given.

Incidence and mortality

In England and Wales there were 1,500 cases of multiple myeloma in men and 1,400 in women in 1997; the crude rates were around 6 and 5 per 100,000 population, respectively (Table 14.1). Although the numbers of cases and the crude rates were only slightly higher in men, all the age-specific rates were markedly higher in men than in women, and the age-standardised rate was almost 50% higher. Only 1% of cases occurred in people under 40 years of age, and 5% in those under 50 (Figure 14.2). Rates rose continuously and steeply with age to reach 50 per 100,000 in men and 30 per 100,000 in women aged 80 and over (Figure 14.1).

Over the period 1971-1997, there were increases in age-specific incidence in both men and women which were larger in the older age groups (Figure 14.3). The age-standardised incidence increased steeply in both men and women up to the mid-1980s; subsequently, the rate of increase slowed markedly in men and levelled off in women (Figure 14.5). Compared with the early 1970s, rates in the mid-1990s were 80% higher in men and 70% in women. Over the years, hospital tests have become more routine, and hospital admission rates have increased (in many countries) particularly in the elderly. The rise in incidence is reflected in the trends by birth cohort, with a levelling off for people born in the 1930s onwards (Figure 14.6).

The patterns in mortality were similar to those in incidence with large increases, particularly in those aged 65 and over, up to the late 1980s, after which rates rose less steeply or levelled off (Figure 14.4). Age-standardised mortality increased from 0.5 per 100,000 in 1950 by a factor of around 8 in men to reach 4 per 100,000 in the late 1980s, and by a factor of 5 in women to around 2.5 per 100,000 (Figure 14.7). Mortality fluctuated around these levels during the 1990s. The pattern in mortality by birth cohort (Figure 14.8) is similar to that for incidence.

There was no pattern in the incidence of multiple myeloma with deprivation in either males or females (Figure 14.9). Mortality, in both males and females, also shows no marked pattern with deprivation, but rates are slightly lower in the most deprived groups (Figure 14.10); however, the numbers of cases in each deprivation group is small, and the confidence intervals around each rate are correspondingly wider than for the more common cancers. There were no consistent geographical patterns in either incidence or mortality in England and Wales (Figures 14.11 and 14.12).

The incidence of multiple myeloma in both males and females in England and Wales is similar to that in the USA, Australia, and other countries in Europe (Figure 14.13). Incidence rates are high in black people in some parts of the world: the USA, the Caribbean (Jamaica, Martinique), Hawaii and New Zealand (Maoris and Polynesians); and low in Chinese, Indians, Japanese and Filipinos. Chinese and Japanese migrants to the USA appear to retain the levels of their country of origin. The international patterns in mortality are similar to those in incidence (Figure 14.14).

Prevalence

Almost 50% of men and women diagnosed with multiple myeloma in the three year period 1990-92 were alive at the beginning of 1993, as were almost 25% of both men and women diagnosed in the 10 years 1983-92 (Figure 14.15).

Table 14.1: Key statistics for multiple myeloma on incidence, mortality and survival by sex, region and country, latest year available

MALES	Incidence Number of cases	CR	ESR*	WSR*	Mortality Number of deaths	CR	ESR*	WSR*	Relative survival (%) patients diagnosed in 1986-90 One-year	Five-year
United Kingdom [a]	1,750	6.1	5.5	3.6	1,210	4.2	3.6	2.3	..	..
Great Britain [a]	1,700	6.1	5.5	3.6	1,180	4.2	3.6	2.3	..	..
England and Wales [b]	1,480	5.8	5.2	3.4	1,100	4.3	3.7	2.3	55	19
England [b]	1,390	5.7	5.1	3.0	1,040	4.3	3.7	2.4	55	19
Northern and Yorkshire [b]	160	5.1	4.5	3.0	120	3.8	3.3	2.2	52	19
Trent [b]	130	5.3	4.6	3.0	120	4.8	4.0	2.5	55	20
Anglia and Oxford [b]	180	6.6	6.1	4.0	110	4.1	3.7	2.4	50	17
North Thames [b]	190	5.4	5.3	3.5	130	3.6	3.5	2.2	67	23
South Thames [b]	190	5.7	5.2	3.5	170	5.1	4.4	2.8	60	24
South and West [b]	240	7.2	5.8	3.8	160	4.7	3.6	2.2	58	18
West Midlands [b]	150	5.7	5.0	3.3	100	3.9	3.4	2.2	52	15
North West [b]	160	4.9	4.6	3.0	140	4.3	3.8	2.4	48	14
Wales [b]	90	6.6	5.6	3.8	60	4.0	3.1	2.0	52	20
Scotland [a]	150	6.0	5.7	3.7	83	3.3	3.0	1.9	50	17
Northern Ireland [a]	50	6.1	6.7	4.3	38	4.6	4.7	3.0	..	..
FEMALES										
United Kingdom [a]	1,540	5.1	3.5	2.3	1,260	4.2	2.6	1.7	..	..
Great Britain [a]	1,500	5.1	3.5	2.3	1,230	4.2	2.6	1.7	..	..
England and Wales [b]	1,390	5.2	3.5	2.3	1,050	3.9	2.5	1.6	56	20
England [b]	1,300	5.2	3.5	2.3	980	3.9	2.5	1.6	56	19
Northern and Yorkshire [b]	160	5.0	3.2	2.1	120	3.7	2.3	1.5	54	19
Trent [b]	140	5.4	3.5	2.3	120	4.8	2.9	1.9	55	18
Anglia and Oxford [b]	150	5.3	3.9	2.6	100	3.7	2.6	1.7	47	17
North Thames [b]	170	4.7	3.4	2.3	110	3.2	2.3	1.5	67	28
South Thames [b]	170	4.8	3.0	2.0	140	4.1	2.4	1.5	64	23
South and West [b]	230	6.8	4.2	2.8	140	4.0	2.1	1.3	57	19
West Midlands [b]	140	5.2	3.6	2.4	120	4.5	3.0	1.9	54	16
North West [b]	150	4.6	3.3	2.2	120	3.6	2.2	1.4	48	14
Wales [b]	90	5.9	3.8	2.5	70	4.5	2.8	1.9	57	28
Scotland [a]	135	5.1	3.7	2.5	104	3.9	2.5	1.6	53	23
Northern Ireland [a]	41	4.8	4.1	2.7	27	3.1	2.4	1.6	..	..

* Directly age-standardised rate
[a] 1996 incidence; 1998 mortality.
[b] 1997 incidence; 1999 mortality.
Figures for England and Wales are provisional: incidence - 1995-1997, mortality - 1999

Survival

The patterns of survival from multiple myeloma are unusual in several ways. For adults diagnosed in 1991-93, one-year survival was 59%, but five-year survival was only 20% (Figure 14.16). There is a long term substantial excess mortality, relative to that of the general population, and the relative survival curves continue to decline up to ten years after diagnosis. For patients diagnosed in 1981-85, ten-year survival was very low: just 7%. Since the early 1970s in England and Wales as a whole, and in each region, survival improved dramatically at one year after diagnosis, and moderately at five years, but did not improve much, if at all, at ten years.

Five-year relative survival declines steadily with age at diagnosis, from 40-50% in young adults (15-39 years) to less than 10% in the elderly (80-99) (Figure 14.17).

For adults diagnosed in 1986-90, one-year survival was 7% points higher in the most affluent group than in the most deprived (Figure 15.18). There was, however, no difference among the deprivation groups in five-year survival. Among the regions, there was wide variation in one-year survival for both men and women, with a range of 20% points around the average of 55%. There were similar, but smaller (absolute) regional differences in five-year survival.

In Scotland, where incidence and mortality both increased slightly during the 1980s, five-year survival from multiple myeloma was better, by around 5%, than in England and Wales in the early 1970s. By the late 1980s, rates were similar. Multiple myeloma was not included in the survival analyses in the first EUROCARE report[3] owing to wide differences in the definition of the disease and in case ascertainment among the participating registries. The results in the second EUROCARE report[4] indicate that one-year survival in England and Scotland was below the European average by around 12% in men and 8% in women; and that five-year survival rates were below average by 11% in men and 5% in women (Figure 14.19). Survival was similar to that in Denmark, Slovenia and Estonia, and well below that in the rest of western Europe. Survival in the USA is close to the European average.

Figure 14.1 **Age-specific incidence, England and Wales, 1997**

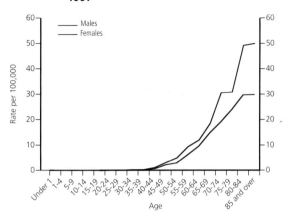

Figure 14.2 **Frequency distribution of new cases by age group, England and Wales, 1997**

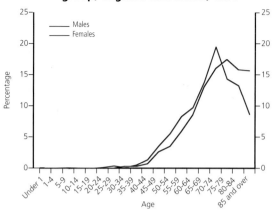

Figure 14.3 **Age-specific incidence, England and Wales, 1971-1997**

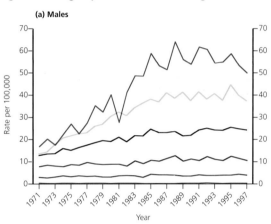

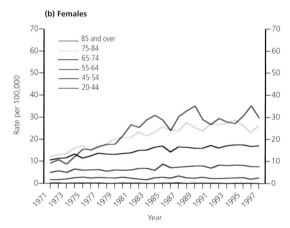

Figure 14.4 **Age-specific mortality, England and Wales, 1950-1999**

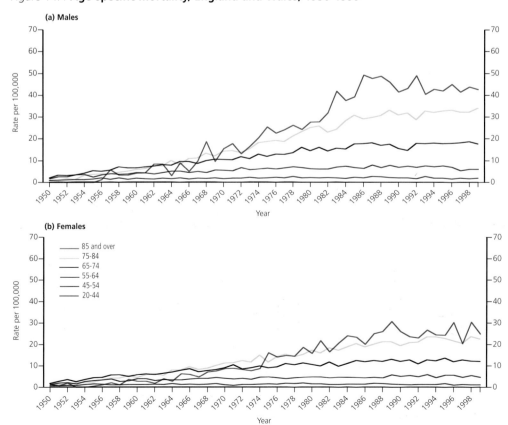

Figure 14.5 **Age-standardised incidence, England and Wales, 1971-1997**

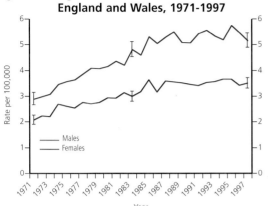

Figure 14.6 **Cohort incidence ratio by sex, England and Wales**

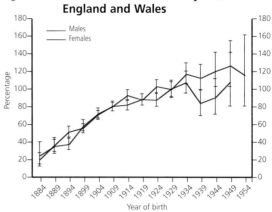

Figure 14.7 **Age-standardised mortality, England and Wales, 1950-1999**

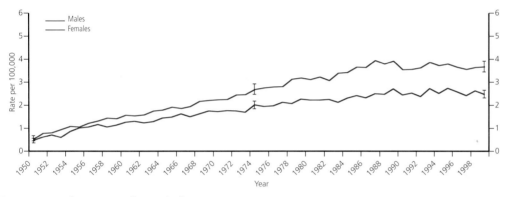

Figure 14.8 **Cohort mortality ratio by sex, England and Wales**

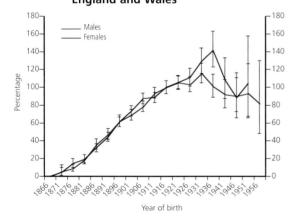

Figure 14.9 **Age-standardised incidence by deprivation category, England and Wales, 1990-93**

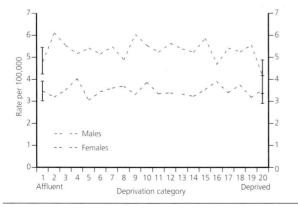

Figure 14.10 **Age-standardised mortality by deprivation category, England and Wales, 1990-93**

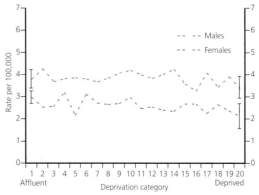

Figure 14.11 **Comparative incidence ratio (CIR) by health region and sex (England and Wales = 100), 1997**

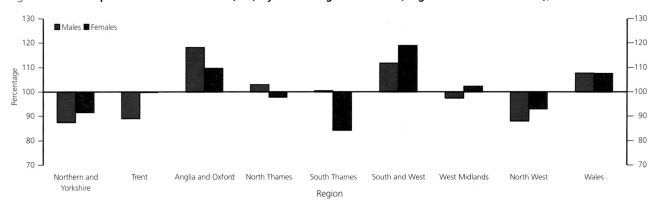

Figure 14.12 **Comparative mortality ratio (CMR) by health region and sex (England and Wales = 100), 1999**

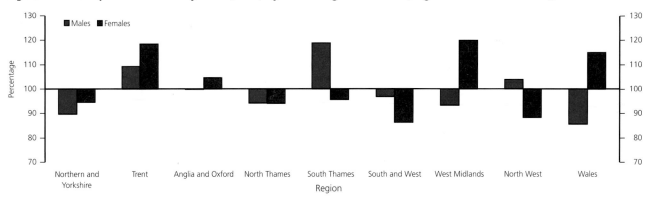

Figure 14.13 **Age-standardised international incidence**

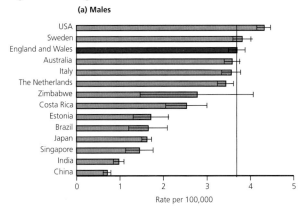

Figure 14.14 **Age-standardised international mortality**

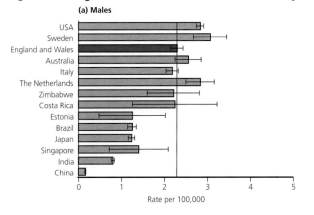
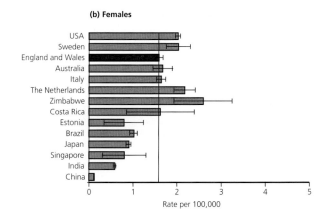

Figure 14.15 **Number and percentage of cases diagnosed in 1990-92 and 1983-92 and alive on 1st January 1993, England and Wales**

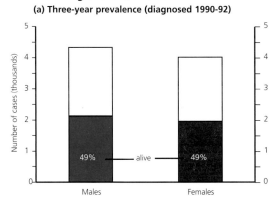

(a) Three-year prevalence (diagnosed 1990-92)

(b) Ten-year prevalence (diagnosed 1983-92)

Figure 14.16 **Relative survival up to ten years, men, England and Wales, 1971-75 to 1991-93**

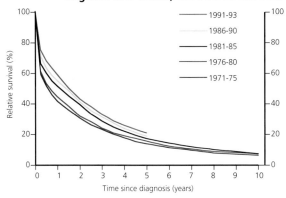

Figure 14.17 **Five-year relative survival by age, England and Wales, adults diagnosed 1991-93**

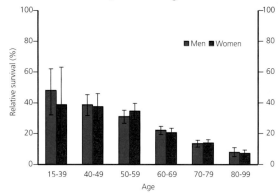

Figure 14.18 **One- and five-year relative survival by deprivation category, England and Wales, adults diagnosed 1986-90**

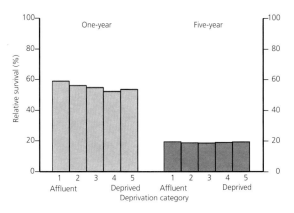

Figure 14.19 **International comparison of five-year relative survival, England and Wales, adults diagnosed 1986-90**

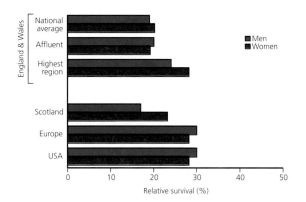

Chapter 15

NON-HODGKIN'S LYMPHOMA

Non-Hodgkin's lymphoma (NHL) is the most common of the leukaemias and lymphomas. It is a collection of malignancies whose clinical behaviour, prognosis, and management vary widely according to the histological subtype, stage and bulk of the disease. Changes have occurred in the registration of NHL over the past thirty years, with a decline in malignancies coded as reticulosarcomas and lymphosarcomas, and a corresponding increase in other types of lymphomas.[1]

Typically NHL arises in lymph node tissue, but in 15-20% of patients, the tumour develops in a site other than a node, for example, in bone, stomach or thyroid, small or large intestine, breast or brain. Little is known about the causes of NHL, but occupational exposure to chemicals appears to increase the risk, particularly exposure to phenoxy herbicides. Other risk factors include family history and immunodeficiency disorders. Kidney transplant patients have also been affected, with a short latent period suggesting the involvement of oncogenic viruses. Incidence has increased in many countries in recent years.[2] Much of the increase of lymphomas in young men can be attributed to AIDS, but this seems unlikely to account for increases in the elderly. Burkitt's lymphoma (BL) is a distinct pathological entity; the areas of high risk coincide with those regions affected by holoendemic malaria, for example, sub-Saharan Africa excepting highland areas. BL probably accounts for 20 to 30% of childhood NHL in most countries, although incidence rates are much higher in Africa.

Until about 25 years ago, most patients died of their disease. Effective combination chemotherapy, initially developed for Hodgkin's disease, has resulted in improved survival and the cure of advanced tumours in some patients with NHL.

Incidence and mortality
The annual number of new cases in England and Wales has risen from 2,200 in 1971 to around 7,600 in 1997 (4,100 in males and 3,500 in females) (Table 15.1). Diffuse (or not otherwise specified) lymphomas account for the majority of cases (86%), while around 8% are nodular or follicular lymphomas. The incidence of NHL increases steeply from middle age, levelling off at ages 75 and over in both males and females (Figures 15.1). About half of all cases occur in the age range 60-79 (Figure 15.2).

The directly age-standardised incidence rate showed a three-fold increase for both sexes, from 5.2 per 100,000 in 1971 to an estimated 14.9 in 1997 in males and 3.5 per 100,000 to 10.1 in females. The rise occurred steadily across the 1970s and 1980s (Figure 15.5). The disease is more common in males than females, with a sex ratio between the age-standardised rates of 1.5. The largest increases in incidence have occurred amongst the elderly, with a five-fold rise in males aged 85 or over (from 15 in 1971 to 77 per 100,000 in 1997) and a four-fold rise for females of the same age (from 11 to almost 50 per 100,000). There was an almost four-fold rise in the rates for males and females aged 65-84 years. Rates doubled in young adults (aged 20-44) but there was little change in the rates for those aged under 20. The trends in incidence by birth

cohort show a steeper rise for both males and females born in 1930s onwards (Figure 15.6).

There were 4,000 deaths from NHL in England and Wales in 1999 (2,100 in males and 1,900 in females) (Table 15.1). The number of deaths more than doubled between 1950 to 1977 after which there was an even steeper rise. Increases in mortality at all ages reflect those occurring in incidence, with particularly large rises in the elderly (Figure 15.4). The age-standardised death rate in males rose from 2.1 (in 1950) to 4.1 per 100,000 (in 1977) and then to 7.6 per 100,000 in 1995 (Figure 15.7). In females, the rise in mortality accelerated slightly in the early 1980s. Overall, there was a four-fold rise in mortality in females, from 1.2 to 4.8 per 100,000, between 1950 and 1999. The cohort pattern for mortality showed the same long term rise as for incidence up to the late 1930s (Figure 15.8). The subsequent levelling off in mortality is consistent with the observed improvements in survival (see below).

There was a slight negative gradient in the incidence of NHL with deprivation category for males (higher rates in the more affluent groups) but not for females (Figure 15.9). There was no significant pattern in mortality by deprivation category for either men or women (Figure 15.10).

There appear to be south-north gradients in both the incidence of and mortality from NHL across England and Wales, with rates above average in the south-east and south-west, and below average in the midlands and the north (Figures 15.11 and 15.12).

The incidence of NHL is generally higher in western Europe and North America than in eastern Europe and Asia, although, in children, it is more common in Africa and the Middle East (Figure 15.13). The highest rates occurred in the USA for both males and females (at 16.0 per 100,000 and 9.8 in males and females, respectively), with the next highest rates in Australia and Italy (at 12-13 in males and 8-9 in females). The rates in England and Wales were similar to those in Sweden and the Netherlands. The lowest rates were in India, China and Estonia (around 4 per 100,000 in males and around 2 in females). The international variation in mortality is similar to that for incidence with the highest rates occurring in the USA and Australia, although survival in Brazil is poor (Figure 15.14); the rates in England and Wales are comparable to those in other European countries.

Prevalence
About 60% of both males and females diagnosed in the three years 1990-92 were still alive at the beginning of 1993, as were about 44% of those diagnosed in the 10 years 1983-92 (Figure 15.15).

Survival
Relative survival from non-Hodgkin's lymphoma was around 66% at one year after diagnosis in both men and women, and around 46% at five years after diagnosis, for patients diagnosed in 1991-93. There was a substantial increase in five-year survival of 14% between the early 1970s and the late 1980s (Figure 15.16).

Table 15.1: Key statistics for non-Hodgkin's lymphoma on incidence, mortality and survival by sex, region and country, latest year available

MALES	Incidence				Mortality				Relative survival (%) patients diagnosed in 1986-90	
	Number of cases	CR	ESR*	WSR*	Number of deaths	CR	ESR*	WSR*	One-year	Five-year
United Kingdom [a]	4,500	15.6	14.7	10.6	2,370	8.2	7.4	5.0	..	..
Great Britain [a]	4,370	15.6	14.6	10.5	2,310	8.2	7.4	5.0	..	..
England and Wales [b]	4,110	16.0	14.9	10.8	2,090	8.1	7.3	4.9	65	45
England [b]	3,730	15.4	14.3	9.6	1,950	8.0	7.2	4.9	66	46
Northern and Yorkshire [b]	470	14.9	13.7	9.6	240	7.8	7.0	4.8	60	41
Trent [b]	360	14.2	13.0	9.3	170	6.7	6.0	4.2	61	41
Anglia and Oxford [b]	390	14.6	13.9	10.1	230	8.6	7.8	5.1	63	44
North Thames [b]	570	16.5	16.5	12.2	270	7.8	7.8	5.3	71	52
South Thames [b]	530	15.7	14.6	10.7	300	8.8	7.8	5.3	71	49
South and West [b]	690	21.0	18.2	13.4	290	8.9	7.2	4.9	69	47
West Midlands [b]	300	11.2	10.3	7.3	220	8.3	7.4	5.0	63	43
North West [b]	440	13.6	12.8	9.5	230	7.1	6.6	4.5	62	44
Wales [b]	230	16.3	14.5	10.5	130	9.3	7.8	5.4	61	41
Scotland [a]	431	17.3	16.5	11.9	182	7.3	6.8	4.6	60	44
Northern Ireland [a]	126	15.4	16.7	12.3	65	7.9	8.3	5.7	..	..
FEMALES										
United Kingdom [a]	3,940	13.1	10.2	7.3	2,110	7.0	4.7	3.2	..	..
Great Britain [a]	3,800	13.1	10.1	7.3	2,040	7.0	4.7	3.1	..	..
England and Wales [b]	3,530	13.3	10.1	7.3	1,930	7.2	4.9	3.4	65	45
England [b]	3,200	12.8	9.7	7.0	1,800	7.2	4.9	3.3	65	45
Northern and Yorkshire [b]	410	12.7	9.5	6.9	240	7.4	5.3	3.6	59	42
Trent [b]	310	11.9	8.8	6.2	180	6.8	4.5	3.1	60	41
Anglia and Oxford [b]	370	13.4	10.8	7.9	180	6.5	4.7	3.2	64	43
North Thames [b]	400	11.4	9.5	6.7	260	7.2	5.3	3.6	71	52
South Thames [b]	500	14.1	10.7	7.6	260	7.4	5.0	3.4	72	50
South and West [b]	560	16.5	11.7	8.4	290	8.6	5.0	3.4	69	48
West Midlands [b]	250	9.4	7.0	5.0	180	6.5	4.5	3.1	61	42
North West [b]	410	12.2	9.4	6.8	230	6.8	4.7	3.2	63	43
Wales [b]	200	13.4	10.0	7.4	120	8.2	5.3	3.8	56	39
Scotland [a]	420	15.9	12.3	8.7	223	8.5	5.7	3.7	60	42
Northern Ireland [a]	134	15.7	13.8	9.9	71	8.2	6.8	4.7	..	..

* *Directly age-standardised rate*
[a] *1996 incidence; 1998 mortality.*
[b] *1997 incidence; 1999 mortality.*
Figures for England and Wales are provisional: incidence - 1995-1997, mortality - 1999

Survival from NHL decreases with age for both sexes. Five-year relative survival was 60-70% for adults aged under 50 at diagnosis, but fell steadily to about 20-30% for those aged 80-99; women have a 5-6% points advantage in survival over men at all ages (Figure 15.17).

There were deprivation gaps in both one- and five-year relative survival of around 7% points for patients diagnosed in 1986-90, with patients in the most deprived group showing a survival disadvantage compared with those in the most affluent group (Figure 15.18). There was some variation in survival across the health regions, with above average survival in the Thames regions and the South and West, and below average survival in Trent, for men in the Northern and Yorkshire and for women in the West Midlands and Wales.

Survival in England and Wales from NHL is similar to that in Scotland, and slightly lower than the average for Europe (by about 3% points) (Figure 15.19). Five-year survival in England and Wales was 4% lower in men than for the USA, but 11% lower for women.

Figure 15.1 **Age-specific incidence, England and Wales, 1997**

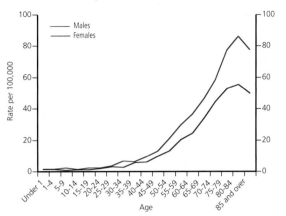

Figure 15.2 **Frequency distribution of new cases by age-group, England and Wales, 1997**

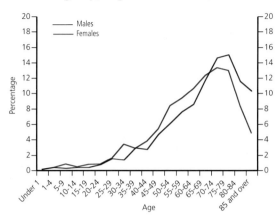

Figure 15.3 **Age-specific incidence, England and Wales, 1971-1997**

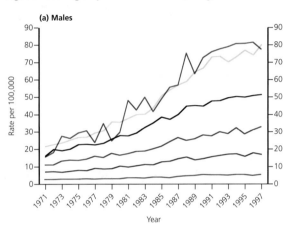

(a) Males

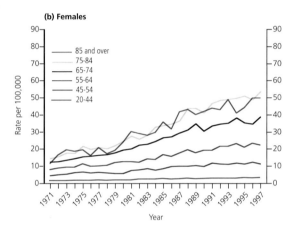

(b) Females

Figure 15.4 **Age-specific mortality, England and Wales, 1950-1999**

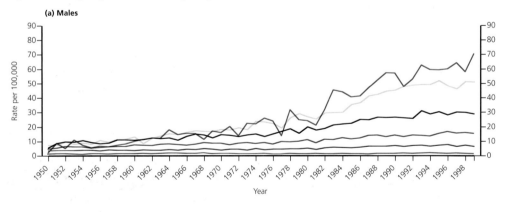

(a) Males

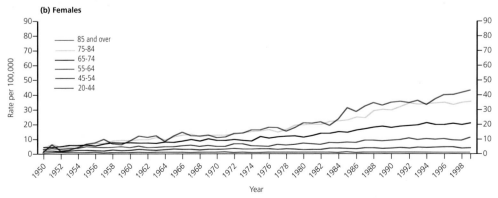

(b) Females

Figure 15.5 **Age-standardised incidence, England and Wales, 1971-1997**

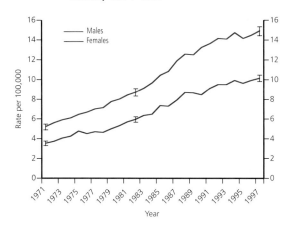

Figure 15.6 **Cohort incidence ratio by sex, England and Wales**

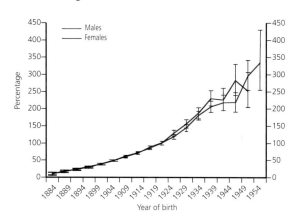

Figure 15.7 **Age-standardised mortality, England and Wales, 1950-1999**

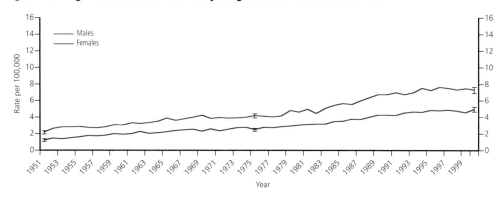

Figure 15.8 **Cohort mortality ratio by sex, England and Wales**

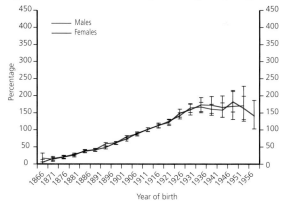

Figure 15.9 **Age-standardised incidence by deprivation category, England and Wales, 1992-93**

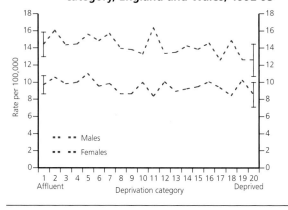

Figure 15.10 **Age-standardised mortality by deprivation category, England and Wales, 1992-93**

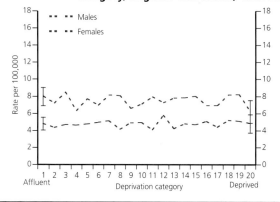

Figure 15.11 **Comparative incidence ratio (CIR) by health region and sex (England and Wales = 100), 1997**

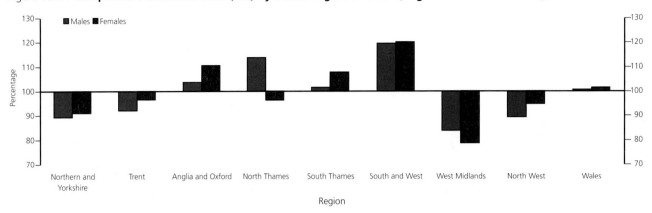

Figure 15.12 **Comparative mortality ratio (CMR) by health region and sex (England and Wales = 100), 1999**

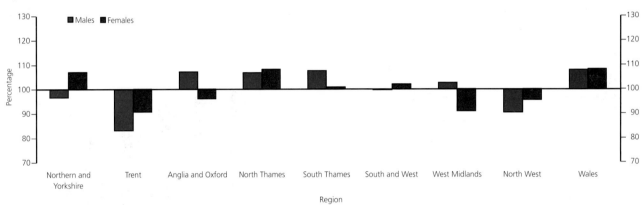

Figure 15.13 **Age-standardised international incidence**

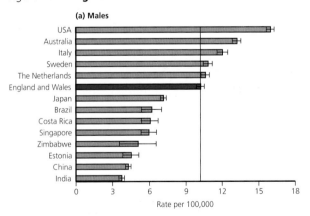
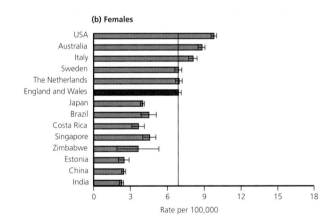

Figure 15.14 **Age-standardised international mortality**

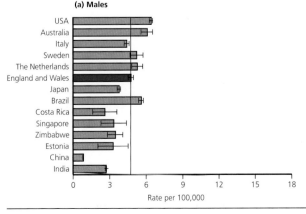
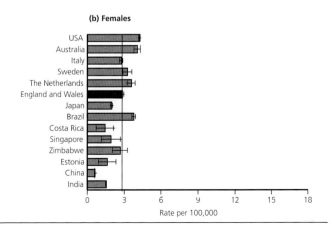

Figure 15.15 **Number and percentage of cases diagnosed in 1990-92 and 1983-92 and alive on 1st January 1993, England and Wales**

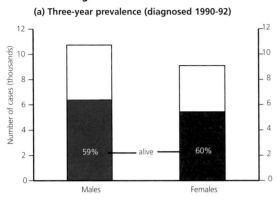

(a) Three-year prevalence (diagnosed 1990-92)

(b) Ten-year prevalence (diagnosed 1983-92)

Figure 15.16 **Relative survival up to ten years, men, England and Wales, 1971-75 to 1991-93**

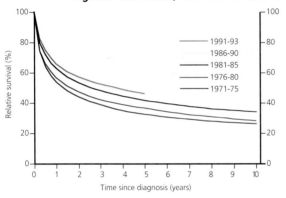

Figure 15.17 **Five-year relative survival by age, England and Wales, patients diagnosed 1991-93**

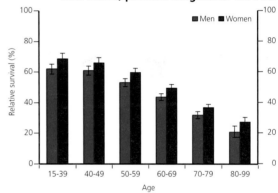

Figure 15.18 **One- and five-year relative survival by deprivation category, England and Wales, adults diagnosed 1986-90**

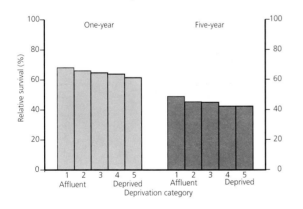

Figure 15.19 **International comparison of five-year relative survival, England and Wales, adults diagnosed 1986-90**

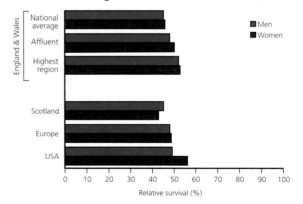

Chapter 16

OESOPHAGUS

Oesophageal cancer is the seventh most common cancer in men and the thirteeth most common in women in England and Wales. Like lung cancer the disease has a strong association with smoking and survival is extremely poor. However, this association does not explain the recent trends in the disease. The incidence of lung cancer has fallen as tobacco smoking has reduced, while the incidence of oesophageal cancer has continued to rise. Alcohol consumption has been suggested as the driving force behind this increase in incidence through its combined (and multiplicative) affect with smoking.[1] Oesophageal cancer is also associated with poverty and nutritional deficiencies, while higher consumption of fruits and vegetables are protective. Each of these risk factors indicates the potential for the prevention of this disease.

Similarities in the incidence of adenocarcinomas of the middle and lower thirds of the oesophagus and of the adjacent part of the stomach, the cardia, have been observed, with rises occurring in a number of countries which appear unrelated to alcohol consumption, smoking, or *Helicobacter pylori* infection.[2,3,4] This disease may be associated with gastro-oesophageal reflux disease and Barrett's oesophagus.[5]

The high risk areas of the world include the 'Asian oesophageal cancer belt' which stretches from the Caspian Sea in northern Iran through the former southern republics of the USSR (Turkmenistan, Kazakhstan and Uzbekistan) to western and northern China; south-eastern Africa; parts of eastern South America; and certain defined areas of western Europe (particularly parts of France and Switzerland). Half of the worldwide cases occur in China. There are also quite marked differences in incidence between ethnic groups – in the USA, rates in black people are four times higher than in white people, and in Singapore rates in Chinese are double those in Indians and ten times those in Malays.

Incidence and mortality

There were 6,100 new cases of oesophageal cancer diagnosed in England and Wales in 1997, 3,600 in males and 2,500 in females (Table 16.1). Cancer of the oesophagus is twice as common in men than in women: the male:female ratio of the age-standardised rates has been around 2:1 since the mid-1980s. The disease is more common at older ages, and is rare below the age of 50 (Figures 16.1 and 16.2). Around two-thirds of oesophageal cancers were registered with a non-specific subsite code. The majority of the remainder (around 25%) were sited in the lower third of the oesophagus, with a further 8% in the middle third and only 3% in the upper third. The morphological distribution of oesophageal cancer varied between the sexes. Over 40% of cases in males were adenocarcinomas, with a further 22% being squamous carcinomas, 3% were mucinous neoplasms and 32% were non-specific epithelial carcinomas. Squamous carcinomas were the most common form of oesophageal cancer in females, accounting for 40% of cases, with 20% being adenocarcinomas. The remainder of cases in females were non-specific epithelial carcinomas.

The incidence of the disease has increased steadily over the past two decades from 7.7 per 100,000 in males in 1971 to 12.7 in 1997, a rise of 62% (Figure 16.5). Incidence in females has also increased but by a lesser degree from 4.2 to 5.8 per 100,000, a rise of 37%. Substantial rises occurred in all age groups from 1971 to 1997 (Figure 16.3). The trends in incidence for successive cohorts of males and females born from the 1880s show a continuous increase for males up to those born in the 1950s, but for females incidence levelled off (possibly declining slightly) for those born in the 1930s onwards (Figure 16.6).

The trends in mortality closely follow those for incidence due to the very poor survival of this disease. The increases in incidence in males since 1970 are reflected in the mortality trends which started to rise in the mid-1960s, reaching almost 13 deaths per 100,000 in 1999 (Figure 16.7). In the 1950s, there was a small decline in mortality in males. In the 1950s and early 1960s the rate for females remained level at just below 4 deaths per 100,000, after which mortality rose very gradually to reach around 5 per 100,000 in the 1990s. Increases occurred in each of the age groups from 45-54 upwards in both males and females (Figure 16.4). The trends in mortality by year of birth were strikingly similar to those for incidence, with increases for males up to the most recent birth cohort while for females, mortality peaked in those born in the late 1920s (Figure 16.8).

The incidence of oesophageal cancer shows a positive gradient with Carstairs deprivation category for both sexes (Figure 16.9), with rates around 30% higher in the most deprived groups. The pattern of mortality with deprivation was closely similar to that for incidence (Figure 16.10).

The incidence of and mortality from oesophageal cancer are above average in the North West region (but not in the rest of the north of England), and generally low in Anglia and Oxford and the south east regions of England (Figures 16.11 and 16.12).

Incidence in England and Wales is around double the rate for a number of other developed countries including the USA, other parts of Europe and Australia (Figure 16.13). The international patterns indicate the relatively high mortality in England and Wales – similar to that in India, Japan and Brazil although well below the noted levels of the disease in China and Zimbabwe (Figure 16.14).

Prevalence

For both males and females, 25% of cases diagnosed with oesophageal cancer in 1990-92 were still alive at the beginning of 1993, as were about 13% of those diagnosed in the 10 years 1983-92 (Figure 16.15).

Survival

The extremely poor five-year relative survival from oesophageal cancers has shown little improvement since the early 1970s when it was 3% for men and 5% for women. For patients diagnosed in

Table 16.1: Key statistics for oesophageal cancer on incidence, mortality and survival by sex, region and country, latest year available

| | Incidence | | | | Mortality | | | | Relative survival (%) patients diagnosed in 1986-90 | |
MALES	Number of cases	CR	ESR*	WSR*	Number of deaths	CR	ESR*	WSR*	One-year	Five-year
United Kingdom [a]	4,100	14.2	13.0	8.7	4,290	14.7	13.2	8.7	..	..
Great Britain [a]	4,010	14.3	13.1	8.7	4,200	14.8	13.2	8.7	..	..
England and Wales [b]	3,560	13.8	12.5	8.3	3,720	14.4	12.8	8.4	23	6
England [b]	3,350	13.8	12.5	8.5	3,500	14.4	12.8	8.4	23	6
Northern and Yorkshire [b]	450	14.5	12.9	8.5	480	15.4	13.6	9.0	21	5
Trent [b]	340	13.4	11.7	7.8	360	14.2	12.4	8.2	21	6
Anglia and Oxford [b]	350	13.1	12.1	8.0	360	13.1	12.0	7.7	22	5
North Thames [b]	370	10.8	10.9	7.2	380	10.9	10.8	7.1	30	8
South Thames [b]	420	12.4	11.1	7.4	430	12.8	11.4	7.5	25	5
South and West [b]	500	15.2	12.7	8.6	550	16.7	13.1	8.4	27	7
West Midlands [b]	400	15.2	13.7	9.1	390	14.6	13.1	8.6	22	6
North West [b]	520	16.0	14.9	10.0	560	17.2	15.7	10.4	20	6
Wales [b]	210	14.6	12.3	8.1	220	15.1	12.5	8.3	22	7
Scotland [a]	484	19.5	18.6	12.6	471	19.0	17.5	11.7	23	5
Northern Ireland [a]	90	11.0	11.9	7.9	93	11.2	12.3	8.3	..	..
FEMALES										
United Kingdom [a]	2,960	9.9	6.2	4.0	2,520	8.4	5.0	3.1	..	..
Great Britain [a]	2,900	10.0	6.2	4.0	2,460	8.4	4.9	3.1	..	..
England and Wales [b]	2,500	9.4	5.8	3.7	2,300	8.7	5.2	3.3	24	7
England [b]	2,330	9.3	5.8	3.7	2,170	8.6	5.2	3.3	24	7
Northern and Yorkshire [b]	310	9.7	5.8	3.7	270	8.3	4.9	3.1	22	6
Trent [b]	240	9.4	5.7	3.6	230	9.0	5.3	3.4	23	8
Anglia and Oxford [b]	190	7.0	4.5	2.8	230	8.2	5.3	3.3	24	9
North Thames [b]	300	8.4	5.6	3.5	210	5.9	3.9	2.4	28	9
South Thames [b]	300	8.5	5.1	3.3	310	8.6	4.8	2.9	28	8
South and West [b]	360	10.5	5.9	3.8	320	9.5	5.1	3.2	25	8
West Midlands [b]	250	9.2	5.8	3.8	250	9.3	5.9	3.8	23	7
North West [b]	380	11.3	7.5	4.9	350	10.4	6.4	4.1	21	7
Wales [b]	170	11.4	6.4	4.0	140	9.2	5.1	3.2	22	8
Scotland [a]	351	13.3	8.8	5.7	259	9.8	6.1	3.8	24	8
Northern Ireland [a]	65	7.6	5.8	3.7	61	7.1	5.2	3.2	..	..

* *Directly age-standardised rate*
[a] *1996 incidence; 1998 mortality.*
[b] *1997 incidence; 1999 mortality.*
Figures for England and Wales are provisional: incidence - 1995-1997, mortality - 1999

1991-93, five-year relative survival was 6% and 7% for men and women, respectively. There was, however, greater improvement in survival within the first few years of diagnosis: one-year relative survival increased by just over 3% every five years between 1971-75 and 1986-90, to reach 21% in men and 25% in women, with a further rise of 2% for men diagnosed in 1991-93 (Figure 16.16).

Women have consistently shown a survival advantage compared with men, by a comparatively large amount given the very low levels of survival for this cancer. Survival at five years after diagnosis was almost twice as high for women aged 50-59 than for men of the same age: 14% compared with 8%. Survival declines with age for both sexes and was only 2% for patients aged 80 or over at diagnosis in 1991-93 (Figure 16.17).

There was little variation in survival between regions. There was also no gradient in five-year survival by deprivation category; however, there was a gap of almost 4% in one-year survival for patients diagnosed in 1986-90, with those in the most deprived group showing worse survival than those in the most affluent group (Figure 16.18). Survival in England and Wales was similar to that in Scotland, although slightly lower than the average for Europe in the EUROCARE II study and rates in the USA (Figure 16.19).

Figure 16.1 **Age-specific incidence, England and Wales, 1997**

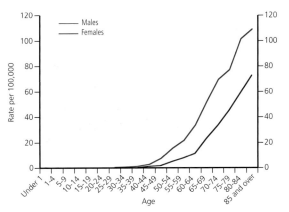

Figure 16.2 **Frequency distribution of new cases by age group, England and Wales, 1997**

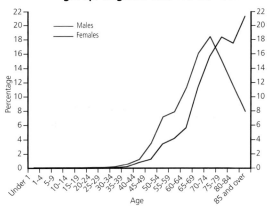

Figure 16.3 **Age-specific incidence, England and Wales, 1971-1997**

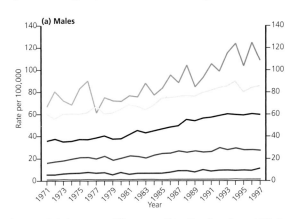

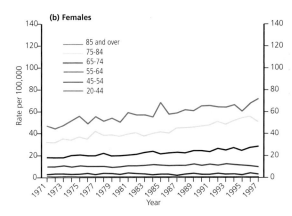

Figure 16.4 **Age-specific mortality, England and Wales, 1950-1999**

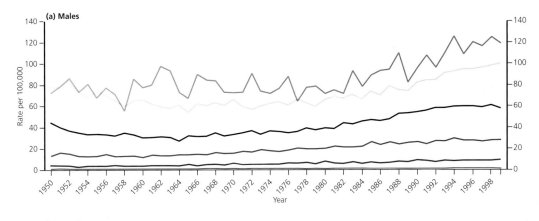

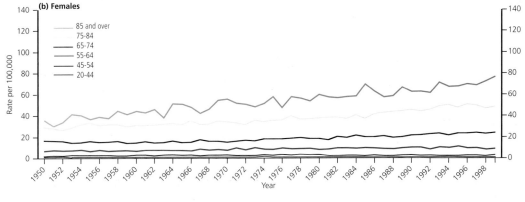

Figure 16.5 **Age-standardised incidence, England and Wales, 1971-1997**

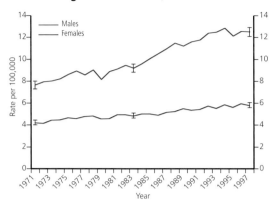

Figure 16.6 **Cohort incidence ratio by sex, England and Wales**

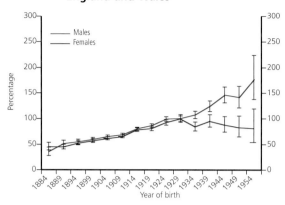

Figure 16.7 **Age-standardised mortality, England and Wales, 1950-1999**

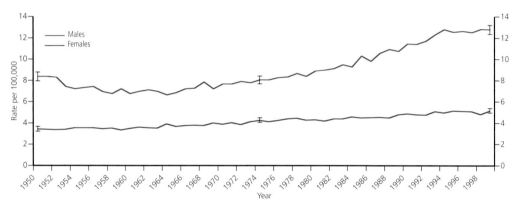

Figure 16.8 **Cohort mortality ratio by sex, England and Wales**

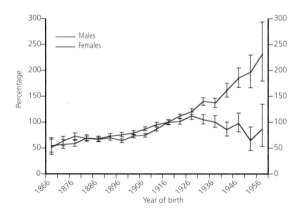

Figure 16.9 **Age-standardised incidence by deprivation category, England and Wales, 1992-93**

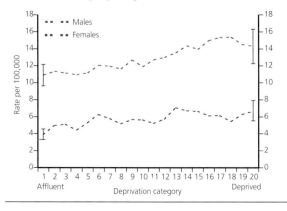

Figure 16.10 **Age-standardised mortality by deprivation category, England and Wales, 1992-93**

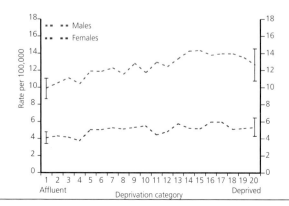

Figure 16.11 **Comparative incidence ratio (CIR) by health region and sex (England and Wales = 100), 1997**

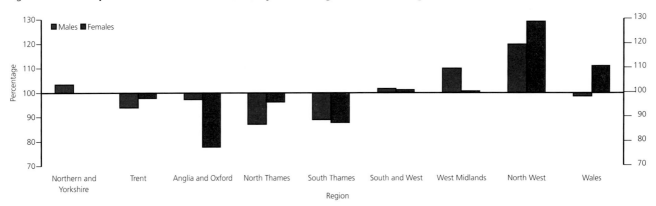

Figure 16.12 **Comparative mortality ratio (CMR) by health region and sex (England and Wales = 100), 1999**

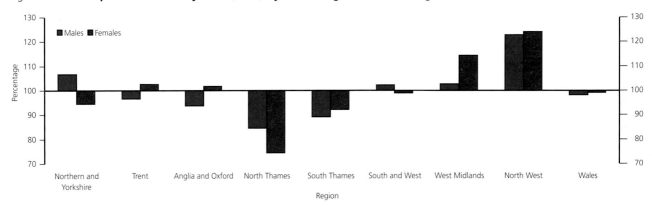

Figure 16.13 **Age-standardised international incidence**

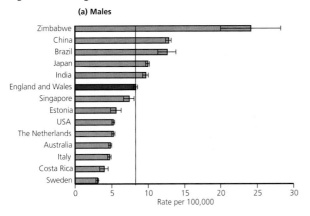

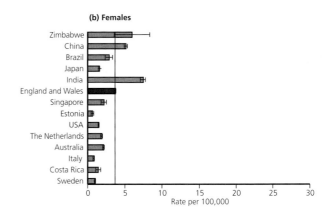

Figure 16.14 **Age-standardised international mortality**

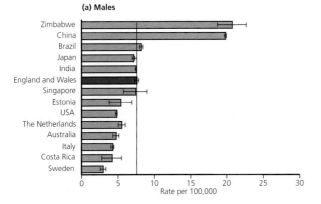

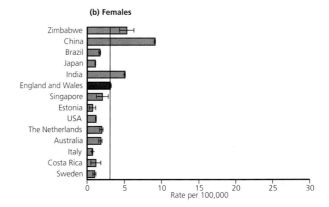

Figure 16.15 **Number and percentage of cases diagnosed in 1990-92 and 1983-92 and alive on 1st January 1993, England and Wales**

(a) Three-year prevalence (diagnosed 1990-92)

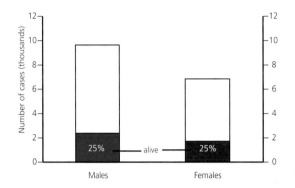

(b) Ten-year prevalence (diagnosed 1983-92)

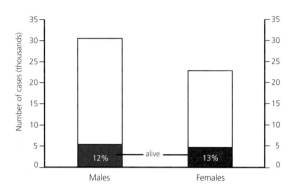

Figure 16.16 **Relative survival up to ten years, men, England and Wales, 1971-75 to 1991-93**

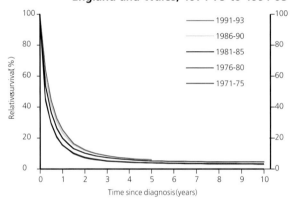

Figure 16.17 **Five-year relative survival by age, England and Wales, patients diagnosed in 1991-93**

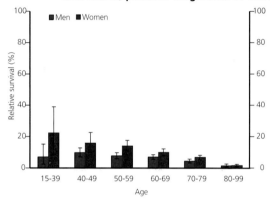

Figure 16.18 **One- and five-year relative survival by deprivation category, England and Wales, adults diagnosed 1986-90**

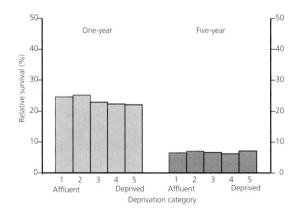

Figure 16.19 **International comparison of five-year relative survival, England and Wales, adults diagnosed 1986-90**

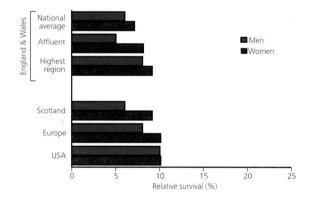

Chapter 17

OVARY

Ovarian cancer is predominantly a disease of older, post-menopausal women. It is an important cause of morbidity and mortality around the world, with an estimated 165,500 cases a year in 1990, making it the sixth most common cancer amongst women worldwide. The majority of ovarian cancers are epithelial. Germ cell tumours account for only 1-2%, but they are more common in young women. Survival is generally poor, partly because the disease has often spread widely before it is diagnosed.

The incidence of ovarian cancer is quite strongly correlated with that of breast and uterine cancers, and like them it is related to hormone levels so it is not surprising that ovarian cancer shares many of their risk factors.[1,2,3] A woman's history of ovulation appears to play a role in the development of the disease. Higher rates occur among women who do not have children, and the risk decreases with the number of pregnancies. A raised risk has also been reported for women who had a late menopause or who are infertile.[4] A lower risk is associated with oral contraceptive use, female sterilisation, and hysterectomy with conservation of the ovaries.[5] About 5% of ovarian cancers may be hereditary. A UK register of familial ovarian cancer was started in 1989. International studies using pooled data from this registry and others in Europe and the USA have helped to identify the genes responsible. In the commonest of the three groups of familial ovarian cancer, it is linked with breast cancer.[6] Rates for Japanese women who migrate to the USA rise to match those in the resident population, suggesting that environmental factors play a role.

Incidence and mortality

In England and Wales there were just over 6,100 new registrations of ovarian cancer in 1997, making it the fourth most common site in women (after breast, colorectal and lung) (Table 17.1). Just under 1% of cancers coded to ovary occur in the Fallopian tubes. About one third of all tumours are adenocarcinomas, one third are cystic, mucinous or serous neoplasms, and the remaining third are poorly specified carcinomas. Ovarian cancer accounts for just over 4% of all cancers in women; there are now about 30% more cases than of uterine cancer and, with the impact of the improved cervical screening programme, almost double the number of cases of cervical cancer. But because of its much poorer survival, it accounts for 6% of all cancer deaths in women, and for more deaths than all the other gynaecological cancers combined.

The disease is rare in pre-menopausal women: under 10% of cases were in women aged under 45. Rates rose steeply after the menopause (Figure 17.1), and the peak in the age distribution was in the 70-74 age group (Figure 17.2). The incidence of ovarian cancer in women aged under 55 has been fairly stable since the early 1970s, but rates in older women have increased steadily (Figure 17.3). The overall age-standardised incidence rate rose steadily from the early 1970s to reach 19 per 100,000 in 1997 (Figure 17.5). The pattern in the risk of ovarian cancer by birth cohort is similar to that for breast cancer, with rates increasing up to the cohorts born around 1930 and then declining – but rates

appear to have increased again for women born from the early 1960s onwards (Figure 17.6).

Mortality from ovarian cancer in women aged 75 and over doubled from 1950 to 1999 and increased by almost 50% for women aged 65-74 (Figure 17.4). Mortality in women aged 55-64 also rose, but less steeply, until the late 1970s, after which the rate stabilised. In younger women (45-54 years) rates were stable until the late 1970s, since when they have fallen by around 40%. From the beginning of this century up to the 1980s, mortality in England and Wales showed continuous marked increases for all women aged over 40 years. Some of this long term increase may have been the result of more accurate diagnosis, especially in the older age groups, but the remainder must be due to an underlying increase in incidence, as there has not been a large improvement in survival rates (see below). The overall age-standardised rate rose by around 20% from 1950 to 1970 and has remained at about 12 per 100,000 since then (Figure 17.7). Mortality by birth cohort, as for incidence, increased up to the 1920s and 1930s, but then stabilised in women born in the 1950s onwards (Figure 17.8).

There were slight *inverse* gradients in both the incidence of and mortality from ovarian cancer with deprivation (although rates fluctuated from category to category) (Figures 17.9 and 17.10). This pattern is similar to, but less clear than, that for breast cancer. There were no strong geographical patterns in ovarian cancer, although both incidence and mortality were below average in Northern and Yorkshire and above average in Anglia and Oxford (Figures 17.11 and 17.12).

Incidence rates vary by a factor of three around the world, with the highest rates in women in northern and western Europe and in North America, and the lowest rates in China, Japan and India (Figure 17.13). Black women in North America have slightly lower rates than white women. Incidence within Europe varies by a factor of two; in England and Wales it was similar to that for Sweden, but higher than the Netherlands and Italy. The international pattern in mortality is similar to that in incidence (Figure 17.14).

Prevalence

About half of women diagnosed with ovarian cancer in the three-year period 1990-92 were alive at the beginning of 1993, as were one-third of those diagnosed in the 10 years 1983-92 (Figure 17.15).

Survival

Survival from ovarian cancer in 1991-93 was the worst of all the gynaecological cancers. One-year relative survival was 58%, and at five years after diagnosis, when women with ovarian cancer were experiencing almost the same mortality as in the general population, survival was 29%, about half that at one year (Figure 17.16). Relative survival at ten years after diagnosis was only about 4% points below five-year survival for women diagnosed in 1981-85. Survival improved during the 1970s and 1980s, from 42% after one year and 21% after five years for cases diagnosed in 1971-75.

Table 17.1: Key statistics for ovarian cancer on incidence, mortality and survival by region and country, latest year available

| | Incidence | | | | Mortality | | | | Relative survival (%) patients diagnosed in 1986-90 | |
| | Number of cases | Rate per 100,000 | | | Number of deaths | Rate per 100,000 | | | | |
		CR	ESR*	WSR*		CR	ESR*	WSR*	One-year	Five-year
United Kingdom [a]	6,570	21.9	18.2	13.2	4,520	15.0	11.6	8.0	..	..
Great Britain [a]	6,430	22.1	18.3	13.2	4,430	15.1	11.6	8.1		
England and Wales [b]	6,130	23.1	19.0	13.7	3,940	14.8	11.2	7.7	56	29
England [b]	5,740	22.9	18.9	13.7	3,680	14.7	11.2	7.7	56	29
Northern and Yorkshire [b]	680	21.1	17.2	12.4	450	13.9	10.3	7.1	53	27
Trent [b]	510	19.8	16.1	11.6	380	14.7	11.3	7.8	57	31
Anglia and Oxford [b]	650	23.9	21.0	15.4	420	15.2	12.4	8.6	56	31
North Thames [b]	720	20.3	18.0	13.0	460	13.0	11.0	7.6	63	34
South Thames [b]	800	22.8	18.3	13.1	530	14.9	11.2	7.7	60	25
South and West [b]	950	28.0	21.2	15.5	550	16.1	11.0	7.6	52	25
West Midlands [b]	640	23.7	19.8	14.3	400	14.7	10.8	7.4	56	31
North West [b]	790	23.5	19.3	13.8	500	14.9	11.4	7.9	52	27
Wales [b]	390	26.4	20.8	15.2	260	17.3	12.2	8.4	53	30
Scotland [a]	622	23.5	18.9	13.5	408	15.5	11.5	8.0	52	28
Northern Ireland [a]	141	16.5	16.0	11.8	94	10.9	9.4	6.5	..	..

* *Directly age-standardised rate*
[a] *1996 incidence; 1998 mortality.*
[b] *1997 incidence; 1999 mortality.*
Figures for England and Wales are provisional: incidence - 1995-1997, mortality - 1999

Although five-year survival for cases diagnosed in 1991-93 in women aged under 40 was around 70%, rates dropped steeply to 45% in women aged 40-49 and to just 12% in the most elderly women (Figure 17.17).

Differences in survival from ovarian cancer between deprivation categories are more pronounced at one year (6% points) than at five years (2% points) (Figure 17.18). This suggests that diagnosis among women in more affluent groups tends to occur earlier. These results are consistent with those from a detailed study of 6,000 women diagnosed in the South Thames Region during 1980-89 which found no consistent difference in survival between deprivation categories.[7] There was no marked variation in survival among the regions.

Survival in England and Wales at both one and five years since diagnosis had been lower than in Scotland in the early 1970s, but in Scotland survival improved by less (by only around 5% for both one-and five-year survival) so the results for the late 1980s were similar. The pattern of survival with age in Scotland in the late 1980s was also similar to that in England and Wales. Results from the second EUROCARE study showed that survival from ovarian cancer in England (and in Scotland) was 6% points below the European average, markedly lower than in several of the other countries – or in parts of them – and only slightly higher than in Slovenia, Poland and Estonia. The European average survival was itself 10% points below that in the USA (Figure 17.19).

Surgery can be curative in early stage disease, with a five-year relative survival of around 80%. Survival is about 60% if growth is limited to the pelvis, but falls to about 25% if growth extends to the abdominal cavity and to only 15% if there are metastases.[8] Unfortunately, 65-75% of women present with advanced stage disease because symptoms are often vague and non-specific. For these cases, where survival rates are much lower, chemotherapy is the main adjuvant treatment; radiotherapy is rarely used because in late stage disease large areas would have to be targeted. Recent evidence from a large randomised controlled trial indicates that earlier detection by screening does reduce mortality.[9]

Figure 17.1 **Age-specific incidence, England and Wales, 1997**

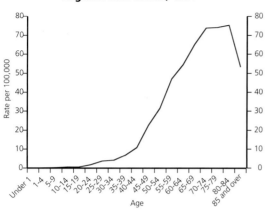

Figure 17.2 **Frequency distribution of new cases by age group, England and Wales, 1997**

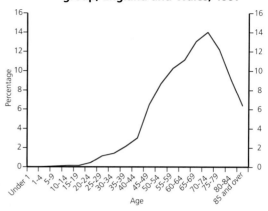

Figure 17.3 **Age-specific incidence, England and Wales, 1971-1997**

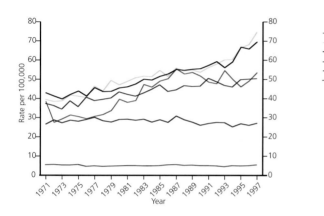

Figure 17.4 **Age-specific mortality, England and Wales, 1950-1999**

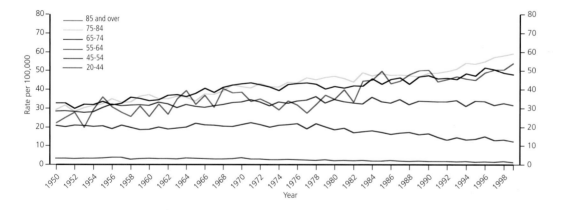

Figure 17.5 **Age-standardised incidence ,
England and Wales, 1971-1997**

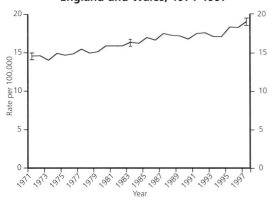

Figure 17.6 **Cohort incidence ratio, England and Wales**

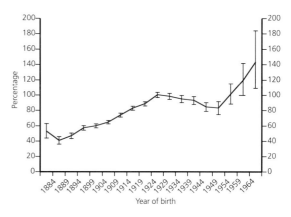

Figure 17.7 **Age-standardised mortality, England and Wales, 1950-1999**

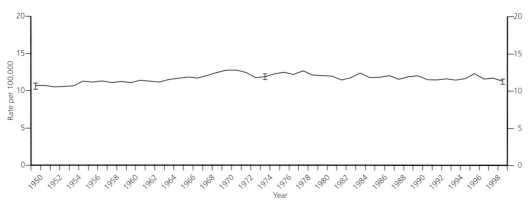

Figure 17.8 **Cohort mortality ratio, England and Wales**

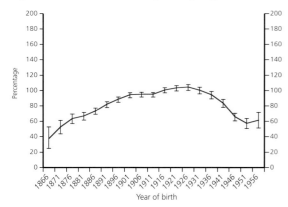

Figure 17.9 **Age-standardised incidence by deprivation
category, England and Wales, 1993**

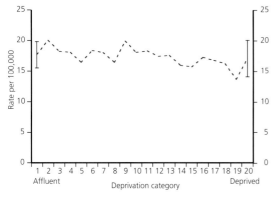

Figure 17.10 **Age-standardised mortality by deprivation
category, England and Wales, 1993**

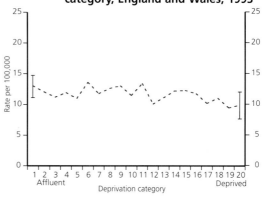

Figure 17.11 **Comparative incidence ratio (CIR) by health region (England and Wales = 100), 1997**

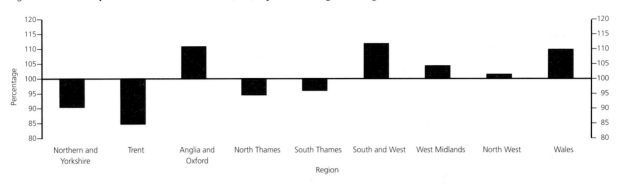

Figure 17.12 **Comparative mortality ratio (CMR) by health region (England and Wales = 100), 1999**

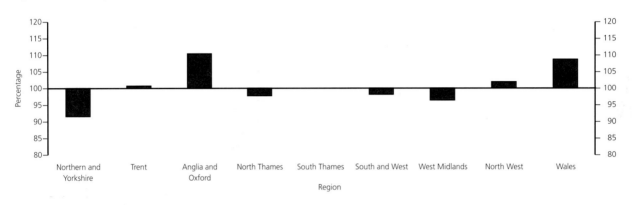

Figure 17.13 **Age-standardised international incidence**

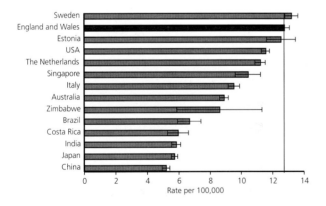

Figure 17.14 **Age-standardised international mortality**

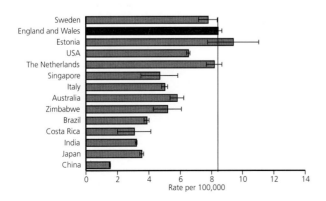

Figure 17.15 **Number and percentage of cases diagnosed in 1990-92 and 1983-92 and alive 1st January 1993, England and Wales**

(a) Three-year prevalence (diagnosed 1990-92)

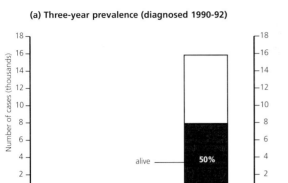

(b) Ten-year prevalence (diagnosed 1983-92)

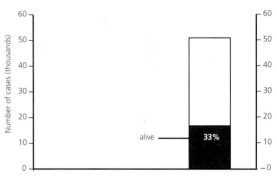

Figure 17.16 **Relative survival up to ten years, England and Wales, 1971-75 to 1991-93**

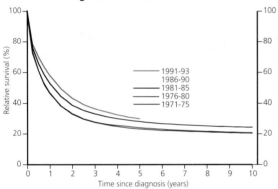

Figure 17.17 **Five-year relative survival by age, England and Wales, 1991-93**

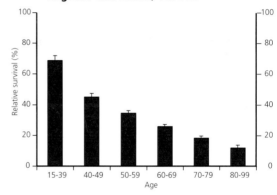

Figure 17.18 **One- and five-year relative survival by deprivation category, England and Wales, 1986-90**

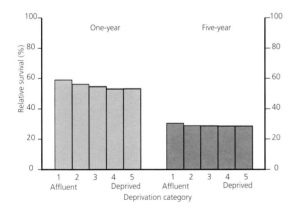

Figure 17.19 **International comparison of five-year relative survival, England and Wales, 1986-90**

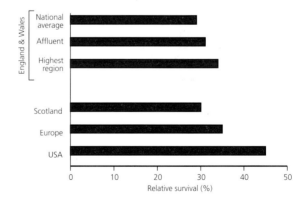

Chapter 18

PANCREAS

Pancreatic cancer is the tenth most common cancer in males and the eleventh most common in females in England and Wales, accounting for around 3% of all cancers. It is rapidly fatal with few patients surviving more than a few years. The only factor to emerge consistently as a risk for pancreatic cancer is tobacco smoking. In major cohort studies of smoking, pancreatic cancer risk has almost always been higher in smokers than in non-smokers.[1] Also, similarities have been observed in the age-specific trends in mortality from cancers of the pancreas and lung in England and Wales, and it has been suggested that the cohort-specific changes were related to differences in the prevalence of smoking among males and females in different age groups. The cohort studies each showed elevated risk for smokers ranging from two- to six-fold with high exposure.[2]

A number of other risk factors also have been suggested – there is some evidence that hereditary pancreatitis (involving pancreatic enzyme deficiencies) and diabetes mellitus are associated with the development of cancer of the pancreas. There is no clear evidence of any specific occupational hazard. Alcohol consumption is only linked to pancreatic cancer via its association with smoking, as is coffee consumption – which formerly had been suggested as a risk factor. There is stronger evidence, however, that a higher intake of plant foods is protective against the disease.[2]

With the exception of the link to smoking, no information is available that can be used to prevent this disease. There are currently no safe and effective means of early diagnosis and the disease is usually in an advanced stage at diagnosis due to the non-specific nature of the symptoms (such as pain and weight loss).[2] Available treatment is not known to be effective and is itself responsible for serious morbidity.

Incidence and mortality
There were around 5,700 cases of pancreatic cancer diagnosed in England and Wales in 1997 – 2,700 in males and 3,000 in females (Table 18.1). The distributions of subsite and histology were similar in males and females. Around 40% of cases were in the head of the pancreas, with 3% in the body and around 2% in the tail of the pancreas; the remainder (55%) had no specific subsite recorded. Over two-thirds of cases were coded as non-specific epithelial neoplasms, while almost 20% were coded as adenocarcinomas, and 2% as mucinous neoplasms.

The incidence of pancreatic cancer has fallen consistently in males since the early 1980s; in females, after a gradual rise in the 1970s and 1980s, it started to decline in the 1990s (Figure 18.5). The age-standardised rate for males was just under 10 per 100,000 in 1997, 18% lower than in 1971. The rate for females in 1997 was just over 7 per 100,000, a fall of 12% from the peak in 1989.

The male:female ratio of the age-standardised rates was 1.3:1 in 1997, compared with 1.7:1 in 1971. This ratio is lower than the ratio of the numbers of new cases due to the high incidence in the

elderly, among whom women greatly out number men. The disease is rare under age 45 in both sexes, but rates subsequently increase steeply with age (Figures 18.1 and 18.2).

The incidence of pancreatic cancer has fallen sharply in successive cohorts of men, starting with those born at the time of the first world war (Figure 18.6). The high risk in those born in the late 19th and early 20th centuries, while not as large, coincides with that for lung cancer and confirms the association with smoking. For women, there has been little variation across successive cohorts.

The trends in mortality of pancreatic cancer closely resemble those in incidence due to its very poor survival, and recent declines in incidence are mirrored in equivalent falls in mortality (Figure 18.7). Prior to this, from 1950 to the late 1960s mortality from pancreatic cancer rose in both sexes, by around 40% in males and 20% in females.

The age-specific mortality for males is similar to that for incidence, but in females mortality in elderly women has not shown the decline observed in incidence. The patterns in mortality across birth cohorts for both males and females were similar to, but less marked than, those for incidence (Figure 18.8).

For cases diagnosed in 1992-93, there was a small positive gradient in the incidence of pancreatic cancer with Carstairs deprivation category for both sexes, with incidence in the most deprived groups some 20% higher than in the most affluent (Figure 18.9). Both sexes show closely similar positive gradients in mortality with deprivation category to those in incidence (Figure 18.10).

There was no consistent geographical variation in pancreatic cancer in England and Wales, with several regions showing different patterns for incidence and mortality for both sexes (Figures 18.11 and 18.12). Around the world, incidence and mortality are generally higher in developed countries such as Japan, parts of Europe (including England and Wales) and the USA (Figures 18.13 and 18.14). Incidence is markedly lower in India and Singapore.

Prevalence
Only 14% of both males and females diagnosed in the three years 1990-92 were still alive at the beginning of 1993, and only 6% of those diagnosed in the ten years 1983-92 (Figure 18.15).

Survival
Survival from pancreatic cancer is extremely poor. One-year relative survival showed a slight improvement from 6-7% for adults diagnosed in 1971-75, to around 10% for those in 1986-90 (Figure 18.16). There was no further improvement for patients diagnosed in 1991-93. Five-year survival showed no improvement for patients diagnosed between 1971 and 1993, remaining at only 2%.

Table 18.1: Key statistics for pancreatic cancer on incidence, mortality and survival by sex, region and country, latest year available

MALES	Incidence				Mortality				Relative survival (%) patients diagnosed in 1986-90	
	Number of cases	CR	ESR*	WSR*	Number of deaths	CR	ESR*	WSR*	One-year	Five-year
United Kingdom [a]	3,230	11.2	10.2	6.7	3,190	10.9	9.7	6.3	..	..
Great Britain [a]	3,150	11.2	10.2	6.7	3,100	11.0	9.7	6.3	..	
England and Wales [b]	2,740	10.7	9.6	6.4	2,790	10.8	9.6	6.3	12	3
England [b]	2,560	10.6	9.5	6.5	2,630	10.8	9.6	6.3	11	3
Northern and Yorkshire [b]	340	11.1	9.9	6.5	340	10.8	9.6	6.4	12	2
Trent [b]	270	10.6	9.3	6.2	310	12.0	10.4	6.9	11	3
Anglia and Oxford [b]	290	10.8	10.0	6.6	260	9.6	8.7	5.7	11	2
North Thames [b]	340	10.0	10.0	6.6	340	9.6	9.6	6.2	17	6
South Thames [b]	370	11.2	10.0	6.6	390	11.4	10.0	6.6	14	3
South and West [b]	350	10.6	8.5	5.6	430	13.0	10.4	6.8	12	3
West Midlands [b]	280	10.7	9.5	6.3	260	10.0	8.8	5.6	8	2
North West [b]	320	9.8	9.0	6.0	320	9.8	9.0	6.0	9	1
Wales [b]	180	12.6	10.5	7.1	160	11.1	9.2	6.2	13	6
Scotland [a]	318	12.8	12.2	8.2	289	11.6	10.6	6.9	10	3
Northern Ireland [a]	82	10.0	10.6	6.6	88	10.6	11.6	7.7	..	..
FEMALES										
United Kingdom [a]	3,630	12.1	7.8	5.1	3,360	11.2	7.1	4.6	..	..
Great Britain [a]	3,560	12.2	7.9	5.1	3,300	11.3	7.1	4.6	..	..
England and Wales [b]	2,990	11.3	7.3	4.8	3,120	11.7	7.4	4.8	10	2
England [b]	2,800	11.2	7.3	4.8	2,920	11.6	7.3	4.7	10	2
Northern and Yorkshire [b]	380	11.9	8.1	5.4	370	11.4	7.1	4.6	9	2
Trent [b]	280	10.7	7.1	4.7	320	12.4	8.0	5.2	10	2
Anglia and Oxford [b]	290	10.6	7.3	4.8	280	10.2	6.9	4.5	9	2
North Thames [b]	360	10.1	7.3	4.8	380	10.5	7.3	4.6	14	3
South Thames [b]	420	12.0	7.5	4.9	450	12.7	7.9	5.1	13	3
South and West [b]	400	11.8	6.8	4.5	480	14.1	7.6	4.9	11	3
West Midlands [b]	300	11.1	7.2	4.7	280	10.4	6.7	4.3	9	2
North West [b]	370	11.1	7.1	4.6	370	10.9	7.1	4.6	9	2
Wales [b]	180	12.1	6.9	4.5	200	13.2	7.9	5.1	12	4
Scotland [a]	335	12.7	8.6	5.6	318	12.1	8.3	5.5	7	2
Northern Ireland [a]	73	8.6	6.7	4.3	66	7.7	5.6	3.6	..	..

* *Directly age-standardised rate*
[a] *1996 incidence; 1998 mortality.*
[b] *1997 incidence; 1999 mortality.*
Figures for England and Wales are provisional: incidence - 1995-1997, mortality - 1999

Survival from pancreatic cancer declines with age (Figure 18.17). While the disease is extremely rare under the age of 40, the five-year survival was about 16-17% for such patients diagnosed in 1991-93; above age 50, survival was less than 5%.

There was no consistent variation in survival by region. There was a slight gradient in one-year survival by deprivation category for patients diagnosed in 1986-90 (Figure 18.18). Survival from pancreatic cancer is equally poor in other countries, including all those in Europe which participated in the EUROCARE study and the USA (Figure 18.19).

Figure 18.1 **Age-specific incidence, England and Wales, 1997**

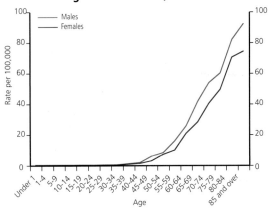

Figure 18.2 **Frequency distribution of new cases by age group, England and Wales, 1997**

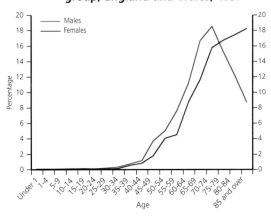

Figure 18.3 **Age-specific incidence, England and Wales, 1971-1997**

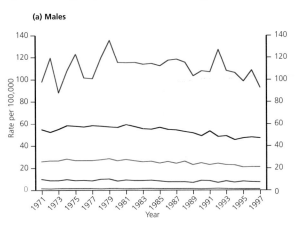

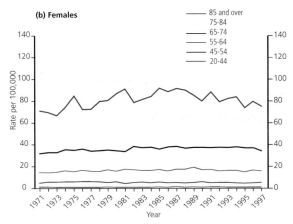

Figure 18.4 **Age-specific mortality, England and Wales, 1950-1999**

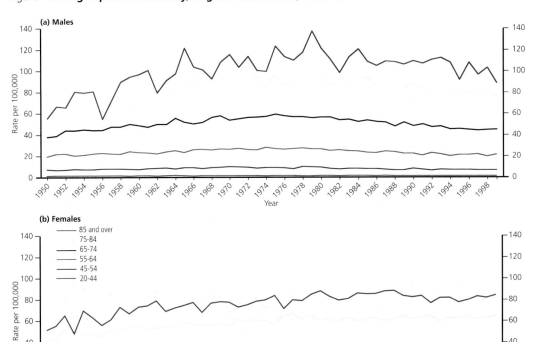

Figure 18.5 **Age-standardised incidence,
England and Wales, 1971-1997**

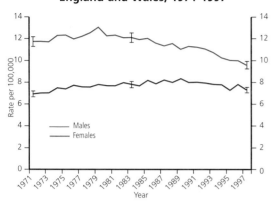

Figure 18.6 **Cohort incidence ratio by sex,
England and Wales**

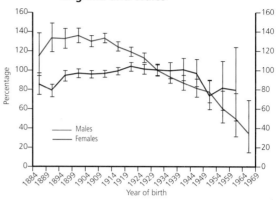

Figure 18.7 **Age-standardised mortality, England and Wales, 1950-1999**

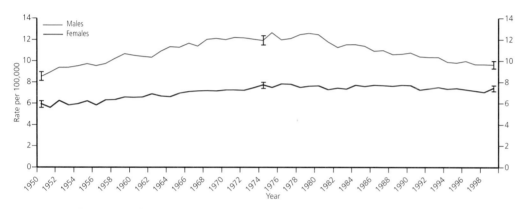

Figure 18.8 **Cohort mortality ratio by sex,
England and Wales**

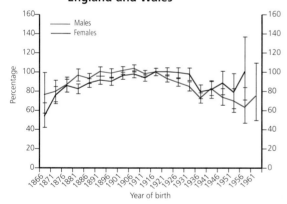

Figure 18.9 **Age-standardised incidence by deprivation
category, England and Wales, 1992-93**

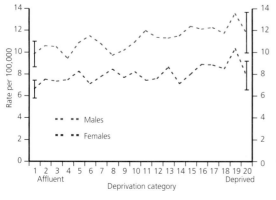

Figure 18.10 **Age-standardised mortality by deprivation
category, England and Wales, 1992-93**

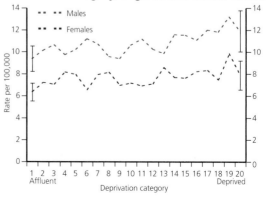

Figure 18.11 **Comparative incidence ratio (CIR) by health region and sex (England and Wales = 100), 1997**

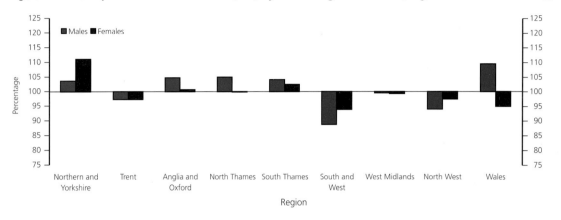

Figure 18.12 **Comparative mortality ratio (CMR) by health region and sex (England and Wales = 100), 1999**

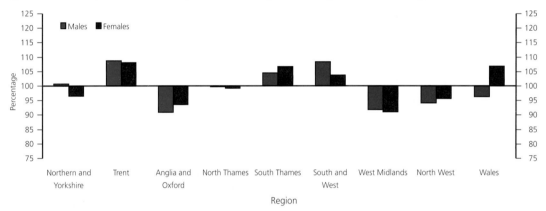

Figure 18.13 **Age-standardised international incidence**

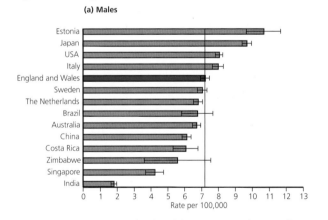
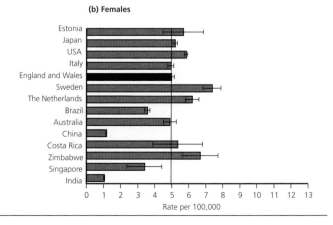

Figure 18.14 **Age-standardised international mortality**

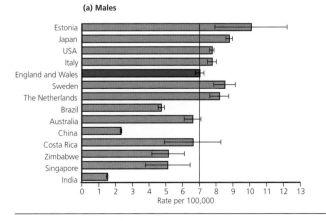

Figure 18.15 **Number and percentage of cases diagnosed in 1990-92 and 1983-92 and alive 1st January 1993, England and Wales**

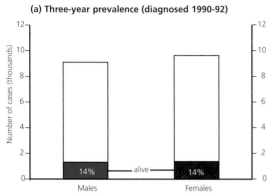

(a) Three-year prevalence (diagnosed 1990-92)

(b) Ten-year prevalence (diagnosed 1983-92)

Figure 18.16 **Relative survival up to ten years, men, England and Wales, 1971-75 to 1991-93**

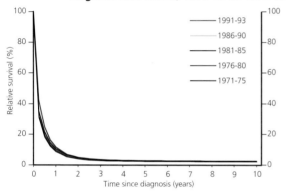

Figure 18.17 **Five-year relative survival by age, England and Wales, patients diagnosed in 1991-93**

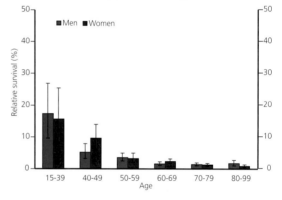

Figure 18.18 **One- and five-year relative survival by deprivation category, England and Wales, adults diagnosed 1986-90**

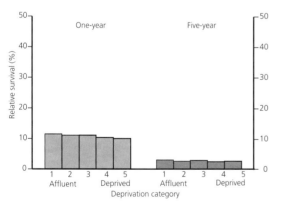

Figure 18.19 **International comparison of five-year relative survival, England and Wales, 1986-90**

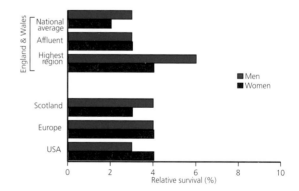

Chapter 19

PROSTATE

Cancer of the prostate is now the second most common cancer in men in England and Wales; the age-standardised rate first exceeded that for colorectal cancer in 1993. Incidence has been increasing at 10-15% every five years in many developed countries, where the lifetime risk up to age 74 for men born around 1940 is already an estimated 3%. Mortality has not been increasing quite as quickly; the corresponding risk of death is 1.3%.

The causes of prostate cancer are essentially unknown, although hormonal factors are involved, and diet may exert an indirect influence. The proportion of cases attributable to genetic factors, ionising radiation, occupational exposure to cadmium or, possibly, vasectomy, appears low.

Urinary symptoms or pain from bone metastases may lead to the diagnosis of prostate cancer. Surgery, radiotherapy and hormone therapy all have roles in treatment. Latent carcinoma of the prostate is common, particularly in old age and such cancers are often diagnosed incidentally on histological examination of a prostate gland removed for benign prostatic hypertrophy, or discovered at post mortem.[1] Most cancer registries record such tumours, and the recorded incidence of prostate cancer can thus be influenced by the frequency of prostatectomy carried out for benign disease.[2,3] Since these cancers may have a better prognosis than cancers diagnosed in men with symptoms due to prostate cancer, their influence on overall trends in population based survival rates must also be borne in mind.[4]

Incidence and mortality
There were 18,300 cases of prostate cancer in England and Wales in 1997, a crude rate of 71 per 100,000 (Table 19.1). Two thirds of cases were coded as adenocarcinomas; the remainder were nearly all poorly specified carcinomas. Incidence rates are very low below age 50, but then rise steeply and continuously with age, reaching almost 1,000 per 100,000 (1%) in men aged 85 and over (Figure 19.1). Less than 0.5% of cases occur in men aged under 50; the peak in the age distribution is in men in their 70s (Figure 19.2).

Incidence rose gradually in all age groups except the youngest from the 1970s to the mid-1980s (Figure 19.3). The steeper increase in rates in the late 1980s may have partly resulted from the detection of latent carcinoma following operations for enlarged prostate or at post mortem in men who had died from other causes.

In 1981, Doll and Peto observed that "if someone ever invents a method of screening apparently healthy men for prostate cancer, the apparent incidence rates may be expected to rise quite quickly by several hundred percent".[5] Their prediction was borne out by the very sharp rise in rates in the early 1990s following the availability of prostate specific antigen (PSA) testing. Similar rises have occurred in Scotland[6] and elsewhere; in the USA the incidence rate rose by 82% between 1986 and 1991.[7] In England and Wales the peak incidence rate in 1994 was over twice the level in the early 1970s (Figure 19.5). Incidence increased in successive birth cohorts, peaking in men born in the mid-1940s (Figure 19.6).

There were 8,500 deaths from prostate cancer in England and Wales in 1999; these formed about 12% of cancer deaths and 3% of all deaths in men (Table 19.1). Trends in mortality by age were slightly different from those in incidence. Following increases during the 1950s, rates reached a plateau during the 1960s and 1970s, only starting to rise again in the mid-1980s (Figure 19.4). Mortality in most age groups peaked in the mid-1990s. Age-standardised mortality reflects the changes described above. After remaining fairly constant at around 20 per 100,000 during the 1960s and 1970s, there was an increase of about 40% between the early 1980s and the early 1990s to just under 30 per 100,000 (Figure 19.7). Mortality in the late 1990s was slightly below the peak in 1993-95. Mortality, like incidence, showed an almost continuous rise across birth cohorts (Figure 19.8).

There was an inverse relationship between the incidence of prostate cancer and deprivation, but the pattern was not linear (Figure 19.9). Rates were similar in the upper six or so deprivation groups but rates in the most deprived groups were some 25% lower. The pattern of mortality with deprivation is similar to that in incidence (Figure 19.10). Incidence was lower than the average for England and Wales in the north of England – Northern and Yorkshire and Trent – and above average in some, but not all, parts of the south (Figure 19.11). There was less regional variation in mortality, and some regions with above average incidence had below average mortality and vice versa (Figure 19.12). The age-standardised incidence rate in Scotland was about 15% higher than in England and Wales, and that in Northern Ireland was slightly lower; but mortality was similar in all countries in the UK (Table 19.1).

Recorded incidence is around twice as high in the USA as in Sweden and Australia, and three times that in England and Wales (Figure 19.13). In contrast, mortality in the USA is not very different from the levels in other developed countries (Figure 19.14) suggesting that a large proportion of prostate cancers registered in the USA have very good prognosis.

Prevalence
About 64% of men diagnosed with prostate cancer in the three-year period 1990–92 were alive at the beginning of 1993, as were 38% of men diagnosed in the 10 years 1983–92 (Figure 19.15).

Survival
Relative survival from prostate cancer for cases diagnosed in England and Wales during 1991-93 was just under 80% at one year and just under 50% at five years, overall increases of 12-16% points since the early 1970s (Figure 19.16). There was no improvement nationally or in any of the regions in the late 1980s; however, there was a rise of 2% in one-year survival and of 7% in five-year survival for men diagnosed in 1991-93. Part of these improvements will be simply due to earlier diagnosis of cases but without any lengthening of life (lead-time bias). Excess mortality continues well beyond five years after diagnosis, and the relative survival curves are still falling at ten years.

Table 19.1: Key statistics for prostate cancer on incidence, mortality and survival by region and country, latest year available

| | Incidence | | | | Mortality | | | | Relative survival (%) patients diagnosed in 1986-90 | |
| | Number of cases | Rate per 100,000 | | | Number of deaths | Rate per 100,000 | | | One-year | Five-year |
		CR	ESR*	WSR*		CR	ESR*	WSR*		
United Kingdom [a]	21,400	74.3	64.4	39.8	9,460	32.5	27.2	15.7	..	..
Great Britain [a]	21,200	75.7	65.4	40.5	9,240	32.6	27.2	15.7	..	..
England and Wales [b]	18,300	71.2	60.7	37.7	8,500	32.9	27.3	15.7	77	42
England [b]	17,200	70.7	60.7	32.9	8,020	32.9	27.3	15.7	78	42
Northern and Yorkshire [b]	1,910	61.3	53.0	32.9	970	31.0	26.4	15.4	78	43
Trent [b]	1,380	54.4	44.9	27.6	830	32.7	26.8	15.5	76	42
Anglia and Oxford [b]	2,050	76.2	68.0	42.5	900	33.3	28.4	16.0	78	43
North Thames [b]	2,190	63.6	61.3	38.2	970	27.7	25.8	14.9	83	46
South Thames [b]	2,470	73.9	62.4	38.7	1,200	35.7	28.4	16.3	83	43
South and West [b]	3,040	93.4	69.7	43.2	1,300	39.6	27.8	16.0	75	40
West Midlands [b]	1,950	74.3	64.1	39.7	870	32.8	28.0	16.0	77	42
North West [b]	2,170	66.9	59.0	36.7	990	30.4	26.7	15.6	75	41
Wales [b]	1,130	78.9	61.5	37.7	480	33.2	26.0	15.0	71	43
Scotland [a]	2,030	81.7	75.3	46.9	677	27.3	25.1	14.6	81	49
Northern Ireland [a]	467	57.2	61.2	37.8	220	26.6	28.0	15.7	..	..

* *Directly age-standardised rate*
[a] *1996 incidence; 1998 mortality.*
[b] *1997 incidence; 1999 mortality.*
Figures for England and Wales are provisional: incidence - 1995-1997, mortality - 1999

The pattern of survival from prostate cancer with age is unusual, and resembles that for breast cancer in women, in that survival is poorer in the younger age groups; the highest survival rates for prostate cancer, however, occur at older ages – 60-69 years – than for breast cancer (50-59 years) (Figure 19.17).

There were significant gradients in survival with deprivation: in 1986-90, one-year survival was higher by 4% points in the most affluent groups than in the most deprived; five-year survival was 3% points higher (Figure 19.18).

Five-year relative survival varies with stage at diagnosis, from 60% or more with malignancy confined to the prostate, to about 20% with bony metastases.[8]

On average, five-year relative survival in England and Wales in the late 1980s was 6% points lower than in Scotland. But survival in England, Wales and Scotland was well below rates in most countries in western Europe where the average level of 56% included countries such as Estonia where the rate was under 40% (Figure 19.19). European differences in survival pale into insignificance, however, when compared with the extraordinary five-year survival rate of 86% in the USA. Regardless of any difference in the efficacy of treatment, there can be little doubt that the prostate cancers registered in the USA are biologically quite different.

Although screening for prostate cancer by either the PSA test or digital rectal examination (DRE) is widespread in the USA and in parts of Europe, there are many arguments against its introduction as a national screening programme in the UK.[9] As already mentioned, high prevalence rates are found at post mortem; as the disease affects predominantly quite elderly men there are few years of life left to gain; there is no clear benefit from treatment; the side effects of prostatectomy are high levels of impotence, and incontinence in around 20% of patients; over-diagnosis leads to significant over-treatment; high rates of biopsy, which is not risk-free, are involved; one third of screen detected cancers are incurable; the costs of screening and the associated follow up and

treatment are high; and no randomised trial has yet shown a reduction in mortality. Large, Europe-wide, randomised trials are, however, in progress.

Figure 19.1 **Age-specific incidence, England and Wales, 1997**

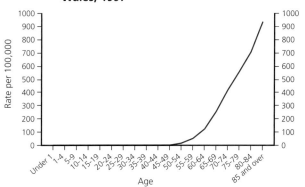

Figure 19.2 **Frequency distribution of new cases by age group, England and Wales, 1997**

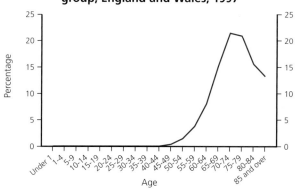

Figure 19.3 **Age-specific incidence, England and Wales, 1971-1997**

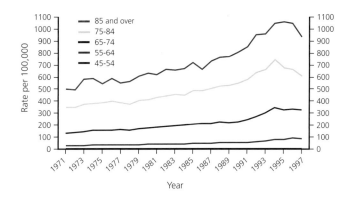

Figure 19.4 **Age-specific mortality, England and Wales, 1950-1999**

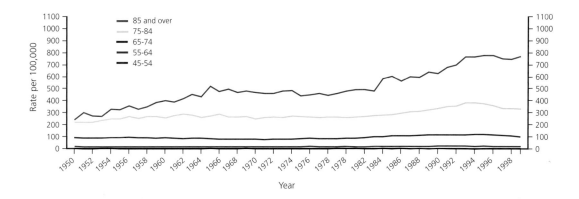

Figure 19.5 **Age-standardised incidence, England and Wales, 1971-1997**

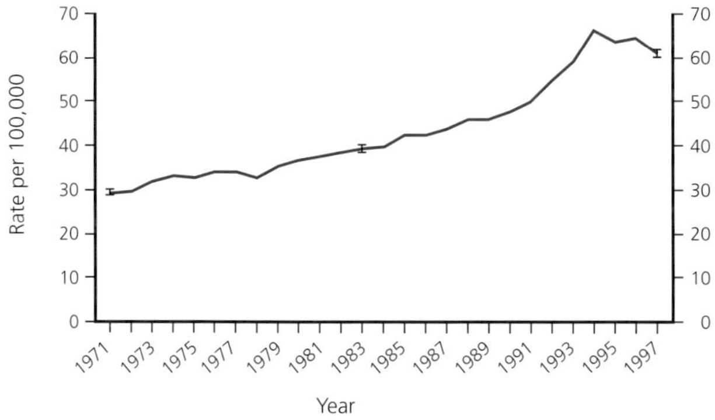

Figure 19.6 **Cohort incidence ratio, England and Wales**

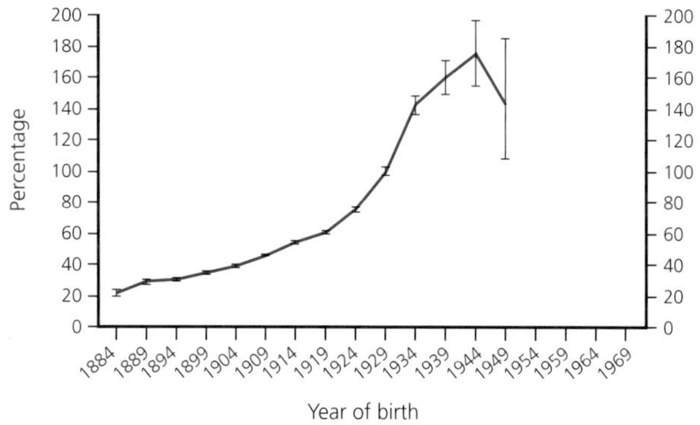

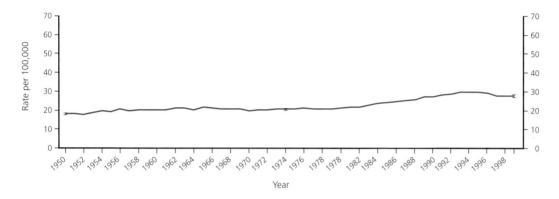

Figure 19.7 **Age-standardised mortality, England and Wales, 1950-1999**

Figure 19.8 **Cohort mortality ratio, England and Wales**

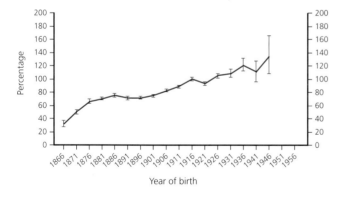

Figure 19.9 **Age-standardised incidence by deprivation category, England and Wales, 1993**

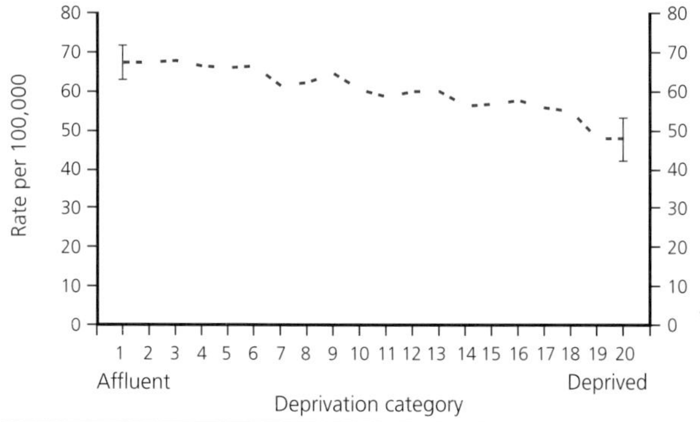

Figure 19.10 **Age-standardised mortality by deprivation category, England and Wales, 1993**

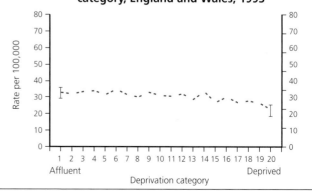

Figure 19.11 **Comparative incidence ratio (CIR) by health region (England and Wales = 100), 1997**

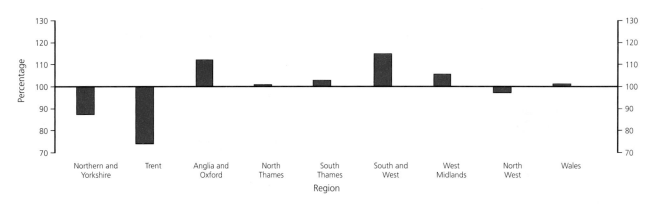

Figure 19.12 **Comparative mortality ratio (CMR) by health region (England and Wales = 100), 1999**

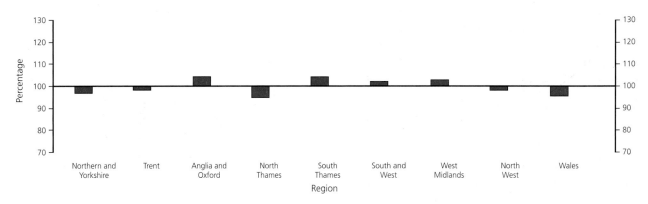

Figure 19.13 **Age-standardised international incidence**

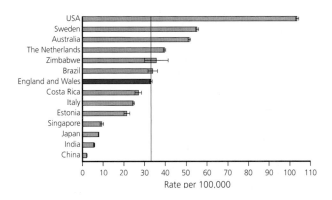

Figure 19.14 **Age-standardised international mortality**

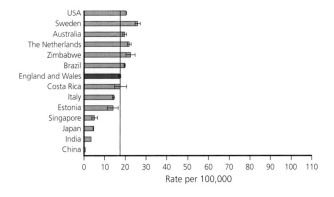

Figure 19.15 **Number and percentage of cases diagnosed in 1990-92 and 1983-92 and alive on 1st January 1993, England and Wales**

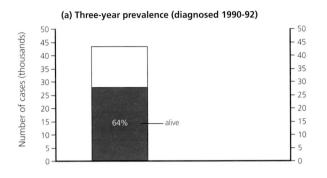

(a) Three-year prevalence (diagnosed 1990-92)

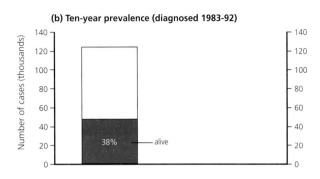

(b) Ten-year prevalence (diagnosed 1983-92)

Figure 19.16 **Relative survival up to ten years, England and Wales, 1971-75 to 1991-93**

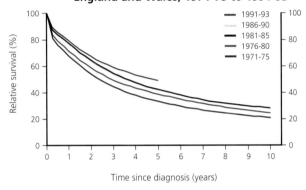

Figure 19.17 **Five-year relative survival by age, England and Wales, 1991-93**

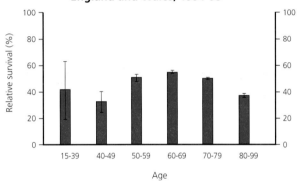

Figure 19.18 **One- and five-year relative survival by deprivation category, England and Wales, 1986-90**

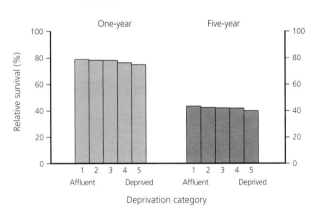

Figure 19.19 **International comparison of five-year relative survival, England and Wales, 1986-90**

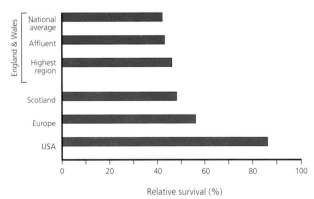

Chapter 20

STOMACH

The most remarkable feature of the epidemiology of cancer of the stomach is the virtually worldwide long-term decline in its incidence and mortality. The decline has been between 2% and 4% per year, but there is still considerable variation between countries and in the timing of its onset. Rates generally appear to be falling more rapidly for men than for women. Stomach cancer was estimated to be the most common cancer in the world in 1980; in 1990 it was the second most common worldwide (after lung cancer) with almost 800,000 new cases each year[1]. Incidence and mortality rates in men are approximately double those in women, but the male:female ratio varies with age.

There are four basic patterns of tumour growth, and these predict how aggressive the growth will be. The most important factor influencing survival is stage of disease at diagnosis: five-year relative survival in the West Midlands in the late 1970s was 77% at stage I, 34% at stage II, 11% at stage III and only 1% at stage IV. Most tumours are not detected until they have metastasised, as symptoms are often vague and non-specific.

Several medical conditions have been studied in relation to cancer of the stomach. Atrophic gastritis has been found to be a predisposing factor; there is controversy over the role of gastric ulcers; intestinal metaplasia has been strongly associated with gastric cancer, and is probably a pre-malignant lesion; and patients with pernicious anaemia have two to three times the general risk. Chronic infection with *Helicobacter pylori* (more prevalent in the lower social classes) is an important risk factor, but its association with duodenal and benign gastric ulcers suggests that its relationship with cancer of the stomach is complex.

Several hypotheses have been proposed to interpret the associations between dietary factors and stomach cancer. Those factors studied most intensively are high intake of salt, and the role of nitrate and nitrite in the endogenous formation of N-nitroso compounds. Vitamin C may have a preventive effect. Case-control studies in several countries have identified fruit and vegetables as major protective factors – daily consumption of several helpings of these foods has been associated with a 50% lower risk of stomach cancer (compared with infrequent consumption).

The results of dietary studies are consistent with the temporal trends in stomach cancer, decreases having occurred in all populations with decreased consumption of salt-preserved and cured meats and increased consumption of fruit and vegetables. Migrant studies have consistently shown that immigrants and their children assume the risk of the host country for stomach cancer. Other risk factors include occupational exposure to dust (for example, the pottery industry), low socio-economic status (involving overcrowded housing, low income and large families), having blood group A, and hereditary non-polyposis colon cancer.

Screening for stomach cancer is a major cancer control measure in Japan, where the incidence of and mortality from the disease are high, but the evidence that this programme is effective in reducing mortality from the disease is slim.

Incidence and mortality

In 1997, there were 5,800 newly diagnosed cases of cancer of the stomach in males in England and Wales, in whom it ranked 5th; and 3,300 cases in women (10th) (Table 20.1). The male:female ratio of the number of cases was 1.8:1. The ratio of the directly age-standardised rates – 20.1 and 7.6 per 100,000 in males and females in 1997, respectively – was much higher at 2.6:1. The lifetime risk of being diagnosed with stomach cancer was 2.3% for males (1 in 44) and 1.2% for females (1 in 86). About 20% of cases of stomach cancer in males and 10% in females were in the cardia; around 60% of cases in males and 70% in females were coded to "stomach, unspecified". About 55% of the tumours were adenomas or adenocarcinomas, and 7% were cystic, mucinous or serous neoplasms; almost 40% were poorly specified carcinomas.

Incidence was low in those aged under 60, but rates then increased steeply with age in both males and females to around 240 and 90 per 100,000, respectively, in those aged 85 and over (Figure 20.1). The peaks in the age distributions of cases occurred in men aged 70-74 and in women aged 75 and over (Figure 20.2). During 1971-1997, the incidence of stomach cancer fell in all age groups in both males and females, but rates in men aged 75 and over and in women aged 85 and over did not start to fall until the early 1980s (Figure 20.3). The age-standardised rates show continuous declines, amounting to nearly 40% in males and to 50% in females (Figure 20.5). There was a substantial fall in the risk of developing stomach cancer for women born in successive cohorts from the late 19th century, up to the 1940s, after which there was little change (Figure 20.6). There was a similar pattern for men, but the fall was – relatively – much smaller, implying that earlier in the last century incidence rates in females were closer to those in males than they were in the 1970s onwards.

Mortality from stomach cancer has been falling steadily since 1950 (and before that) in all age groups and in both males and females (Figure 20.4). Age-standardised mortality in males is now only 30%, and that in females only 20%, of the levels 50 years ago. The birth cohort pattern of mortality is similar to that for incidence with a relatively larger fall for females than for males, starting with those born in the 1880s (Figure 20.8).

In 1992-93 there were very strong positive gradients in the incidence of stomach cancer by Carstairs deprivation category (Figure 20.9). For both males and females, the rates in the most deprived groups were about twice those in the most affluent. The pattern in mortality by deprivation category was closely similar to that in incidence (Figure 20.10). There were marked north-south patterns in both incidence and mortality, with higher than average rates in the north and Wales, and lower than average rates in the south (Figure 20.11 and 20.12).

Around the world, the highest recorded incidence rates occur in Japan and in other areas of eastern Asia, including Korea and parts of China (Figure 20.13). High rates have also been reported from the former Soviet Union, eastern Europe and parts of Latin

Table 20.1: Key statistics for stomach cancer on incidence, mortality and survival by sex, region and country, latest year available

| | Incidence | | | | Mortality | | | | Relative survival (%) patients diagnosed in 1986-90 | |
MALES	Number of cases	CR	ESR*	WSR*	Number of deaths	CR	ESR*	WSR*	One-year	Five-year
United Kingdom [a]	6,610	22.9	20.5	13.3	4,450	15.3	13.3	8.5	..	..
Great Britain [a]	6,440	23.0	20.4	13.3	4,330	15.3	13.2	8.4	..	..
England and Wales [b]	5,840	22.7	20.1	13.1	3,810	14.8	12.8	8.1	26	10
England [b]	5,350	22.1	19.6	14.2	3,520	14.4	12.6	7.9	26	10
Northern and Yorkshire [b]	760	24.4	21.8	14.2	540	17.3	14.9	9.5	24	10
Trent [b]	570	22.4	19.2	12.3	460	18.2	15.4	9.7	25	9
Anglia and Oxford [b]	470	17.3	15.7	10.0	330	12.3	11.0	6.9	25	8
North Thames [b]	630	18.3	18.0	11.8	410	11.7	11.4	7.2	34	13
South Thames [b]	690	20.7	18.1	11.7	370	11.0	9.4	5.9	29	10
South and West [b]	680	20.9	16.6	10.9	440	13.4	10.2	6.3	28	11
West Midlands [b]	640	24.4	21.7	14.2	420	16.0	14.1	8.9	24	10
North West [b]	920	28.3	25.9	17.0	550	16.8	15.1	9.7	22	8
Wales [b]	490	34.0	27.7	18.1	290	20.3	16.1	10.2	23	9
Scotland [a]	580	23.3	22.0	14.4	387	15.6	14.3	9.4	23	9
Northern Ireland [a]	167	20.5	22.4	14.7	121	14.6	16.1	10.8	..	..
FEMALES										
United Kingdom [a]	3,740	12.5	7.7	5.0	2,880	9.6	5.6	3.5	..	..
Great Britain [a]	3,630	12.5	7.7	4.9	2,790	9.5	5.5	3.5	..	..
England and Wales [b]	3,280	12.4	7.6	4.9	2,310	8.7	5.0	3.2	25	11
England [b]	3,010	12.0	7.4	4.8	2,140	8.5	5.0	3.1	26	11
Northern and Yorkshire [b]	490	15.2	9.6	6.3	340	10.5	6.0	3.7	25	10
Trent [b]	320	12.4	7.3	4.6	220	8.6	5.0	3.2	26	12
Anglia and Oxford [b]	230	8.5	5.5	3.5	190	6.9	4.2	2.6	23	8
North Thames [b]	360	10.2	7.2	4.8	270	7.6	5.2	3.3	31	13
South Thames [b]	370	10.6	6.2	3.9	260	7.5	4.1	2.5	28	12
South and West [b]	420	12.3	6.8	4.5	250	7.4	3.8	2.4	27	13
West Midlands [b]	340	12.5	7.7	4.9	260	9.7	5.8	3.6	25	10
North West [b]	480	14.2	8.7	5.6	340	10.2	6.0	3.8	23	10
Wales [b]	270	18.1	10.4	6.7	170	11.1	5.8	3.5	24	14
Scotland [a]	398	15.1	10.1	6.7	293	11.1	7.1	4.6	23	10
Northern Ireland [a]	116	13.6	10.5	7.0	94	10.9	7.7	4.9	..	..

* Directly age-standardised rate
[a] 1996 incidence; 1998 mortality.
[b] 1997 incidence; 1999 mortality.
Figures for England and Wales are provisional: incidence - 1995-1997, mortality - 1999

America. Stomach cancer is generally rare in Africa. Rates in white people in the USA are about half those in black people. Incidence is generally 2 to 2½ times higher in males than in females. The pattern in the rates across the various countries shown in Figure 20.13 was broadly similar in males and females. The international patterns in mortality (Figure 20.14) are broadly similar to those in incidence with the exception of Costa Rica – where incidence is much lower than in Japan but mortality is markedly higher (implying poorer survival).

Prevalence

About 25% of both males and females diagnosed in the three years 1990-92 were still alive at the beginning of 1993, as were about 13% of those diagnosed in the 10 years 1983-92 (Figure 20.15).

Survival

Survival from stomach cancer in England and Wales as a whole for patients diagnosed in 1991-93 was poor – just 28% after one year and 10% after five years in both men and women (Figure 20.16). Survival within the first year after diagnosis has improved, particularly between the late 1970s and early 1980s, but showed little improvement thereafter. The age-standardised trends show an average increase of 3-4% points in one-year survival in both sexes between successive five-year periods (from 1971-75 to 1986-90). Improvements in five-year survival have been smaller – about 2% points every five years. Regional variation in survival was not marked.

For patients diagnosed in 1991-93, five-year survival declined continuously with age from around 20% in both men and women aged under 50 to only 5-6% in the elderly (80 and over) (Figure

20.17). Women have a small but consistent 2% points advantage over men at both one and five years after diagnosis. One-year survival was slightly higher in the more affluent groups, but five-year survival differed very little between the deprivation categories (Figure 20.18).

The results from the second EUROCARE study showed that in the late 1980s, the participating countries could be divided into three groups. Survival was above the European average in Iceland, France, Germany, Switzerland, Italy and Spain; about average in Finland, Sweden and the Netherlands; and below average in Denmark, Scotland, England, Slovakia, Slovenia, Poland and Estonia. Rates of both one- and five-year survival in England were around 11% points lower than the European average, and lower than in the USA by a similar amount.

There were no major improvements in the treatment of stomach cancer in the period covered by the survival analyses. Surgery is the primary form of treatment, but late diagnosis means that the large majority (80%) of patients present too late for curative resection. Laser therapy is increasingly used for palliative treatment, and radiotherapy may be used to relieve chronic bleeding and pain from bone metastases. Trials are being carried out of conservative against more radical surgery, of different combinations of chemotherapy, of adjuvant chemotherapy to destroy metastatic disease, and to determine whether chemotherapy can increase the percentage of patients undergoing surgery. All the results described above support other evidence that operative mortality has fallen in the past 20 years, possibly due to better staging and selection procedures, and improved operating techniques and patient care.

Figure 20.1 **Age-specific incidence, England and Wales, 1997**

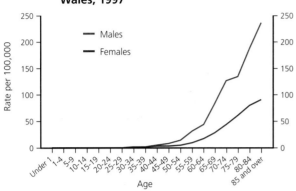

Figure 20.2 **Frequency distribution of new cases by age group, England and Wales, 1997**

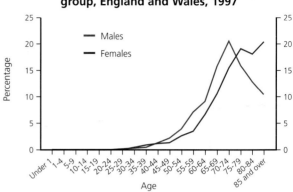

Figure 20.3 **Age-specific incidence, England and Wales, 1971-1997**

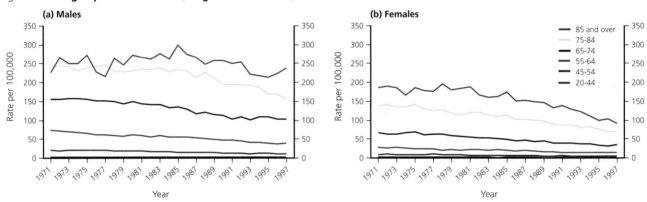

Figure 20.4 **Age-specific mortality, England and Wales, 1950-1999**

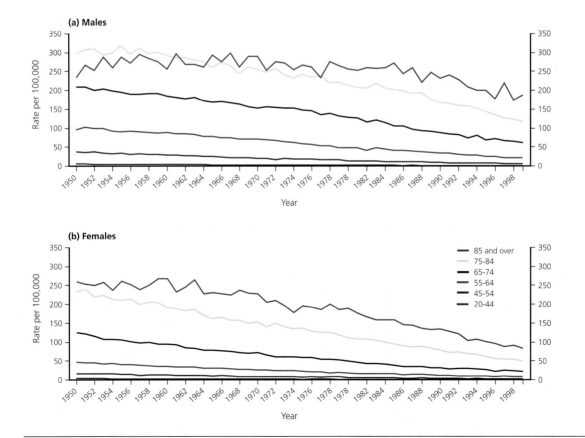

Figure 20.5 **Age-standardised incidence, England and Wales, 1971-1997**

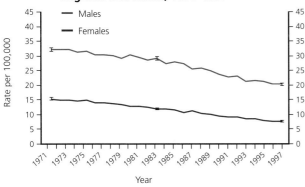

Figure 20.6 **Cohort incidence ratio by sex, England and Wales**

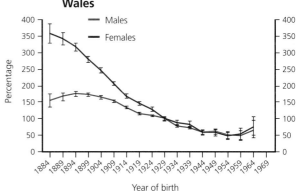

Figure 20.7 **Age-standardised mortality by sex, England and Wales, 1950-1999**

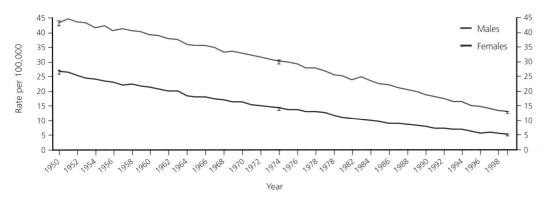

Figure 20.8 **Cohort mortality ratio by sex, England and Wales**

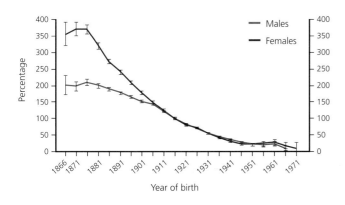

Figure 20.9 **Age-standardised incidence by deprivation category, England and Wales, 1992-93**

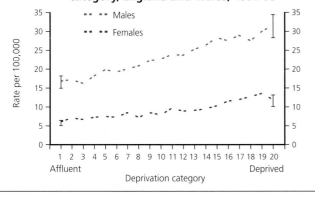

Figure 20.10 **Age-standardised mortality by deprivation category, England and Wales, 1992-93**

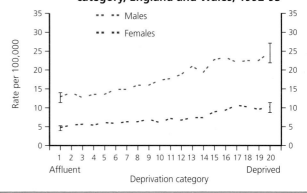

Figure 20.11 **Comparative incidence ratio (CIR) by health region and sex (England and Wales = 100), 1997**

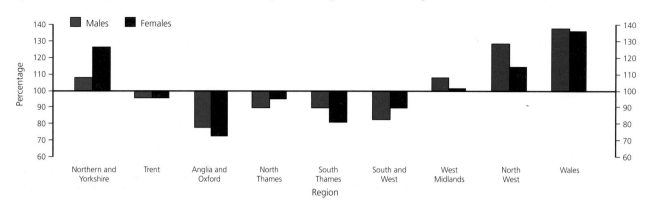

Figure 20.12 **Comparative mortality ratio (CMR) by health region and sex (England and Wales = 100), 1999**

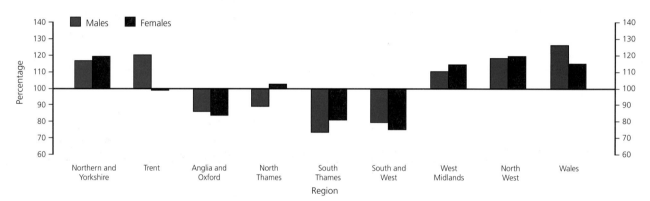

Figure 20.13 **Age-standardised international incidence**

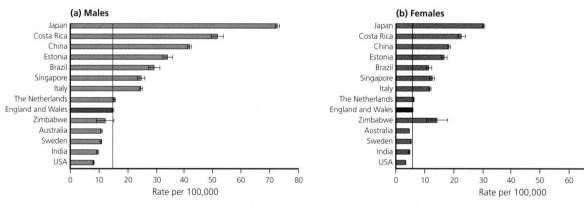

Figure 20.14 **Age-standardised international mortality**

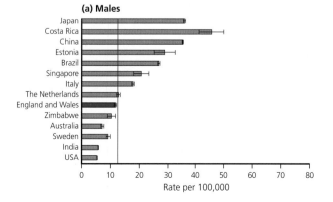
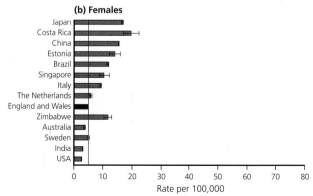

Figure 20.15 **Number and percentage of cases diagnosed in 1990-92 and 1983-92 and alive on 1st January 1993, England and Wales**

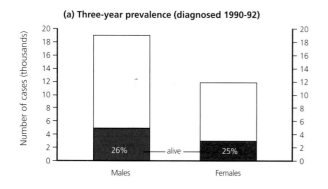

(a) Three-year prevalence (diagnosed 1990-92)

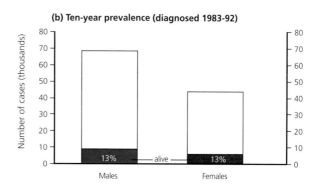

(b) Ten-year prevalence (diagnosed 1983-92)

Figure 20.16 **Relative survival up to ten years, men, England and Wales, 1971-75 to 1991-93**

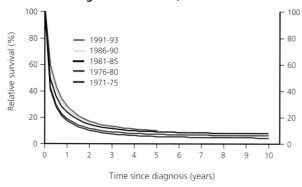

Figure 20.17 **Five-year relative survival by age, England and Wales, patients diagnosed 1991-93**

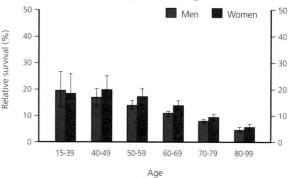

Figure 20.18 **One- and five-year relative survival by deprivation category, England and Wales, adults diagnosed 1986-90**

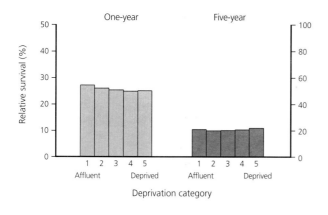

Figure 20.19 **International comparison of five-year relative survival, England and Wales, 1986-90**

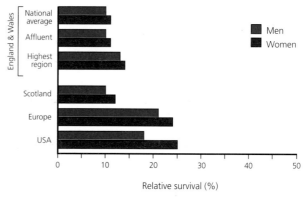

Chapter 21

TESTIS

Cancer of the testis is relatively uncommon, accounting for about 1% of all cancers in men worldwide. It has a highly unusual age distribution, being primarily a disease of younger men. Incidence is low in all parts of the world, but incidence and mortality rates rose during the last century, and the age distribution has shifted towards the younger age groups.

Germ cell tumours account for the vast majority (around 90%) of testicular cancers. Teratomas tend to occur in younger men, with a peak in incidence in those aged 20-29; the incidence of seminomas, which are slower growing, peaks in men aged 30-49.[1]

The clinical literature suggests that the incidence rate is 20 times higher than normal in men with a history of persistently undescended testis, but such cases account for only a very small proportion of the total. The risk is increased if both testes are undescended. A family history of the disease and certain rare familial syndromes also carry a higher risk, but trauma, including vasectomy, have not been proved to be a cause. Testicular cancer is positively associated with inguinal hernia, testicular torsion and exposure to oestrogen in utero. Carcinoma in situ is a risk factor, as 50% of such cases progress to invasive cancer within five years, and almost all cases do so within ten years. If a man has had one testicular cancer, the other testicle is at greater risk. Single men have higher incidence than married men at all ages. And there is a social class gradient, with the highest rates occurring in professional and non-manual workers.

There have been two major advances since the 1960s in the treatment of testicular cancer, the first of which was the use of two serum markers for its detection. The other was the introduction of platinum based drugs which have proved highly successful for patients with metastatic disease – Cisplatin from the mid-1970s, and Carboplatin (which unlike Cisplatin does not damage the kidney or nerves and has also been used to treat lung and ovarian cancer) from the mid-1980s.[1]

The standard treatment for the 50-60% of patients with seminoma who have early stage disease is surgical removal of the affected testicle, with adjuvant radiotherapy. Chemotherapy is effective for metastatic disease. The 30% of patients with teratoma who have early stage disease are also treated surgically; patients with widespread metastases, however, do not respond so well to chemotherapy.[2]

Incidence and mortality

There were over 1,400 cases of testicular cancer in England and Wales in 1997, accounting for just over 1% of all cancers in men (Table 21.1). Over 90% of tumours were specified as germ cell neoplasms; most of the remainder being poorly specified carcinomas. One-third of carcinomas were teratomas and just over a half were seminomas. Incidence rose with age to a peak of around 13 per 100,000 in men aged 30-34 (Figure 21.1). Testicular cancer was by far the most common cancer in men aged 20-34, accounting for almost a third of all cases in that age group. Just

over 50% of cases occurred in men aged under 35, and around 80% in men under 45 (Figure 21.2). Incidence has increased in all age groups from 20 to 59, but not in men age 60 or over (Figure 21.3 – NB. as the age distribution of testicular cancer is so very different from all other cancers, the age groups illustrated are 0-19, 20-29, 30-39, 40-49, 50-59, and 60 and over). Overall, incidence was 5.4 per 100,000 in 1997 compared with 2.9 in 1971 (Figure 21.5). There was no change in the risk of developing testicular cancer for cohorts of men born up to the early 1920s but rates increased continuously in subsequent cohorts (Figure 21.6).

Age-standardised mortality was around 1 per 100,000 in the 1950s and rose slightly during the 1960s and 1970s (Figure 21.7). But as a result of the advances in treatment described above, the trends in incidence and mortality have diverged over the past 20 years. Mortality declined by over 75% between 1971 and 1999 to less than 0.3 per 100,000; most of this decrease took place in the late 1970s and early 1980s. Despite the increasing risk of developing the disease, men born from the early 1950s (and particularly those born in the 1960s onwards) experienced substantially lower mortality than those in earlier birth cohorts (Figure 21.8).

There was an inverse relationship in the incidence of testicular cancer with higher rates in the more affluent groups (Figure 21.9). Data on mortality by deprivation group have not been shown because the total number of deaths is very low, around 70 each year.

There is some consistency between the pattern of incidence with socio-economic deprivation and with the pattern of regional variation – substantially higher than average incidence is suggested for Anglia and Oxford and the South and West, and below average incidence in Northern and Yorkshire and in Trent (Figure 21.11).

Around the world, there is considerable geographical variation in testicular cancer with the highest incidence rates occurring in northern Europeans and the disease being rare in non-Caucasian populations, eg in Africa, Asia and China. This is reflected in Figure 21.13 with incidence in England and Wales amongst the highest. The pattern in mortality among the selected countries illustrated in Figure 21.14 is very different from that for incidence: the four countries with the highest incidence (USA, Sweden, Australia and England and Wales) have relatively low mortality – implying that survival is far worse in countries such as Brazil, Estonia and Costa Rica.

Prevalence

Ninety per cent of men diagnosed with testicular cancer in the three-year period 1990-92 were alive at the beginning of 1993, as were 88% of men diagnosed in the 10 years 1983-92 (Figure 21.15).

Survival

Survival from testicular cancer in England and Wales in the early 1970s was good – 82% one year after diagnosis and 69% five years after. As a result of the major advances in treatment described

Table 21.1: Key statistics for testicular cancer on incidence, mortality and survival by region and country, latest year available

| | Incidence | | | | Mortality | | | | Relative survival (%) patients diagnosed in 1986-90 | |
| | Number of cases | Rate per 100,000 | | | Number of deaths | Rate per 100,000 | | | | |
		CR	ESR*	WSR*		CR	ESR*	WSR*	One-year	Five-year
United Kingdom [a]	1,770	6.1	5.8	5.4	90	0.3	0.3	0.2	..	..
Great Britain [a]	1,710	6.1	5.7	5.3	90	0.3	0.3	0.3		
England and Wales [b]	1,440	5.6	5.4	5.0	70	0.3	0.3	0.2	96	91
England [b]	1,370	5.6	5.4	4.6	70	0.3	0.3	0.2	96	91
Northern and Yorkshire [b]	160	5.0	4.9	4.6	10	0.3	0.2	0.2	94	89
Trent [b]	100	4.1	3.8	3.5	10	0.3	0.3	0.3	95	89
Anglia and Oxford [b]	200	7.3	7.0	6.5	0	0.1	0.1	0.1	95	90
North Thames [b]	170	4.8	4.4	4.1	10	0.1	0.1	0.1	98	95
South Thames [b]	200	6.1	5.6	5.2	10	0.2	0.2	0.1	96	93
South and West [b]	220	6.6	6.5	6.2	10	0.3	0.3	0.3	96	91
West Midlands [b]	140	5.3	5.2	5.0	10	0.4	0.4	0.3	94	91
North West [b]	180	5.6	5.5	5.1	20	0.5	0.5	0.4	96	93
Wales [b]	80	5.4	5.4	4.9	0	:	:	:	91	87
Scotland [a]	169	6.8	6.3	5.9	11	0.4	0.4	0.4	95	92
Northern Ireland [a]	56	6.9	6.8	6.2	0	:	:	:	..	..

*   *Directly age-standardised rate*
[a] *1996 incidence; 1998 mortality.*
[b] *1997 incidence; 1999 mortality.*
Figures for England and Wales are provisional: incidence - 1995-1997, mortality - 1999

above, one- and five-year survival rose to 94% and 88%, respectively, in the early 1980s; there were only small further improvements in the late 1980s; however, for men diagnosed in 1991-93 one-year survival increased to almost 98% and five-year survival to almost 95% (Figure 21.16). Survival is now the highest of any cancer in men. Three years after diagnosis, excess mortality is very low – the relative survival curves are almost flat.

Five-year relative survival falls sharply with age: for men diagnosed under the age of 50 in 1991-93 it was 95%, but fell to around 60% for men aged 70-79 and further to under 30% in elderly men (80-99 years) in whom the cancer is rare (Figure 21.17).

For cases diagnosed in 1986-90, survival was significantly better for men in the most affluent group: the gaps between them and men in the most deprived groups was over 3% points at one year after diagnosis and over 6% five years after (Figure 21.18). These are substantial differences, given the high overall survival rates. There was little variation in survival among the regions.

In Scotland, overall relative survival at all periods up to five years from diagnosis, the pattern of survival with age, and improvements in survival since the early 1970s were all closely similar to the results for England and Wales. Testicular cancer is one of only a very small number of cancers for which survival in the UK was similar to – or better than – most other countries which participated in the EUROCARE studies. The highest regional five-year survival rate, and survival in the most affluent group, in England and Wales (93-95%) were similar to the survival rate in the USA (Figure 21.19).

Survival is sharply dependent on stage at diagnosis: five-year survival is over 95% for stages I and II, about 85% for stage III, and only 70% for stage IV.[3] Thus although survival from testicular cancer is now very good, the overall rate could be improved still further if a higher proportion of patients were diagnosed at an early stage of their disease when treatment is most effective and least toxic.[4]

Figure 21.1 **Age-specific incidence, England and Wales, 1997**

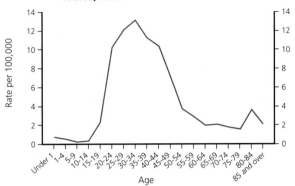

Figure 21.2 **Frequency distribution of new cases by age group, England and Wales, 1997**

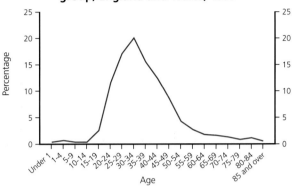

Figure 21.3 **Age-specific incidence, England and Wales, 1971-1997**

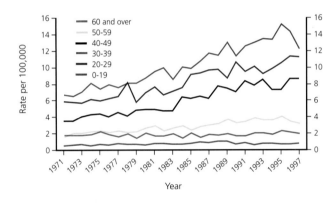

Figure 21.5 **Age-standardised incidence, England and Wales, 1971-1997**

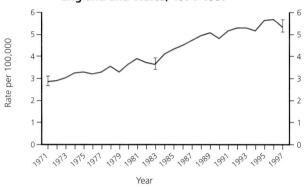

Figure 21.6 **Cohort incidence ratio, England and Wales**

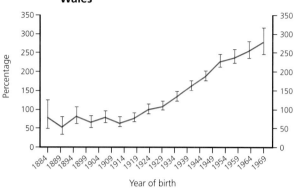

Figure 21.7 **Age-standardised mortality, England and Wales, 1950-1999**

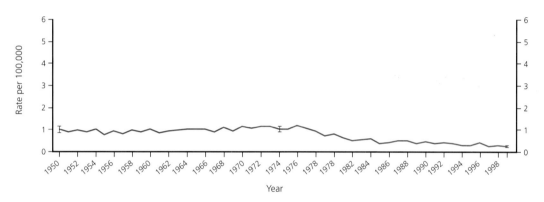

Figure 21.8 **Cohort mortality ratio, England and Wales**

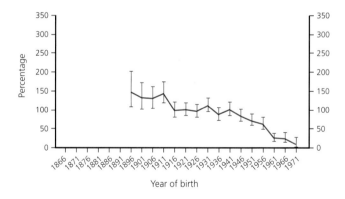

Figure 21.9 **Age-standardised incidence by deprivation category, England and Wales, 1991-93**

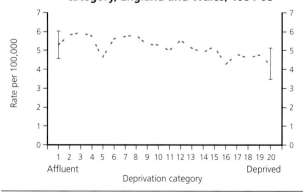

Figure 21.11 **Comparative incidence ratio (CIR) by health region (England and Wales = 100), 1997**

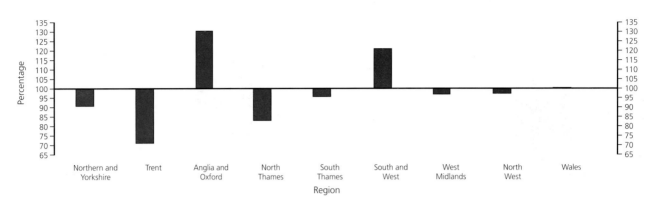

Figure 21.13 **Age-standardised international incidence**

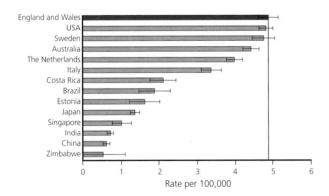

Figure 21.14 **Age-standardised international mortality**

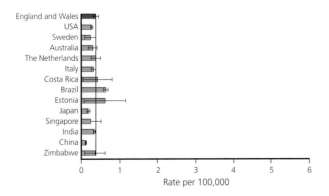

Figure 21.15 **Number and percentage of cases diagnosed in 1990-92 and 1983-92 and alive on 1st January 1993, England and Wales**

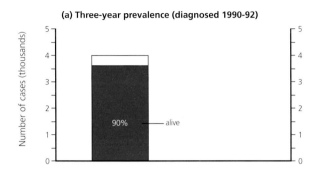

(a) Three-year prevalence (diagnosed 1990-92)

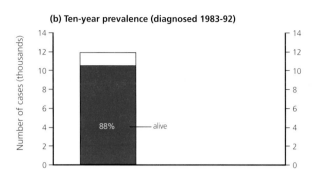

(b) Ten-year prevalence (diagnosed 1983-92)

Figure 21.16 **Relative survival up to ten years, England and Wales, 1971-75 to 1991-93**

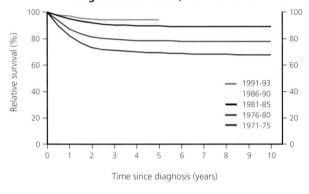

Figure 21.17 **Five-year relative survival by age, England and Wales, 1991-93**

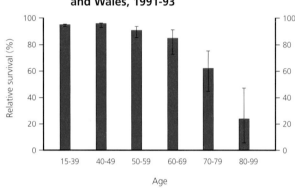

Figure 21.18 **One- and five-year relative survival by deprivation category, England and Wales, 1986-90**

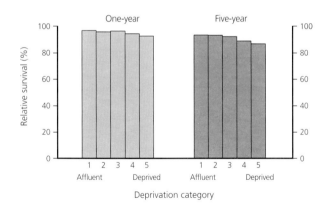

Figure 21.19 **International comparison of five-year relative survival, England and Wales, 1986-90**

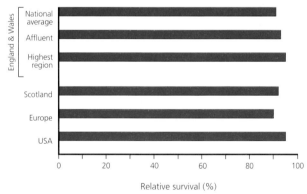

Chapter 22

UTERUS

There were an estimated 142,000 new cases of cancer of the body of the uterus worldwide in 1990, making it just over one third as frequent as cancer of the cervix. In developed countries, however, it is now slightly more common. Incidence varies widely, is higher among more affluent populations, and increases with the adoption of a more westernised lifestyle.

About 90% of cancers of the body of the uterus occur in the inner lining of the womb (endometrium); these are mostly adenocarcinomas, sometimes mixed with squamous elements. The rest occur in the muscle that forms the wall (myometrium) and are sarcomas.

Endometrial cancer shares many epidemiological features and risk factors with breast and ovarian cancer.[1,2,3] It too is hormonally dependent, with risk factors of early age at menarche, low parity, and late age at menopause. There appears to be no association with age at first birth, however. Obesity has consistently been found to be a risk factor, probably a causal one. Combination-type oral contraceptives are protective. This has led to the prescription of progestogens together with oestrogens for menopausal symptoms, an approach that may reduce the risk of endometrial cancer but may have adverse effects on the incidence of breast cancer, ischaemic heart disease and osteoporosis. There is a slightly increased risk of endometrial cancer in women treated with tamoxifen for breast cancer.

Assessing international trends in mortality presents problems because of variation between countries and over time in the use in death certification of the code "malignant neoplasm of uterus, part unspecified", and the effect of trends in hysterectomy. Nevertheless, it appears that mortality from uterine cancer in many countries has declined steadily for women in all age groups. The trends in incidence are more variable – in the USA (and to a lesser extent Canada) uterine cancer increased substantially among postmenopausal women (particularly at ages 55-64) in the 1970s. The increased risk was ascribed to the use of postmenopausal oestrogens; after this was recognised, and the use of these drugs fell, incidence rates declined to previous levels and even below. The increased risk in Europe was much smaller. In the 1980s, there were no marked trends in incidence in women under 65, but slight increases in women aged 65-74. Changes in the prevalence of the main aetiological factors over time may be responsible for much of the observed change in incidence around the world.

Cancers of the uterus should not normally be registered without sufficient information being available to classify them to the cervix or body of the uterus.[4] In practice, around 8% of uterine cancers registered in England and Wales in 1994 were assigned to the non-specific code for uterus (179 in ICD9) – but the proportion of such registrations varied widely among the regions (see below). The results for incidence described below relate only to cancers coded to 182 in ICD9 and C54 in ICD10. For mortality, however, deaths coded to the non-specific code 179 make up about one third of the total for codes 179 and 182. The results for mortality have therefore been based on records for these codes combined.

Incidence and mortality

There were 4,000 cases of cancer of the body of the uterus in England and Wales in 1997, representing almost 4% of all cancer cases in women, in whom it ranked fifth (Table 22.1). Almost all tumours were coded to the body of the uterus. About 70% of cases were adenocarcinomas; 5% were complex mixed and stromal tumours; and around 20% were poorly specified carcinomas. The cancer was rare in women below the age of 35, but incidence rates rose through the 40s and 50s to reach a plateau of almost 50 per 100,000 population in women in all age groups over 60. Cases in women under 50 years of age accounted for only 7% of the total; there was a peak in the 55-69 age group, which formed almost 45% of the total cases.

In women aged 45-54, the incidence of cancer of the uterus declined steadily from 22 per 100,000 in the mid-1970s to just over 15 in 1997 (Figure 22.3). In women aged 55-74, rates were fairly stable up to the late 1980s but then increased slightly, while the rise in older women just occurred in the 1970s and 1980s. The overall age-standardised rate has remained close to 12 per 100,000 since the early 1970s (Figure 22.5). Incidence increased in successive birth cohorts from the 1880s up to the late 1920s and early 1930s and then declined (Figure 22.6).

In contrast to the trends in incidence, there have been long term declines in mortality from cancer of the uterus in all age groups except the very elderly (Figure 22.4); in women aged 45-64, mortality in 1999 was only half that in the early 1950s. The age-standardised rate has halved – from 6 per 100,000 in 1950 to 3 per 100,000 in 1999 (Figure 22.7). Cohort mortality has declined more or less continuously since the 1890s (Figure 22.8).

There was no marked pattern in the incidence of cancer of the uterus with deprivation, although there was a slight decreasing trend in the rates in the most deprived categories (Figure 22.9). There was no variation in mortality with deprivation (Figure 22.10).

There were no consistent geographical patterns in either the incidence of or mortality from cancer of the uterus among the regions (Figures 22.11 and 22.12). Any real variation in incidence is obscured by the varying proportions of cases assigned to the non-specific code 179 in ICD9: virtually zero in Anglia and Oxford and in the Thames regions; around 10% in Trent, South and West, West Midlands and North West and around 15% in Northern and Yorkshire and in Wales.

Around the world, high incidence rates are found in the USA in white women, and in Canada and western Europe, and low rates in Middle Eastern and Asian populations (except for Jews in Israel, and Chinese in Singapore and Hong Kong) (Figure 22.13). Relative to incidence, mortality is high in Estonia and Brazil (implying worse survival) and low in the USA (Figure 22.14).

Table 22.1: Key statistics for uterine cancer on incidence, mortality and survival by region and country, latest year available

| | Incidence | | | | Mortality | | | | Relative survival (%) patients diagnosed in 1986-90 | |
	Number of cases	CR	ESR*	WSR*	Number of deaths	CR	ESR*	WSR*	One-year	Five-year
		Rate per 100,000				Rate per 100,000				
United Kingdom [a]	4,520	15.1	12.6	8.8	930	3.1	2.0	1.4	..	..
Great Britain [a]	4,420	15.2	12.6	8.8	910	3.1	2.1	1.4	..	..
England and Wales [b]	4,000	15.1	12.6	8.9	1,230	4.6	3.0	2.0	85	73
England [b]	3,780	15.1	12.7	8.9	1,150	4.6	3.0	2.0	85	73
Northern and Yorkshire [b]	380	11.7	9.9	7.1	130	3.9	2.5	1.6	85	74
Trent [b]	410	15.7	12.7	8.8	110	4.2	2.7	1.8	84	71
Anglia and Oxford [b]	470	17.4	15.6	11.1	120	4.4	3.0	2.0	83	71
North Thames [b]	470	13.2	12.0	8.3	140	3.9	3.0	2.0	88	73
South Thames [b]	530	15.1	12.4	8.8	170	4.9	3.0	2.0	87	75
South and West [b]	600	17.7	13.7	9.6	210	6.2	3.8	2.5	88	76
West Midlands [b]	440	16.3	13.8	9.7	120	4.5	3.0	1.9	85	74
North West [b]	480	14.3	12.0	8.5	160	4.7	3.0	2.0	83	69
Wales [b]	220	14.7	12.2	8.7	80	5.2	3.2	2.1	83	68
Scotland [a]	423	16.0	13.9	9.9	92	3.5	2.3	1.5	86	75
Northern Ireland [a]	101	11.8	11.4	8.1	13	1.5	1.3	0.9	..	..

* *Directly age-standardised rate*
[a] *1996 incidence; 1998 mortality.*
[b] *1997 incidence; 1999 mortality.*
Figures for England and Wales are provisional: incidence - 1995-1997, mortality - 1999

Prevalence

About 76% of women diagnosed with cancer of the uterus in the three-year period 1990-92 were alive at the beginning of 1993, as were 64% of women diagnosed in the 10 years 1983-92 (Figure 22.15).

Survival

During the 1970s and 1980s, one-year relative survival from cancer of the uterus improved by 6% points to 84%, and five-year survival by 9% points to 70% (Figure 22.16). Survival improved only slightly for women diagnosed in 1991-93. One-year survival is about 5% points lower than for breast cancer, but five-year survival is similar.

For cases diagnosed in 1991-93 there was little difference in five-year relative survival with age in women under 60, with rates of 80-85%, but rates then declined with age to 46% in elderly women (aged 85 and over) (Figure 22.17).

There were gaps of 3-4% points in both one- and five-year relative survival between the most affluent and the most deprived groups in 1986-90 (Figure 22.18). There was only small variation in survival in absolute terms among the regions, but five-year survival was significantly above average in South Thames and South West, and below average in the North West and in Wales.

Survival from cancer of the uterus varied widely across Europe in the late 1980s. The patterns in survival at one year and five years were similar: in England and Wales, survival was markedly better than in Poland and Estonia, similar to that in Scotland, Spain, Slovakia and Slovenia, and only slightly below the European average. Survival was, however, well below that in Iceland, Finland, Sweden, Denmark, the Netherlands, Germany, Switzerland, France and Italy; and more than 10% below rates in the USA (Figure 22.19).

Survival is worse for later stage disease: five-year relative survival rates are around 30% for stage IV, compared with around 90% for stage I. If detected at an early stage, both endometrial cancer and pre-cancerous changes that carry a high risk of progressing to cancer are curable in most cases – usually by surgery, or (if this is

not feasible) by hormone treatment. There have been no recent advances in the treatment of cancer of the uterus that can be expected to increase survival and reduce mortality further.

Figure 22.1 **Age-specific incidence, England and Wales, 1997**

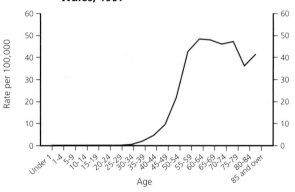

Figure 22.2 **Frequency distribution of new cases by age group, England and Wales, 1997**

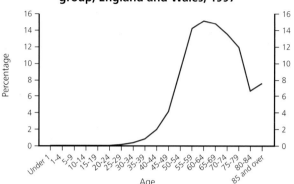

Figure 22.3 **Age-specific incidence, England and Wales, 1971-1997**

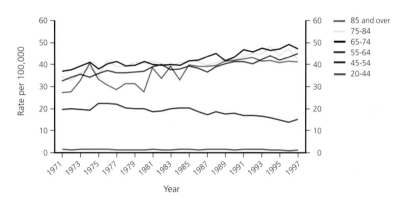

Figure 22.4 **Age-specific mortality, England and Wales, 1950-1999**

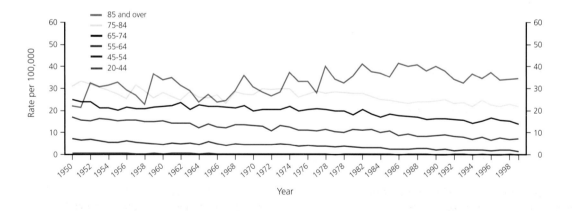

Figure 22.5 **Age-standardised incidence, England and Wales, 1971-1997**

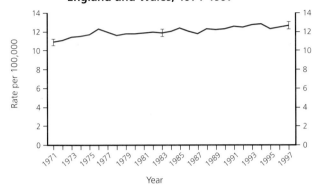

Figure 22.6 **Cohort incidence ratio, England and Wales**

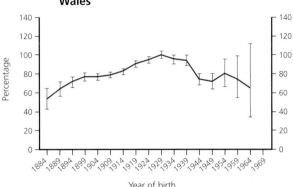

Figure 22.7 **Age-standardised mortality, England and Wales, 1950-1999**

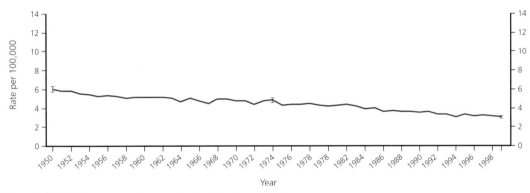

Figure 22.8 **Cohort mortality ratio, England and Wales**

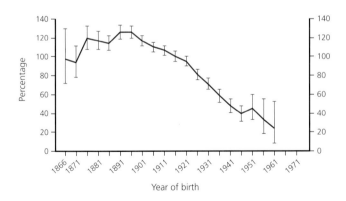

Figure 22.9 **Age-standardised incidence by deprivation category, England and Wales, 1989-93**

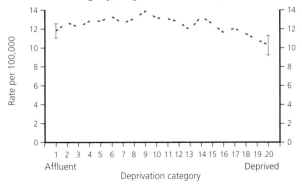

Figure 22.10 **Age-standardised mortality by deprivation category, England and Wales, 1989-93**

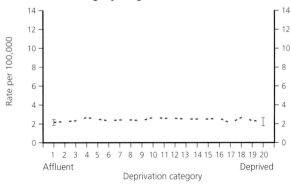

Figure 22.11 **Comparative incidence ratio (CIR) by health region (England and Wales = 100), 1997**

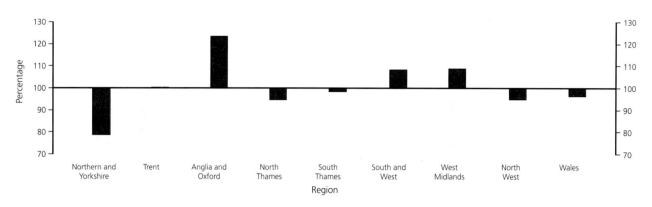

Figure 22.12 **Comparative mortality ratio (CMR) by health region (England and Wales = 100), 1999**

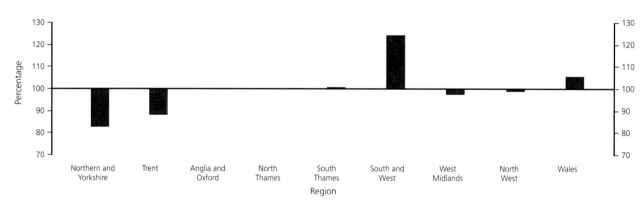

Figure 22.13 **Age-standardised international incidence**

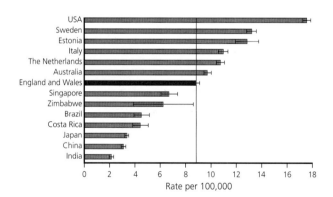

Figure 22.14 **Age-standardised international mortality**

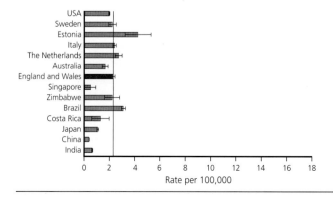

Figure 22.15 **Number and percentage of cases diagnosed in 1990-92 and 1983-92 and alive on 1st January 1993, England and Wales**

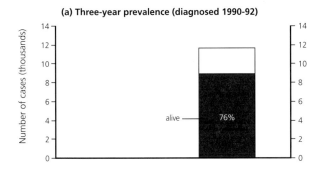

(a) Three-year prevalence (diagnosed 1990-92)

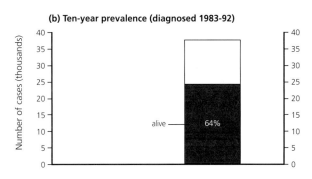

(b) Ten-year prevalence (diagnosed 1983-92)

Figure 22.16 **Relative survival up to ten years, England and Wales, 1971-75 to 1991-93**

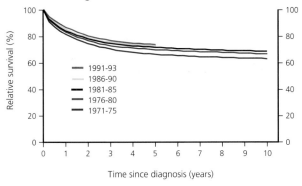

Figure 22.17 **Five-year relative survival by age, England and Wales, 1991-93**

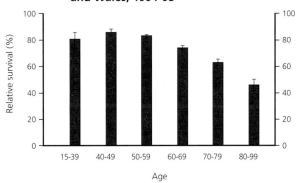

Figure 22.18 **One- and five-year relative survival by deprivation category, England and Wales, 1986-90**

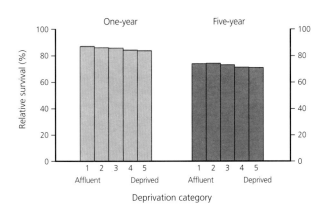

Figure 22.19 **International comparison of five-year relative survival, England and Wales, 1986-90**

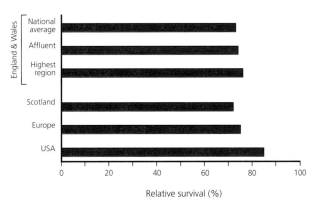

Chapter 23

CHILDHOOD CANCERS

This chapter describes the trends in incidence, mortality and survival for the six most common cancers in children: leukaemia, brain and spinal cord tumours, lymphomas, soft tissue sarcomas, neuroblastoma and Wilms' tumour.

Only just over $\frac{1}{2}$% of all cancers in England and Wales occur in children (aged under 15). As only small numbers of even the major childhood cancers are diagnosed each year, the following commentary is based on the numbers of new cases in 1993 and 1994 combined. There were on average 1,290 new cases of cancer diagnosed in children aged under 15 in England and Wales in 1993-94. Of these, leukaemia accounted for 32%, brain and spinal cord tumours 21%, lymphomas 10%, soft tissue sarcomas 7%, neuroblastoma 6%, and Wilms' tumour 5% (Table 23.1).

Leukaemia

Leukaemia accounts for one third of all childhood cancers; the main form in children is acute lymphoblastic leukaemia (ALL), accounting for four out of five leukaemia cases. Childhood leukaemia has been consistently associated with higher socio-economic status.[1] Pre- and post-natal radiation exposure are known risk factors, but evidence of paternal preconception exposure to radiation leading to leukaemia in children has not been established, while viral infection in (mainly rural) areas with population mixing has been argued as an alternative explanation for childhood leukaemia clusters.[2,3] A fairly weak risk has been found with a pre-natal exposure to influenza. Although the risk is small, 'flu' is very common, and suggests one avenue for prevention.

Leukaemia in children occurs mainly at ages 2-3 (Table 23.1). The age-specific incidence was around 85 per million for children aged 2-3, with on average 120 new cases diagnosed in 1993-94; rates for children aged under 2 were less than half this – around 40 per million and rates in older children over 7 were 25-30 per million (Figure 23.1.1). There was some fluctuation in the overall incidence of leukaemia between 1971 and 1994, with rises of 7% in the number of cases and of 25% in the age-standardised rate (world) (from 34 per million to 42 per million) (Figure 23.2.1). The greatest increase in incidence from 1971 to 1994 was in the children aged under 1, with a doubling of the rate. There are around 100 deaths from leukaemia in children each year. Mortality from leukaemia fell steeply during the 1970s and 1980s, from just under 30 deaths per million to around 10 per million in 1992 (Figure 23.3.1). It remained at around this level throughout the 1990s. This fall reflects the major improvements in treatment and consequent large gains in survival.

One-year relative survival from leukaemia was 86% for children diagnosed in 1986-90, and five-year survival was 69%. Five-year survival from leukaemia has shown a substantial improvement since the early 1970s of around 12% points every five years, with an overall rise of over 35% points (Figure 23.4.1).

Brain and spinal cord

Cancers of the brain and spinal cord are the most common type of solid tumour in children, and second in incidence only to leukaemia. They account for a fifth of all malignant neoplasms in children in England and Wales. Astrocytomas are the most common form of malignant brain tumours in children, accounting for 30-50% of all cases. A number of causes of brain and spinal cord cancer have been suggested including ionising radiation (particularly when children are exposed post-natally); electromagnetic fields; maternal use of marijuana during pregnancy; and pre-natal exposure to barbituates, particularly sodium pentothal. Brain tumours also have been associated with the occupational exposure of parents to hydrocarbons.

There were on average around 270 new cases of brain and spinal cord tumours diagnosed in 1993-94 in children aged under 15 (Table 23.1; Figure 23.1.2). The overall incidence of these cancers rose gradually from 19 per million in 1971 to 26 per million in 1994, an increase of around 40% (Figure 23.2.2). The greatest increase in incidence from 1971 occurred in the 5-9 age group, with the rate doubling between 1971 and 1994. There were 100 deaths from tumours of the brain and spinal cord in children in 1999. Mortality fell from around 17 per million in the 1970s to around 10 per million from the early 1980s onwards (Figure 23.3.2).

Survival from cancers of the brain and spinal cord increased by around 6% every five years since the early 1970s, to reach 76% one year after diagnosis and 60% after five years for children diagnosed in 1986-90 (Figure 23.4.2). Survival from brain tumours is relatively poor compared with the other cancers in this group.

Lymphomas

There are around 130 new cases of lymphoma (non-Hodgkin's lymphoma (NHL) and Hodgkin's disease) each year in children in England and Wales. They are a collection of malignancies whose clinical behaviour, prognosis, and management vary widely according to the histological subtype, stage and bulk of the disease. Typically lymphomas arise in lymph node tissue, but may also develop in other sites such as, bone, stomach, small or large intestine.[4] Little is known about the causes of lymphomas, but risk factors include family history and immunodeficiency disorders. Lymphomas are more common in Africa and Papua New Guinea, due to the high incidence of Burkitt's lymphoma (BL). BL accounts for up to 30% of childhood non-Hodgkin's lymphoma in many countries, although incidence rates are very much higher in Africa. The risk is highest in areas with endemic malaria, suggesting an aetiological role for an insect-borne oncogenic virus and the influence of malaria.[1]

Lymphomas occur more frequently in older children, with an average of around 130 cases diagnosed in 1993-94 in children aged 10-14 years (Table 23.1). Incidence rises with age to around 20 per million in 13 and 14 year olds (Figure 23.1.3). Since the mid-1970s the incidence of childhood lymphomas has remained at 11-15 per

Table 23.1 Major cancers in children (under 15), England and Wales, 1993 and 1994 combined

| | Numbers of new cases | | | | | | | |
Age	Leukaemia	Brain and spinal cord	Lymphomas	Soft tissue sarcoma	Neuroblastoma	Wilms' tumour	Other	Total
0	47	29	12	18	46	9	97	**258**
1	58	36	4	15	24	27	55	**219**
2	117	36	8	21	22	22	41	**267**
3	122	58	18	20	21	28	16	**283**
4	86	43	18	12	11	19	16	**205**
5	71	35	15	14	10	12	21	**178**
6	45	50	13	13	3	10	17	**151**
7	45	40	11	4	9	2	12	**123**
8	36	36	21	11	4	2	18	**128**
9	33	30	18	9	0	4	24	**118**
10	35	31	18	5	3	2	27	**121**
11	37	24	29	6	2	2	18	**118**
12	34	30	17	9	1	0	29	**120**
13	30	30	29	16	4	0	33	**142**
14	37	23	26	12	1	2	52	**153**
Total under 15	**833**	**531**	**257**	**185**	**161**	**141**	**476**	**2,584**
Annual average	417	266	129	93	81	71	238	**1,292**
% all cases	32.2	20.5	9.9	7.2	6.2	5.5	18.4	**100.0**

| | Rates per million | | | | | | | |
Age	Leukaemia	Brain and spinal cord	Lymphomas	Soft tissue sarcoma	Neuroblastoma	Wilms' tumour	Other	Total
0	35.1	21.7	9.0	13.4	34.3	6.7	72.4	**192.7**
1	41.8	26.0	2.9	10.8	17.3	19.5	39.7	**157.9**
2	83.5	25.7	5.7	15.0	15.7	15.7	29.2	**190.5**
3	88.8	42.2	13.1	14.6	15.3	20.4	11.6	**205.9**
4	63.0	31.5	13.2	8.8	8.1	13.9	11.7	**150.1**
5	51.4	25.3	10.8	10.1	7.2	8.7	15.2	**128.8**
6	33.6	37.3	9.7	9.7	2.2	7.5	12.7	**112.7**
7	34.0	30.2	8.3	3.0	6.8	1.5	9.1	**92.8**
8	27.3	27.3	15.9	8.3	3.0	1.5	13.6	**97.1**
9	26.1	23.7	14.2	7.1	0.0	3.2	19.0	**93.2**
10	27.6	24.4	14.2	3.9	2.4	1.6	21.3	**95.2**
11	29.2	18.9	22.9	4.7	1.6	1.6	14.2	**93.1**
12	26.4	23.3	13.2	7.0	0.8	0.0	22.5	**93.2**
13	23.1	23.1	22.3	12.3	3.1	0.0	25.4	**109.3**
14	29.6	18.4	20.8	9.6	0.8	1.6	41.5	**122.2**
Total under 15	41.9	26.7	12.9	9.3	8.1	7.1	23.9	**130.0**

million (Figure 23.2.3). Deaths from lymphomas have fallen sharply since 1971. There were only 16 deaths in children in 1999, compared with over 80 deaths in 1971. The age-standardised rate fell from 7 per million to 1 per million over the same period (Figure 23.3.3).

Survival is presented separately for non-Hodgkin's lymphoma (NHL) and Hodgkin's disease in Figures 23.4.3(a) and 23.4.3(b). Five-year survival from Hodgkin's disease has been high – at least 80% – since the early 1970s, and was 92% for children diagnosed in 1986-90. Survival from NHL is somewhat lower – five-year relative survival was 76% for children diagnosed in 1986-90. However, there were substantial improvements from the early 1970s, with an average rise of 17% points every five years – the greatest increase in survival to occur for a childhood cancer.

Soft tissue sarcomas

Soft tissue sarcomas account for 7% of all malignant neoplasms in children and are the fourth most common form of childhood cancer in England and Wales. They occur in muscle, fat, fibrous tissues and blood vessels throughout the body. While they may arise in any site, most soft tissue sarcomas occur in the head and neck region and the genitourinary tract. The most common form in children is rhabdomyosarcoma (striated muscle). Several risk factors are suspected in the aetiology of the disease, including radiation therapy, exposure to pesticides, and paternal smoking.[1] The disease is also associated with a number of genetic disorders. Kaposi's sarcoma, a further form of the disease, is associated with AIDS. This has been demonstrated by the very large increase in Uganda, a country severely effected by AIDS.[5] A higher risk of soft tissue sarcoma was also observed in children not immunised for small pox, or measles, mumps and rubella.[6]

There are around 90 new cases of soft tissue sarcoma (STS) diagnosed each year. STS occur most frequently in children aged under 4 with around 40 such cases diagnosed annually (Table 23.1). Incidence generally declined with age up to 10-11 years, but then increased (Figure 23.1.4). The incidence of STS increased over the 1970s and early 1980s from 4 cases per million to around 8 per million (Figure 23.2.4). There are around 15 deaths from STS each year and mortality has fluctuated mostly between 1 and 2 deaths per million since the early 1970s (Figure 23.3.4).

One-year relative survival from soft tissue sarcomas was 86% and five-year survival was 66% for children diagnosed in 1986-90. There was a rise of around 25% points in five-year survival compared with children diagnosed in 1971-75, an average rise of 8-9% every five years (Figure 23.4.5).

Neuroblastoma

Neuroblastoma is an embryonal tumour, which accounts for 6% of all cancers diagnosed in children in England and Wales. It is a disease of young children, with around three-quarters of cases occurring in children aged under 4. Around two-thirds of neuroblastoma tumours occur in the abdomen, with 60% of these in the adrenal gland.[7] The remainder occur in the chest, pelvis, and the head and neck region. Little is know about the cause of neuroblastoma; however, there is an association with neurofibromatosis, a genetic syndrome.[7] There is also some evidence of a link with *in utero* exposures to alcohol, diuretic drugs, non-prescriptive pain relievers and hair-colouring products, and exposure to maternal smoking.[1]

There were on average 80 new cases of neuroblastoma diagnosed in 1971 to 1994. Neuroblastomas occur most frequently in the youngest ages and are extremely rare over the age of 10 (Table 23.1). There were on average 23 cases in 1993-94 in children aged under 1, and 39 cases in those aged 1-4. In children under 1, the incidence rate was 34 per million in 1993-94 (Figure 23.1.5) but there has been some variability in the rate over time mainly in the range 20-28 per million in the late 1980s and early 1990s. Incidence in children aged 1-4 was 16 per million in 1994, compared with 4 per million in the 5-9 age group and less than 2 per million in those aged 10-14. The overall incidence of the disease has fluctuated mostly between 6 and 9 cases per million since 1971 (Figure 23.2.5). Mortality data are not available for neuroblastoma since this disease is defined through morphology information which is not available on death certificates.

Survival from neuroblastoma is amongst the lowest in children: five-year survival (using the actuarial method) was less than 40% for children diagnosed in 1986-88.[8] This cancer was not included in *Cancer Survival Trends*, and so the data are not shown in Figure 23.4.

Renal cancers – Wilms' tumour

The great majority of renal cancers are Wilms' tumours (nephroblastoma). They are usually diagnosed in children between the ages of one and three, and are associated with congenital anomalies, familial (hereditary) conditions and other genetic disorders, as well as high birth weight.[1,9] Wilms' tumour has a high tendency to spread to other parts of the body, generally via the bloodstream rather than the lymph nodes, and give rise to metastases – most commonly in the lungs. The prognosis for localised disease is very good, as Wilms' tumour is both radio- and chemosensitive.

During the 1980s, there were 50-60 cases of the renal cancer, Wilms' tumour, each year, but there were on average 70 cases in 1993-94. Wilms' tumour occurs most frequently in children aged 1-4 – there were on average 48 such cases diagnosed in 1993-94 (Table 23.1), and an incidence rate of almost 20 per million across that age range (Figure 23.1.6). The disease is extremely rare in children aged over 6, with incidence of around 1 per million. The overall incidence of the disease has remained relatively stable since 1971, at 6-8 new cases per million (Figure 23.2.6) with stable rates in each age group. In the absence of morphology information at death registration, it is not possible to present mortality figures for Wilms' tumours. However, since Wilms' tumours account for the majority of renal cancers in children, the figures for deaths from cancers of the kidney (ICD9 189.0) have been given. There are around 15 deaths from kidney cancer in children. Mortality declined from 4 per million in 1971 to between 1 and 2 per million in the 1990s (Figure 23.3.6).

Relative survival from Wilms' tumour for children diagnosed in 1986-90 in England and Wales was 94% at one year and 84% at five years. Survival rose by around 20% points since the early 1970s (Figure 23.4.6). Ten-year survival is very similar to survival at five years. The plateau in the relative survival curves reflects very low excess mortality among children who survived three years or more and can effectively be considered as cured.

International variations in incidence

The incidence of childhood cancers in England and Wales tends to be moderate compared with other countries and the rate is particularly low for lymphomas (Figure 23.5). The rates in the USA, in contrast, tend to be higher than for England and Wales, particularly for brain tumours and lymphomas. The extremely high incidence of soft tissue sarcomas in Zimbabwe reflects the incidence of AIDs-related Kaposi sarcoma.

Figure 23.1 **Age-specific incidence of the major cancers in children, England and Wales, 1993-94**

Figure 23.1.1 **Leukaemia***

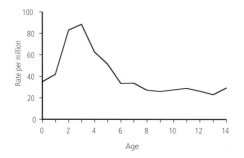

Figure 23.1.2 **Brain and spinal cord**

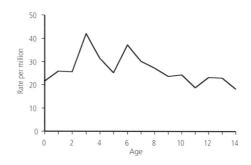

Figure 23.1.3 **Lymphomas**

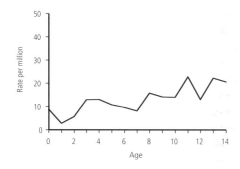

Figure 23.1.4 **Soft tissue sarcoma**

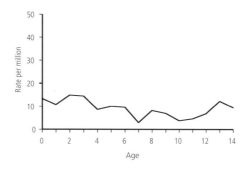

Figure 23.1.5 **Neuroblastoma**

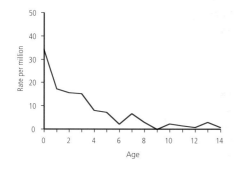

Figure 23.1.6 **Wilms' tumour**

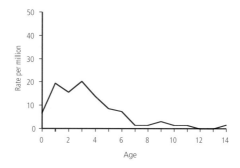

*Note different vertical scale

Figure 23.2 **Age-standardised incidence* of the major cancers in children, England and Wales, 1971-1994**

Figure 23.2.1 **Leukaemia**

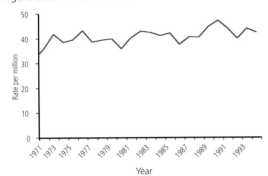

Figure 23.2.2 **Brain and spinal cord**

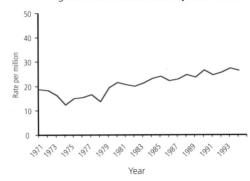

Figure 23.2.3 **Lymphomas**

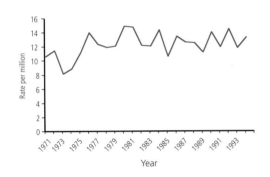

Figure 23.2.4 **Soft tissue sarcoma**

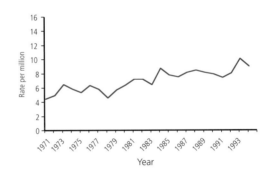

Figure 23.2.5 **Neuroblastoma**

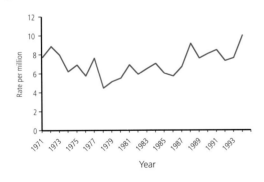

Figure 23.2.6 **Wilms' tumour**

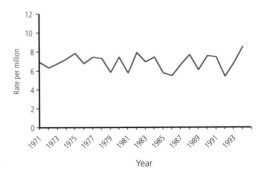

* Directly age-standardised using the World standard population
Note different vertical scales

Figure 23.3 **Mortality* from the major cancers in children, England and Wales, 1971-1999**

Figure 23.3.1 **Leukaemia**

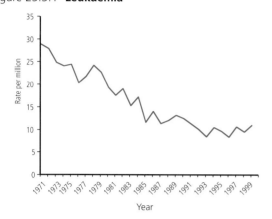

Figure 23.3.2 **Brain and spinal cord**

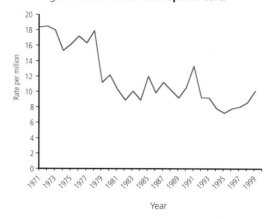

Figure 23.3.3 **Lymphomas**

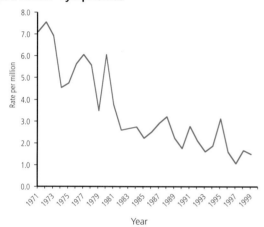

Figure 23.3.4 **Soft tissue sarcoma**

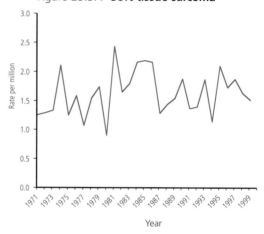

Figure 23.3.5 **Neuroblastoma**

Mortality data for neuroblastoma are not available since this disease is defined through morphology information which is not available on death certificates.

Figure 23.3.6 **Kidney**

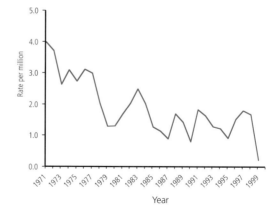

* Directly age-standardised using the World standard population
Note different vertical scales

Figure 23.4 **Survival up to ten years after diagnosis in children (aged under 15), by calendar period, England and Wales, 1971-1990**

Figure 23.4.1 **Leukaemia**

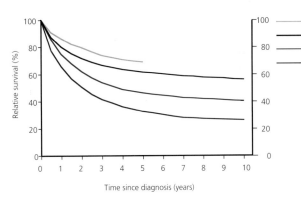

Figure 23.4.2 **Brain and spinal cord**

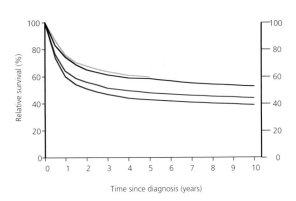

Figure 23.4.3(a) **Hodgkin's Disease**

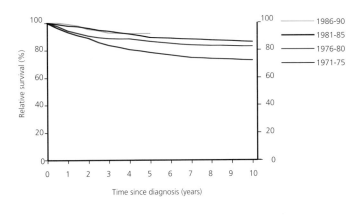

Figure 23.4.4 **Soft tissue sarcoma**

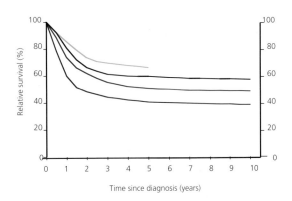

Figure 23.4.3(b) **Non-Hodgkin's lymphoma**

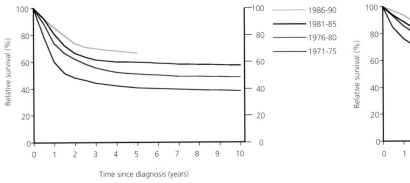

Figure 23.4.6 **Wilms' tumour**

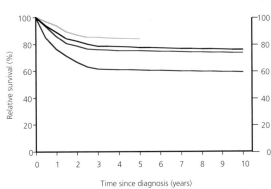

Note: Figure 23.4.5 is not presented as neoblastoma survival has not been calculated using the national cancer database

Figure 23.5 **Age-standardised international incidence***

Figure 23.5.1 **Leukaemia**

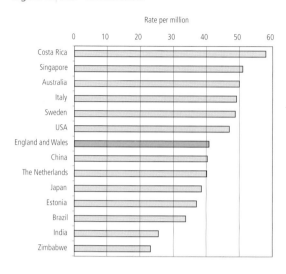

Figure 23.5.2 **Brain and spinal cord**

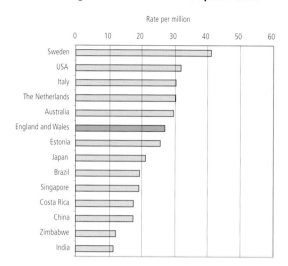

Figure 23.5.3 **Lymphomas**

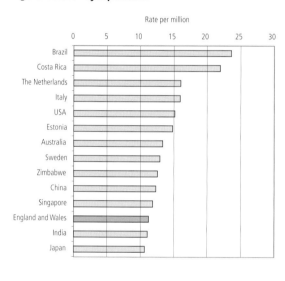

Figure 23.5.4 **Soft tissue sarcoma**

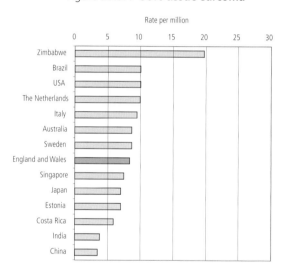

Figure 23.5.5 **Neuroblastoma**

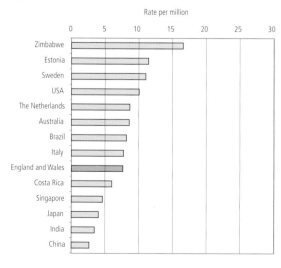

Figure 23.5.6 **Wilms' tumour**

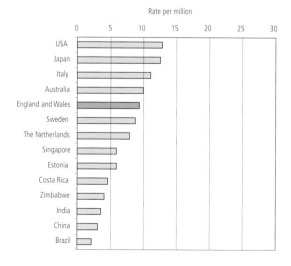

* Directly age-standardised using the World standard population

Appendix A

KEY CANCER STATISTICS, UNITED KINGDOM

Males

ICD10	Site	Incidence (1996) Number of cases	CR	ESR*	WSR*	Rank	Lifetime risk	Odds	Mortality (1998) Number of deaths	CR	ESR*	WSR*	One-year	Five-year	ICD10
C00-C97 xC44	All cancers excluding NMSC^a	126,200	437.4	395.8	265.9	:	35.1	1 in 3	80,200	275.3	242.5	156.8	::	::	C00-C97 xC44
C00-C14	Lip, mouth and pharynx	2,940	10.2	9.9	7.0	11	0.8	1 in 126	1,080	4.4	4.2	2.9	74+	44+	C00-C14
C15	Oesophagus	4,100	14.2	13.0	8.7	7	1.3	1 in 75	4,290	14.7	13.2	8.7	25	6	C15
C16	Stomach	6,610	22.9	20.5	13.3	5	2.3	1 in 44	4,450	15.3	13.3	8.5	28	10	C16
C18-C21	Colorectal	17,500	60.6	55.0	36.3	3	5.7	1 in 18	8,910	30.6	26.9	17.2	64**	42**	C18-C21
C25	Pancreas	3,230	11.2	10.2	6.7	10	1.0	1 in 96	3,190	10.9	9.7	6.3	11	2	C25
C32	Larynx	1,870	6.5	6.2	4.3	14	0.5	1 in 182	700	2.4	2.3	1.5	83	64	C32
C33-C34	Lung	25,700	88.9	79.6	52.1	1	8.0	1 in 13	21,800	74.9	65.8	42.6	20	5	C33-C34
C43	Melanoma of skin	2,240	7.7	7.5	5.6	12	0.7	1 in 147	800	2.7	2.5	1.8	91	72	C43
C61	Prostate	21,400	74.3	64.4	39.8	2	7.3	1 in 14	9,460	32.5	27.2	15.7	79	49	C61
C62	Testis	1,770	6.1	5.8	5.4	15	0.4	1 in 259	90	0.3	0.3	0.2	97	94	C62
C64-C66,C68	Kidney	3,450	11.9	11.2	7.9	8	1.1	1 in 89	1,880	6.5	5.9	4.0	59	41	C64-C66,C68
C67	Bladder	9,710	33.7	30.1	19.5	4	3.3	1 in 30	3,300	11.3	9.8	5.9	82	66	C67
C71	Brain	2,170	7.5	7.3	5.8	13	0.7	1 in 147	1,790	6.1	5.9	4.4	30	13	C71
C82-C85,C91.4,C96	Non-Hodgkin's lymphoma	4,500	15.6	14.7	10.6	6	1.4	1 in 69	2,370	8.2	7.4	5.0	66	46	C82-C85,C91.4,C96
C88,C90	Multiple myeloma	1,750	6.1	5.5	3.6	16	0.6	1 in 177	1,210	4.2	2.3	3.6	59	20	C88,C90
C91-C95 xC91.4	Leukaemia	3,380	11.7	10.8	8.3	9	1.0	1 in 95	2,140	7.3	6.5	4.6	56	30	C91-C95 xC91.4
	Other	13,900	48.4	44.3	31.0	:	:	:	12,700	42.9	39.3	23.7	:	:	

Females

ICD10	Site	Incidence (1996) Number of cases	CR	ESR*	WSR*	Rank	Lifetime risk	Odds	Mortality (1998) Number of deaths	CR	ESR*	WSR*	One-year	Five-year	ICD10
C00-C97 xC44	All cancers excluding NMSC^a	128,700	429.9	329.5	234.0	:	33.1	1 in 3	74,000	245.9	166.5	111.6	::	::	C00-C97 xC44
C00-C14	Lip, mouth and pharynx	1,620	5.4	4.3	3.1	16	0.5	1 in 215	710	2.4	1.7	1.1	73+	52+	C00-C14
C15	Oesophagus	2,960	9.9	6.2	4.0	13	1.1	1 in 95	2,520	8.4	5.0	3.1	25	7	C15
C16	Stomach	3,740	12.5	7.7	5.0	11	1.2	1 in 86	2,880	9.6	5.6	3.5	27	11	C16
C18-C21	Colorectal	16,800	56.0	37.5	24.8	2	4.9	1 in 20	8,200	27.2	16.5	10.5	61**	40**	C18-C21
C25	Pancreas	3,630	12.1	7.8	5.1	10	1.1	1 in 95	3,360	11.2	7.1	4.6	11	2	C25
C33-C34	Lung	15,200	50.9	37.4	25.4	3	4.3	1 in 23	13,100	43.5	30.3	20.2	20	5	C33-C34
C43	Melanoma of skin	3,110	10.4	9.1	7.1	8	0.9	1 in 117	840	2.8	2.1	1.5	96	86	C43
C50	Breast	36,100	120.5	104.3	75.9	1	10.9	1 in 9	13,200	43.8	32.6	22.6	92	74	C50
C53	Cervix	3,320	11.1	10.0	8.0	7	0.9	1 in 116	1,330	4.4	3.5	2.6	83	64	C53
C54	Uterus	4,520	15.1	12.6	8.8	5	1.4	1 in 73	930	3.1	2.0	1.4	87	73	C54
C56,C57	Ovary	6,570	21.9	18.2	13.2	4	2.1	1 in 48	4,520	15.0	11.6	8.0	58	29	C56,C57
C64-C66,C68	Kidney	2,160	7.2	5.5	4.0	14	0.6	1 in 162	1,230	4.1	2.8	1.9	54	37	C64-C66,C68
C67	Bladder	3,890	13.0	8.5	5.6	9	1.3	1 in 79	1,710	5.7	3.2	2.0	71	58	C67
C71	Brain	1,780	5.9	5.2	4.3	15	0.5	1 in 207	1,380	4.6	3.9	3.0	30	14	C71
C82-C85,C91.4,C96	Non-Hodgkin's lymphoma	3,940	13.1	10.2	7.3	6	1.2	1 in 83	2,110	7.0	4.7	3.2	67	48	C82-C85,C91.4,C96
C88,C90	Multiple myeloma	1,540	5.1	3.5	2.3	17	0.5	1 in 204	1,260	4.2	2.6	1.7	56	18	C88,C90
C91-C95 xC91.4	Leukaemia	2,850	9.5	7.0	5.6	12	0.8	1 in 127	1,800	6.0	4.0	2.8	51	30	C91-C95 xC91.4
	Other	15,000	50.1	34.5	24.4	:	:	:	12,900	43.2	27.3	17.9	:	:	

Relative survival (%) for patients diagnosed in 1991-93#

a Non-melanoma skin cancer

* Directly age-standardised rate per 100,000, using the European (E) and World (W) standard populations

England and Wales

+ Cancer of the oral cavity (patients diagnosed 1986-90)

** Colon cancer

INCIDENCE

Appendix B1(a) Number of new cases, England and Wales, 1994 - males

ICD10	Site	All ages	Under 1	1-4	5-9	10-14	15-19	20-24	25-29	30-34	35-39	40-44	45-49	50-54	55-59	60-64	65-69	70-74	75-79	80-84	85 and over	ICD10
	All registrations	137,593	94	270	206	212	282	559	849	1,218	1,449	2,262	4,136	6,021	9,574	14,344	20,647	27,173	20,349	17,268	10,680	
C00-C97	All cancers	131,549	78	249	193	181	253	493	761	1,100	1,341	2,075	3,849	5,680	9,080	13,695	19,785	26,129	19,548	16,689	10,370	C00-C97
C00-C97 xC44	All cancers excluding NMSC	112,145	76	249	193	176	244	468	699	966	1,127	1,685	3,162	4,714	7,643	11,686	17,110	22,412	16,684	14,125	8,726	C00-C97 xC44
C00-C14	Lip, mouth and pharynx	2,364	-	3	-	-	7	3	13	16	47	93	175	248	292	312	344	348	209	157	96	C00-C14
C00	Lip	202	-	-	-	-	-	-	1	-	2	4	8	16	19	22	37	38	23	19	13	C00
C01-C02	Tongue	510	-	-	-	1	1	1	5	4	10	22	45	54	69	67	79	71	46	18	18	C01-C02
C03-C06	Mouth	614	-	-	-	-	-	-	2	2	19	21	39	67	69	89	89	74	58	47	26	C03-C06
C07-C08	Salivary glands	236	-	-	-	-	-	-	3	5	4	10	18	15	26	24	26	37	23	20	25	C07-C08
C09-C10	Oropharynx	311	-	-	-	-	-	-	-	1	4	18	30	43	41	53	36	49	20	12	4	C09-C10
C11	Nasopharynx	122	-	3	-	-	6	1	3	4	5	7	9	11	13	11	15	17	9	4	1	C11
C12-C13	Hypopharynx	244	-	-	-	-	-	-	-	-	1	6	18	22	36	24	36	46	22	27	6	C12-C13
C14	Pharynx unspecified	125	-	-	-	-	-	-	-	-	2	5	8	8	18	17	26	16	8	10	3	C14
C15	Oesophagus	3,537	-	-	-	-	-	-	4	8	20	53	114	188	278	450	567	687	481	402	285	C15
C16	Stomach	6,115	-	-	-	-	1	2	8	15	26	69	135	226	374	655	984	1,300	945	881	494	C16
C17	Small intestine	241	-	-	-	-	-	1	1	3	5	7	16	17	13	36	36	38	37	23	9	C17
C18-C21	Colorectal	14,811	1	-	-	2	2	4	22	44	84	165	409	649	1,104	1,715	2,388	3,007	2,204	1,865	1,147	C18-C21
C18	Colon	8,729	1	-	1	1	-	3	16	25	41	85	220	331	565	941	1,399	1,791	1,372	1,185	753	C18
C19-C21	Rectum	6,082	1	-	-	-	1	1	6	19	43	80	189	318	539	774	989	1,216	832	680	394	C19-C21
C22	Liver	942	4	4	1	1	1	1	2	10	8	13	25	50	83	120	161	190	132	77	60	C22
C23-C24	Gallbladder	468	-	1	-	-	-	-	-	-	2	7	11	15	25	40	80	104	78	67	38	C23-C24
C25	Pancreas	2,846	-	-	-	-	-	-	3	6	14	34	92	129	233	325	409	546	444	370	241	C25
C26	Digestive organs etc. unspecified	239	-	-	-	-	-	1	1	1	-	3	5	11	13	21	33	33	30	42	46	C26
C30-C31	Nasal cavities	199	1	2	-	-	-	2	-	5	4	5	14	20	12	17	23	40	25	17	10	C30-C31
C32	Larynx	1,714	-	-	-	-	-	-	1	1	9	38	76	149	178	246	297	344	176	134	65	C32
C33-C34	Lung	23,314	1	-	-	-	1	4	10	25	49	173	485	907	1,622	2,607	4,153	5,307	3,709	2,750	1,511	C33-C34
C37-C39,C45	Respiratory system	1,129	-	-	1	3	3	2	4	4	11	23	37	75	123	175	222	199	131	71	48	C37-C39,C45
C40-C41	Bone	221	1	-	10	15	22	20	9	6	11	10	10	14	10	15	19	20	11	13	5	C40-C41
C43	Melanoma of skin	1,909	-	-	4	4	11	29	48	87	100	127	184	177	208	191	202	204	145	112	76	C43
C44	Non-melanoma of skin	19,404	2	1	5	5	9	25	62	134	214	390	687	966	1,437	2,009	2,675	3,717	2,864	2,564	1,644	C44
C47,C49	Connective tissue	577	5	14	7	13	14	15	13	22	24	26	37	35	47	48	60	78	54	42	23	C47,C49
C48	Retroperitoneum and peritoneum	149	1	-	1	-	2	2	2	3	-	6	6	14	16	17	25	20	22	4	7	C48
C50	Breast	189	-	-	-	-	-	-	-	3	1	4	6	13	18	24	27	28	31	21	15	C50
C60,C63	Penis	359	-	4	-	-	-	-	4	4	5	7	20	17	36	33	37	62	48	42	31	C60,C63
C61	Prostate	19,399	1	4	-	-	-	1	-	-	4	6	42	211	633	1,457	2,777	4,472	3,823	3,558	2,408	C61
C62	Testis	1,381	4	8	1	4	29	140	247	307	207	146	110	50	49	15	20	15	10	13	6	C62
C64-C66,C68	Kidney	2,985	5	22	8	3	-	7	7	17	41	78	146	201	302	412	503	518	328	254	132	C64-C66,C68
C67	Bladder	8,516	-	2	-	4	4	9	12	14	40	77	167	320	573	911	1,355	1,756	1,319	1,207	749	C67
C69	Eye	204	13	12	1	1	-	-	2	3	6	5	12	9	18	20	28	31	22	16	7	C69
C70,C72	Other nervous system	85	-	6	1	2	-	-	2	-	8	4	5	9	8	9	10	7	5	4	-	C70,C72
C71	Brain	1,986	6	39	48	33	30	38	46	71	84	101	152	161	191	244	244	247	142	78	31	C71
C73	Thyroid	259	-	-	-	3	3	6	10	20	10	16	27	19	21	16	40	29	24	9	8	C73
C74-C75	Other endocrine glands	116	11	19	4	2	6	2	3	-	2	2	5	3	7	9	14	11	4	8	4	C74-C75
C81	Hodgkin's disease	674	1	2	17	15	32	84	91	67	52	52	40	43	36	40	39	32	16	9	4	C81
C82-C85,C91.4,C96	Non-Hodgkin's lymphoma	3,919	3	11	16	27	26	43	65	110	147	151	248	298	372	436	471	590	415	306	184	C82-C85,C91.4,C96
C88,C90	Multiple myeloma	1,442	-	-	-	1	-	-	5	1	9	16	55	74	119	152	216	293	197	177	126	C88,C90
C91-C95,xC91.4	Leukaemia	3,053	13	98	72	52	45	37	51	55	54	60	114	125	181	268	368	464	365	397	234	C91-C95,xC91.4
C91,xC91.4	Lymphoid leukaemia	1,482	4	83	60	38	24	15	15	13	17	20	44	53	85	134	174	229	171	189	116	C91,xC91.4
C92	Myeloid leukaemia	1,401	6	12	8	12	21	34	24	41	35	37	68	66	88	126	174	206	175	174	94	C92
C93	Monocytic leukaemia	29	2	-	-	1	-	-	-	-	1	3	-	2	3	1	1	6	3	4	4	C93
C94-C95	Other leukaemia	141	3	3	3	1	-	2	2	1	1	3	2	4	5	7	19	23	16	30	20	C94-C95
C76-C80	Other and unspecified	6,803	3	2	1	-	2	14	14	31	44	94	181	245	444	651	958	1,392	1,102	999	626	C76-C80

Appendix B1(b) Number of new cases, England and Wales, 1994 - females

ICD10	Site	All ages	Under 1	1-4	5-9	10-14	15-19	20-24	25-29	30-34	35-39	40-44	45-49	50-54	55-59	60-64	65-69	70-74	75-79	80-84	85 and over	ICD10
	All registrations	159,106	71	233	191	200	553	3,691	6,521	6,447	5,653	6,058	8,570	9,295	10,730	13,480	15,988	20,938	17,122	16,954	16,411	C00-C97
C00-C97	All cancers	129,558	60	219	168	164	197	483	958	1,788	2,606	4,121	6,955	7,966	9,515	12,243	14,862	19,579	16,127	15,964	15,583	C00-C97 xC44
C00-C97 xC44	All cancers excluding NMSC	112,175	60	219	164	156	183	459	891	1,642	2,381	3,711	6,206	7,187	8,545	10,918	12,995	16,897	13,507	13,276	12,778	C00-C14
C00-C14	Lip, mouth and pharynx	1,306				4	3	6	10	13	24	55	76	63	108	138	174	176	150	155	149	C00
C00	Lip	78			2	1			1			2	3	1	6	4	6	14	6	19	13	C01-C02
C01-C02	Tongue	277			1	1		2	1	3	3	14	19	19	20	35	29	44	33	23	30	C03-C06
C03-C06	Mouth	378					1		2	1	8	18	23	8	36	35	54	53	44	48	47	C07-C08
C07-C08	Salivary glands	189				1	1		3	5	4	9	9	12	14	21	17	25	18	23	26	C09-C10
C09-C10	Oropharynx	142						3			3	4	4	14	12	17	21	18	17	23	13	C11
C11	Nasopharynx	61			1		1	3		4	4	4	4	2	3	4	9	4	5	6	5	C12-C13
C12-C13	Hypopharynx	123					1			4	5		6	7	3	13	27	10	22	17	8	C14
C14	Pharynx unspecified	58										2	4	7	8	9	11	8	5	4	7	C15
C15	Oesophagus	2,444						1		2	4	22	38	73	118	176	263	426	411	450	459	C16
C16	Stomach	3,599								16	21	42	50	71	129	249	366	565	625	704	759	C17
C17	Small intestine	227						1	1	5	1	6	11	9	26	18	20	32	37	32	29	C18-C21
C18-C21	Colorectal	14,093				2	1	10	17	31	87	138	312	510	826	1,207	1,727	2,401	2,048	2,357	2,419	C18
C18	Colon	9,575			2	2	1	7	10	17	54	82	192	318	543	798	1,151	1,584	1,425	1,653	1,738	C19-C21
C19-C21	Rectum	4,518						3	7	14	33	56	120	192	283	409	576	817	623	704	681	C22
C22	Liver	633	5	2	2		3	2	2	3	9	10	19	20	23	45	72	108	93	110	105	C23-C24
C23-C24	Gallbladder	737		2				1	1	2	1	4	11	25	40	60	91	116	115	132	139	C25
C25	Pancreas	3,170				2	2		1	2	9	29	54	93	128	281	389	567	481	561	571	C26
C26	Digestive organs etc. unspecified	305									3	3	2	4	8	15	23	39	60	59	90	C30-C31
C30-C31	Nasal cavities	143		1		2	2	1	2	4	4	3	2	8	13	17	11	22	19	15	19	C32
C32	Larynx	362		2	1			2		1	4	2	15	18	40	39	84	72	33	25	27	C33-C34
C33-C34	Lung	12,297	1		1	1	1	2	4	18	44	122	312	477	728	1,287	2,144	2,819	1,875	1,439	1,022	C37-C39 C45
C37-C39 C45	Respiratory system	281	2			1		4	5		5	4	10	19	29	35	31	61	32	17	12	C40-C41
C40-C41	Bone	151		1	11	15	16	6	9	8	3	7	6	6	5	10	2	9	14	11	12	C43
C43	Melanoma of skin	2,816			5	3	13	79	148	169	156	194	301	231	212	214	244	256	211	202	178	C44
C44	Non-melanoma of skin	17,383		7	5	8	14	24	67	146	225	410	749	779	970	1,325	1,867	2,682	2,620	2,688	2,805	C47,C49
C47,C49	Connective tissue	496	10	13	5	8	11	11	18	20	23	25	23	29	26	34	55	55	53	36	55	C48
C48	Retroperitoneum and peritoneum	135		9		1		1	1	2	5	6	6	7	17	17	10	18	12	17	12	C50
C50	Breast	31,671	3	2	1		5	22	163	538	1,070	1,899	3,255	3,499	3,411	3,518	3,062	3,629	2,622	2,515	2,459	C51-C52
C51-C52	Other female genital	1,089	1	2		1	1	2	4	16	21	46	50	42	50	92	101	162	150	170	206	C53
C53	Cervix	3,173				3	15	55	190	369	356	329	292	220	204	213	209	271	191	160	112	C54
C54	Uterus	4,044						6	6	5	36	70	173	340	510	639	596	580	444	356	289	C55
C55	Uterus unspecified	333					2	1		2	2	10	25	14	17	28	34	52	38	44	66	C56-C57
C56-C57	Ovary	5,349		1	5	7	10	27	53	78	112	195	353	465	549	653	724	774	522	476	345	C58
C58	Placenta	12					2		4	2	3											C64-C66,C68
C64-C66,C68	Kidney	1,781	3	35	10	5		5	4	18	21	33	83	94	139	210	240	293	216	206	165	C67
C67	Bladder	3,286		2	2		1	3	1	8	9	27	51	96	141	266	391	599	541	574	578	C69
C69	Eye	183	7	13	2	6	1	2	3	2	7	5	11	10	12	4	2	30	6	19	13	C70,C72
C70,C72	Other nervous system	101	1	9	7	6	1	1	2	5	3	6	8	10	19	12	15	13	6	8	5	C71
C71	Brain	1,516	6	33	47	30	24	24	39	58	54	59	103	91	128	166	170	195	147	96	46	C73
C73	Thyroid	786		12		3	15	48	42	72	63	74	65	45	42	48	55	73	52	51	38	C74-C75
C74-C75	Other endocrine glands	115	4		5	4	2	5	5	4	5	8	6	6	8	6	8	8	10	5	4	C81
C81	Hodgkin's disease	509		2	4	8	38	76	78	46	41	29	22	20	30	25	22	26	9	18	15	C82-C85,C91.4,C96
C82-C85,C91.4,C96	Non-Hodgkin's lymphoma	3,352	2	7	12	11	11	24	40	50	70	91	162	207	256	340	388	578	447	376	280	C88,C90
C88,C90	Multiple myeloma	1,436			1				1	3	5	13	36	47	80	139	177	264	229	253	188	C91-C95,xC91.4
C91-C95,xC91.4	Leukaemia	2,506	12	85	43	41	24	30	24	39	35	63	88	93	115	167	244	326	321	348	408	C91,xC91.4
C91,xC91.4	Lymphoid leukaemia	1,130	6	75	34	30	12	6	6	6	11	21	30	54	77	101	148	134	159	216		C92
C92	Myeloid leukaemia	1,210	3	8	9	9	10	22	17	32	31	50	64	57	54	81	132	156	162	156	157	C93
C93	Monocytic leukaemia	20						1	1		2				1		2	4	3	3	5	C94-C95
C94-C95	Other leukaemia	146	3	2		2	2				1	2	3	8	4	9	18	22	3	30	30	C76-C80
C76-C80	Other and unspecified	7,738	2	4	1	1	3	9	10	28	67	108	178	228	366	550	851	1,282	1,278	1,279	1,492	D05
D05	Breast in situ	1,715								27	54	113	197	400	310	308	111	92	41	41	15	D06
D06	Cervix in situ	19,611					270	3,045	5,307	4,344	2,685	1,549	1,001	486	406	289	128	57	28	14	1	

Appendix B2(a) Incidence rates per 100,000 population, England and Wales, 1994 - males

ICD10	Site	CR	ESR	WSR	Under 1	1-4	5-9	10-14	15-19	20-24	25-29	30-34	35-39	40-44	45-49	50-54	55-59	60-64	65-69	70-74	75-79	80-84	85 and over	ICD10
	All registrations	**543.8**	**492.7**	**329.8**	**27.3**	**19.2**	**11.9**	**12.9**	**18.3**	**30.1**	**40.0**	**58.0**	**81.2**	**133.7**	**229.0**	**420.9**	**728.7**	**1,186.7**	**1,860.6**	**2,720.0**	**3,500.6**	**4,274.3**	**4,665.8**	
C00-C97	**All cancers**	**519.9**	**470.6**	**314.2**	**22.7**	**17.7**	**11.2**	**11.0**	**16.5**	**26.5**	**35.8**	**52.4**	**75.2**	**122.7**	**213.1**	**397.1**	**691.1**	**1,133.0**	**1,782.9**	**2,615.5**	**3,362.8**	**4,130.9**	**4,530.4**	**C00-C97**
C00-C97 xC44	**All cancers excluding NMSC**	**443.2**	**400.8**	**268.2**	**22.1**	**17.7**	**11.2**	**10.7**	**15.9**	**25.2**	**32.9**	**46.0**	**63.2**	**99.6**	**175.1**	**329.5**	**581.7**	**966.8**	**1,541.9**	**2,243.4**	**2,870.1**	**3,496.3**	**3,812.1**	**C00-C97 xC44**
C00-C14	Lip, mouth and pharynx	**9.3**	**9.1**	**6.4**	-	0.2	-	0.1	0.5	0.2	0.6	0.8	2.6	5.5	9.7	17.3	22.2	25.8	31.0	34.8	36.0	38.9	41.9	**C00-C14**
C00	Lip	**0.8**	**0.7**	**0.5**	-	-	-	-	-	0.1	-	-	0.1	0.2	0.4	1.1	1.4	1.8	3.3	3.8	4.0	4.7	5.7	**C00**
C01-C02	Tongue	**2.0**	**2.0**	**1.4**	-	-	-	-	0.1	0.1	0.2	0.2	0.6	1.3	2.5	3.8	5.3	5.5	7.1	7.1	7.9	4.5	7.9	**C01-C02**
C03-C06	Mouth	**2.4**	**2.4**	**1.7**	-	-	-	-	-	-	0.1	0.1	1.1	1.2	2.2	5.5	5.3	7.4	8.0	7.4	10.0	11.6	11.4	**C03-C06**
C07-C08	Salivary glands	**0.9**	**0.9**	**0.6**	-	-	-	-	-	-	0.1	0.2	0.2	0.6	1.0	1.0	2.0	2.0	2.3	3.7	4.0	5.0	10.9	**C07-C08**
C09-C10	Oropharynx	**1.2**	**1.2**	**0.9**	-	-	-	-	-	-	-	0.0	0.2	1.1	1.7	3.0	3.1	4.4	3.2	4.9	3.4	3.0	1.7	**C09-C10**
C11	Nasopharynx	**0.5**	**0.5**	**0.4**	-	0.2	-	0.1	0.4	0.1	0.1	0.2	0.3	0.4	0.5	0.9	1.0	0.9	1.4	1.7	1.5	1.0	0.4	**C11**
C12-C13	Hypopharynx	**1.0**	**0.9**	**0.6**	-	-	-	-	-	-	-	-	0.1	0.4	1.0	1.5	2.7	2.0	3.2	4.6	3.8	6.7	2.6	**C12-C13**
C14	Pharynx unspecified	**0.5**	**0.5**	**0.3**	-	-	-	-	-	-	-	-	0.1	0.3	0.4	0.5	1.4	1.8	2.3	1.6	1.4	2.5	1.3	**C14**
C15	Oesophagus	**14.0**	**12.8**	**8.5**	-	-	-	-	-	-	0.2	0.4	1.1	3.1	6.3	13.1	21.2	37.2	51.1	68.8	82.7	99.5	124.5	**C15**
C16	Stomach	**24.2**	**21.6**	**14.0**	-	-	-	-	0.1	0.1	0.4	0.7	1.5	4.1	7.5	15.8	28.5	54.2	88.7	130.1	162.6	218.1	215.8	**C16**
C17	Small intestine	**1.0**	**0.9**	**0.6**	-	-	-	-	-	-	0.0	0.1	0.3	0.4	0.9	1.2	1.0	3.0	3.2	3.8	6.4	5.7	3.9	**C17**
C18-C21	Colorectal	**58.5**	**53.0**	**34.9**	0.3	-	-	0.1	0.1	0.2	1.1	2.1	4.7	9.8	22.6	45.4	84.0	141.9	215.2	301.0	379.2	461.6	501.1	**C18-C21**
C18	Colon	**34.5**	**31.0**	**20.2**	-	-	-	0.1	0.1	0.2	0.8	1.2	2.3	5.0	12.2	23.1	43.0	77.9	126.1	179.3	236.0	293.3	329.0	**C18**
C19-C21	Rectum	**24.0**	**22.0**	**14.7**	0.3	-	-	-	0.1	0.1	0.3	0.9	2.4	4.7	10.5	22.2	41.0	64.0	89.1	121.7	143.1	168.3	172.1	**C19-C21**
C22	Liver	**3.7**	**3.4**	**2.4**	1.2	0.3	0.1	-	0.1	0.1	0.1	0.5	0.4	0.8	1.4	3.5	6.3	9.9	14.5	19.0	22.7	19.1	26.2	**C22**
C23-C24	Gallbladder	**1.8**	**1.6**	**1.1**	-	0.1	-	-	-	-	-	-	0.1	0.4	0.6	1.0	1.9	3.3	7.2	10.4	13.4	16.6	16.6	**C23-C24**
C25	Pancreas	**11.2**	**10.2**	**6.7**	-	-	-	-	-	-	0.1	0.3	0.8	2.0	5.1	9.0	17.7	26.9	36.9	54.7	76.4	91.6	105.3	**C25**
C26	Digestive organs etc. unspecified	**0.9**	**0.9**	**0.5**	-	-	-	-	-	-	0.0	0.0	-	0.2	0.3	0.8	1.0	1.7	3.0	3.3	5.2	10.4	20.1	**C26**
C30-C31	Nasal cavities	**0.8**	**0.7**	**0.5**	0.3	0.1	-	-	0.1	0.1	0.0	0.2	0.2	0.3	0.8	1.4	0.9	1.4	2.1	4.0	4.3	4.2	4.4	**C30-C31**
C32	Larynx	**6.8**	**6.4**	**4.4**	-	-	-	-	-	-	0.0	0.0	0.5	2.2	4.2	10.4	13.5	20.4	26.8	34.4	30.3	33.2	28.4	**C32**
C33-C34	Lung	**92.1**	**82.6**	**54.2**	0.3	-	-	-	0.1	0.2	0.5	1.2	2.7	10.2	26.9	63.4	123.4	215.7	374.2	531.2	638.1	680.7	660.1	**C33-C34**
C37-C39, C45	Respiratory	**4.5**	**4.2**	**2.9**	-	-	-	0.1	0.2	0.1	0.2	0.2	0.6	1.4	2.0	5.2	9.4	14.5	20.0	19.9	22.5	17.6	21.0	**C37-C39, C45**
C40-C41	Bone	**0.9**	**0.9**	**0.8**	0.3	-	0.6	0.9	1.4	1.1	0.4	0.3	0.6	0.6	0.6	1.0	0.8	1.2	1.7	2.0	1.9	3.2	2.2	**C40-C41**
C43	Melanoma of skin	**7.5**	**7.3**	**5.5**	-	-	0.2	0.2	0.7	1.6	2.3	4.1	5.6	7.5	10.2	12.4	15.8	15.8	18.2	20.4	24.9	27.7	33.2	**C43**
C44	Non-melanoma of skin	**76.7**	**69.7**	**46.0**	0.6	-	-	0.3	0.6	1.3	2.9	6.4	12.0	23.1	38.0	67.5	109.4	166.2	241.1	372.1	492.7	634.7	718.2	**C44**
C47, C49	Connective tissue	**2.3**	**2.2**	**1.7**	1.5	1.0	0.4	0.8	0.9	0.8	0.6	1.0	1.3	1.5	2.0	2.4	3.6	4.0	5.4	7.8	9.3	10.4	10.0	**C47, C49**
C48	Retroperitoneum and peritoneum	**0.6**	**0.6**	**0.4**	0.3	-	0.1	-	0.1	0.1	0.1	0.1	0.1	0.4	0.3	1.0	1.2	1.4	2.3	2.0	3.8	1.0	3.1	**C48**
C50	Breast	**0.7**	**0.7**	**0.5**	-	-	0.1	-	-	-	-	-	0.1	0.2	0.3	0.9	1.4	2.0	2.4	2.8	5.3	5.2	6.6	**C50**
C60, C63	Penis	**1.4**	**1.3**	**0.9**	-	-	-	-	-	-	0.2	0.2	0.4	1.2	0.9	1.2	2.7	2.7	3.3	6.2	8.3	10.4	13.5	**C60, C63**
C61	Prostate	**76.7**	**66.1**	**40.4**	0.3	0.3	-	-	-	0.1	0.0	0.0	0.2	0.4	2.3	14.8	48.2	120.5	250.2	447.6	657.7	880.7	1,052.0	**C61**
C62	Testis	**5.5**	**5.2**	**4.8**	1.2	0.6	0.1	0.2	1.9	7.5	11.6	14.6	11.6	8.6	6.1	3.5	3.7	1.2	1.8	1.5	1.7	3.2	2.6	**C62**
C64-C66, C68	Kidney	**11.8**	**11.1**	**7.8**	1.5	1.6	0.5	0.2	0.1	0.4	0.3	0.8	2.3	4.6	8.1	14.1	23.0	34.1	45.3	51.9	56.4	62.9	57.7	**C64-C66, C68**
C67	Bladder	**33.7**	**30.2**	**19.6**	0.3	0.1	-	-	0.3	0.5	0.6	0.7	2.2	4.6	9.2	22.4	43.7	75.4	122.1	175.8	226.9	298.8	327.2	**C67**
C69	Eye	**0.8**	**0.8**	**0.6**	3.8	0.9	0.1	0.1	-	-	-	0.1	0.3	0.3	0.7	0.6	1.4	1.7	2.5	3.1	3.8	4.0	3.1	**C69**
C70, C72	Other nervous system	**0.3**	**0.3**	**0.3**	0.3	0.4	0.1	0.1	0.1	0.1	0.0	0.4	0.2	0.3	0.5	0.4	0.4	0.7	0.9	0.7	0.9	1.0	-	**C70, C72**
C71	Brain	**7.8**	**7.7**	**6.1**	1.7	2.8	2.8	2.0	2.0	2.0	2.2	3.4	4.7	6.0	8.4	11.3	14.5	20.2	22.0	24.7	24.4	19.3	13.5	**C71**
C73	Thyroid	**1.0**	**1.0**	**0.7**	-	-	-	0.1	0.2	0.3	0.5	1.0	0.6	0.9	1.5	1.3	1.6	1.3	3.6	2.9	4.1	2.2	3.5	**C73**
C74-C75	Other endocrine glands	**0.5**	**0.5**	**0.5**	3.2	1.3	0.2	0.1	0.4	0.1	0.1	-	0.1	0.1	0.3	0.2	0.5	0.7	1.3	1.1	0.7	2.0	1.7	**C74-C75**
C81	Hodgkin's disease	**2.7**	**2.6**	**2.4**	0.3	0.1	1.0	0.9	2.1	4.5	4.3	3.2	2.9	3.1	2.2	2.7	3.3	3.3	3.5	3.2	2.8	2.2	1.7	**C81**
C82-C85, C91.4, C96	Non-Hodgkin's lymphoma	**15.5**	**14.7**	**10.7**	0.9	0.8	0.9	1.6	1.7	2.3	3.1	5.2	8.2	8.9	13.7	20.8	28.3	36.1	42.4	59.1	71.4	75.7	80.4	**C82-C85, C91.4, C96**
C88, C90	Multiple myeloma	**5.7**	**5.2**	**3.4**	-	-	-	0.1	0.0	0.1	0.2	0.0	0.5	0.9	3.0	5.2	9.1	12.6	19.5	29.3	33.9	43.8	55.0	**C88, C90**
C91-C95, xC91.4	Leukaemia	**12.1**	**11.1**	**8.4**	3.8	7.0	4.2	3.2	2.9	2.0	2.4	2.6	3.0	3.5	6.3	8.7	13.8	22.2	33.2	46.4	62.8	98.3	102.2	**C91-C95, xC91.4**
C91, xC91.4	Lymphoid leukaemia	**5.9**	**5.5**	**4.3**	1.2	5.9	3.5	2.3	1.6	0.7	0.7	0.6	1.0	1.2	2.4	3.7	6.5	11.1	15.7	22.9	29.4	46.8	50.7	**C91, xC91.4**
C92	Myeloid leukaemia	**5.5**	**5.1**	**3.7**	1.7	0.9	0.5	0.7	1.4	1.3	1.6	2.0	2.0	2.2	3.8	4.6	6.7	10.4	15.7	20.6	30.1	43.1	41.1	**C92**
C93	Monocytic leukaemia	**0.1**	**0.1**	**0.1**	0.6	-	0.1	0.1	-	-	-	-	0.1	-	-	0.1	0.2	0.1	0.1	0.6	0.5	1.0	1.7	**C93**
C94-C95	Other leukaemia	**0.6**	**0.5**	**0.3**	0.3	0.2	0.2	0.1	-	-	0.1	0.0	0.1	0.2	0.1	0.3	0.4	0.6	1.7	2.3	2.8	7.4	8.7	**C94-C95**
C76-C80	Other and unspecified	26.9	**24.1**	**15.6**	0.9	0.1	0.1	-	0.1	0.8	0.7	1.5	2.5	5.6	10.0	17.1	33.8	53.9	86.3	139.3	189.6	247.3	273.5	**C76-C80**

Appendix B2(b) Incidence rates per 100,000 population, England and Wales, 1994 - females

ICD10	Site	CR	ESR	WSR	Under 1	1-4	5-9	10-14	15-19	20-24	25-29	30-34	35-39	40-44	45-49	50-54	55-59	60-64	65-69	70-74	75-79	80-84	85 and over	ICD10
	All registrations	604.6	477.3	358.9	21.7	17.4	11.6	12.9	38.2	208.9	319.1	318.3	320.5	359.4	474.8	648.7	807.0	1,061.8	1,275.6	1,633.1	1,921.2	2,215.6	2,383.9	
C00-C97	All cancers	492.3	372.6	264.0	18.4	16.3	10.2	10.5	13.6	27.3	46.9	88.3	147.8	244.5	385.3	556.0	715.6	964.3	1,185.7	1,527.1	1,809.7	2,086.3	2,263.7	C00-C97
C00-C97,xC44	All cancers excluding NMSC	426.3	327.0	233.2	18.4	16.3	10.0	10.0	12.6	26.0	43.6	81.1	135.0	220.1	343.8	501.6	642.7	860.0	1,036.8	1,317.9	1,515.7	1,735.0	1,856.2	C00-C97,xC44
C00-C14	Lip, mouth and pharynx	5.0	3.8	2.7			0.1	0.3	0.2	0.3	0.5	0.6	1.4	3.3	4.2	4.4	8.1	10.9	13.9	13.7	16.8	20.3	21.6	C00-C14
C00	Lip	0.3	0.2	0.1			0.1	0.1						0.1	0.1	0.2	0.1	0.5	0.3	0.5	1.1	0.7	2.5	1.9	C00
C01-C02	Tongue	1.1	0.8	0.6						0.1	0.1			0.2	0.8	1.1	1.3	1.5	2.8	2.3	3.4	3.7	3.0	4.4	C01-C02
C03-C06	Mouth	1.4	1.1	0.8				0.1				0.1	0.1	0.5	1.1	1.3	0.6	2.7	2.8	4.3	4.1	4.9	6.3	6.8	C03-C06
C07-C08	Salivary glands	0.7	0.6	0.4				0.1		0.1	0.1	0.2		0.5	0.5	0.8	1.0	1.1	1.7	1.4	1.9	2.0	3.0	3.8	C07-C08
C09-C10	Oropharynx	0.5	0.4	0.3										0.2	0.2	0.4	1.0	0.9	1.3	1.7	1.4	1.9	2.0	1.9	C09-C10
C11	Nasopharynx	0.2	0.2	0.2							0.2		0.2		0.1	0.2	0.1	0.2	0.3	0.7	0.3	0.6	0.8	0.7	C11
C12-C13	Hypopharynx	0.5	0.4	0.3				0.1	0.1						0.1	0.3	0.5	0.7	1.0	2.2	0.8	2.5	2.2	1.2	C12-C13
C14	Pharynx unspecified	0.2	0.2	0.1												0.2		0.6	0.7	0.9	0.6	0.6	0.5	1.0	C14
C15	Oesophagus	9.3	5.9	3.8	0.3					0.1				0.2	1.3	2.1	5.1	8.9	13.9	21.0	33.2	46.1	58.8	66.7	C15
C16	Stomach	13.7	8.3	5.3								0.1	0.8	1.2	2.5	2.8	5.0	9.7	19.6	29.2	44.1	70.1	92.0	110.3	C16
C17	Small intestine	0.9	0.6	0.4								0.0	0.2	0.1	0.4	0.6	2.0	2.0	1.4	1.6	2.5	4.2	4.2	4.2	C17
C18-C21	Colorectal	53.6	35.6	23.6					0.1	0.6	0.8	1.5	4.9	8.2	17.3	35.6	62.1	95.1	137.8	187.3	229.9	308.2	351.2	C18-C21
C18	Colon	36.4	23.8	15.7					0.1	0.4	0.5	0.8	3.1	4.9	10.6	22.2	40.8	62.9	91.8	123.5	159.9	216.0	252.5	C18
C19-C21	Rectum	17.2	11.8	7.9					0.2	0.2	0.3	0.7	1.9	3.3	6.6	13.4	21.3	32.2	46.0	63.7	69.9	92.0	98.9	C19-C21
C22	Liver	2.4	1.6	1.1	1.5		0.1		0.2	0.1	0.1	0.1	0.5	0.6	1.1	1.4	1.7	3.5	5.7	8.4	10.4	14.4	15.3	C22
C23-C24	Gallbladder	2.8	1.8	1.2							0.0	0.1	0.1	0.2	0.6	1.7	3.0	4.7	7.3	9.0	12.9	17.3	20.2	C23-C24
C25	Pancreas	12.0	7.7	5.1							0.1	0.1	0.5	1.7	3.0	6.5	9.6	22.1	31.0	44.2	54.0	73.3	82.9	C25
C26	Digestive organs etc. unspecified	1.2	0.6	0.4									0.1	0.2	0.2	0.3	0.6	1.2	1.8	3.0	6.7	7.7	13.1	C26
C30-C31	Nasal cavities	0.5	0.4	0.3		0.1	0.1	0.1		0.1	0.1	0.2	0.2	0.2	0.1	0.6	1.0	1.3	0.9	1.7	2.1	2.0	2.8	C30-C31
C32	Larynx	1.4	1.1	0.8							0.1	0.0	0.2	0.1	0.8	1.3	3.0	3.1	6.7	5.6	3.7	3.3	3.9	C32
C33-C34	Lung	46.7	33.7	23.0	0.3					0.1	0.2	0.9	2.5	7.2	17.3	33.3	54.8	101.4	171.1	219.9	210.4	188.1	148.5	C33-C34
C37-C39,C45	Respiratory	1.1	0.9	0.6	0.6	0.1		0.1	0.1	0.2	0.2		0.3	0.2	0.6	1.3	2.2	2.8	2.5	4.8	3.6	2.2	3.5	C37-C39,C45
C40-C41	Bone	0.6	0.5	0.5		0.1	0.7	1.0	1.1	0.3	0.4	0.4	0.2	0.4	0.3	0.4	0.4	0.8	0.2	0.7	1.6	1.4	1.7	C40-C41
C43	Melanoma of skin	10.7	9.4	7.4			0.3	0.2	0.9	4.5	7.2	8.3	8.8	11.5	16.7	16.1	15.9	16.9	19.5	20.0	23.7	26.4	25.9	C43
C44	Non-melanoma of skin	66.1	45.6	30.9			0.2	0.5	1.0	1.4	3.3	7.2	12.8	24.3	41.5	54.4	73.0	104.4	149.0	209.2	294.0	351.3	407.5	C44
C47,C49	Connective tissue	1.9	1.6	1.3	3.1	0.4	0.3	0.5	0.3	0.6	0.9	1.0	1.3	1.5	1.3	2.0	2.0	2.7	4.4	4.3	5.9	4.7	8.0	C47,C49
C48	Retroperitoneum and peritoneum	0.5	0.4	0.3		0.2		0.1			0.0	0.2	0.2	0.3	0.3	0.5	1.3	1.3	0.8	1.4	1.3	2.2	1.7	C48
C50	Breast	120.3	104.7	76.5	0.9				0.3	1.2	8.0	26.6	60.7	112.6	180.3	244.2	256.5	277.1	244.3	283.1	294.2	328.7	357.2	C50
C51-C52	Other female genital	4.1	2.8	1.9	0.3				0.1	0.1	0.2	0.8	1.3	1.2	2.5	2.9	3.8	7.2	8.1	12.6	16.8	22.2	29.9	C51-C52
C53	Cervix	12.1	11.0	8.8					0.1	3.1	9.3	18.2	20.2	19.5	16.2	15.4	15.3	16.8	16.7	21.1	21.4	20.9	16.3	C53
C54	Uterus	15.4	12.8	9.0							0.3	0.2	2.0	4.2	9.6	23.7	38.4	50.3	47.6	45.2	49.8	46.5	42.0	C54
C55	Uterus unspecified	1.3	0.9	0.6						0.1			0.1	0.6	1.4	1.0	1.3	2.2	2.7	4.1	4.3	5.8	9.6	C55
C56-C57	Ovary	20.3	17.0	12.4			0.3	0.4	0.7	1.5	2.6	3.9	6.4	11.6	19.6	32.5	41.3	51.4	57.8	60.4	58.6	62.2	50.1	C56-C57
C58	Placenta	0.0	0.0	0.0				0.1	0.1	0.3	0.2	0.1	0.2											C58
C64-C66,C68	Kidney	6.8	5.2	3.9	0.9	2.6	0.6	0.3	0.1	0.3	0.2	0.9	1.2	2.0	4.6	6.6	10.5	16.5	19.1	22.9	24.2	26.9	24.0	C64-C66,C68
C67	Bladder	12.5	8.0	5.2		0.1						0.4	0.5	1.6	2.8	6.7	10.6	21.0	31.2	46.7	60.7	75.0	84.0	C67
C69	Eye	0.7	0.6	0.5	2.1	1.0	0.1			0.1	0.1	0.1	0.1	0.4	0.3	0.6	0.3	1.4	0.9	1.2	2.3	1.7	1.9	1.9	C69
C70,C72	Other nervous system	0.4	0.4	0.3	0.3	0.7	0.4	0.4	0.1	0.1	0.1	0.2	0.2	0.4	0.4	0.7	0.3	0.3	0.2	1.0	0.7	1.0	0.7	C70,C72
C71	Brain	5.8	5.1	4.3	1.8	2.5	2.9	1.9	1.7	1.4	1.9	2.9	3.1	3.5	5.7	6.4	9.6	13.1	13.6	15.2	16.5	12.5	6.7	C71
C73	Thyroid	3.0	2.7	2.2				0.2	1.0	2.7	2.1	3.6	3.6	4.4	3.6	3.1	3.2	3.8	4.4	5.7	5.8	6.7	5.5	C73
C74-C75	Other endocrine glands	0.4	0.4	0.4	1.2	0.9	0.3	0.3	0.1	0.3	0.2	0.2	0.3	0.5	0.3	0.4	0.6	0.5	0.6	0.6	1.1	0.7	0.6	C74-C75
C81	Hodgkin's disease	1.9	1.9	1.8			0.2	0.5	2.6	4.3	3.8	2.3	2.3	1.7	1.2	1.4	2.3	2.0	1.8	2.0	1.0	2.4	2.2	C81
C82-C85,C91.4,C96	Non-Hodgkin's lymphoma	12.7	9.9	7.1	0.6		0.7	0.7	0.8	1.4	2.0	2.5	4.0	5.4	9.0	14.4	19.3	26.8	31.0	45.1	50.2	49.1	40.7	C82-C85,C91.4,C96
C88,C90	Multiple myeloma	5.5	3.7	2.4							0.0	0.1	0.3	0.8	2.0	3.3	6.0	10.9	14.1	20.6	25.7	33.1	27.3	C88,C90
C91-C95,xC91.4	Leukaemia	9.5	7.0	5.5	3.7	6.3	2.6	2.6	1.7	1.7	1.2	1.9	2.0	3.7	4.9	6.5	8.6	13.2	19.5	25.4	36.0	45.5	59.3	C91-C95,xC91.4
C91,xC91.4	Lymphoid leukaemia	4.3	3.1	2.6	1.8	5.6	2.1	1.9	0.8	0.8	0.3	0.3	0.2	0.7	1.2	2.1	4.1	6.1	8.1	11.5	15.0	20.8	31.4	C91,xC91.4
C92	Myeloid leukaemia	4.6	3.4	2.6	0.9	0.6	0.5	0.6	0.7	1.2	0.8	1.6	1.8	3.0	3.5	3.8	4.3	6.4	10.5	12.2	18.2	20.4	22.8	C92
C93	Monocytic leukaemia	0.1	0.0	0.0												0.1		0.1	0.2	0.3	0.3	0.4	0.7	C93
C94-C95	Other leukaemia	0.6	0.4	0.3	0.9			0.1	0.1	0.1	0.0	0.0		0.1	0.2	0.6	0.3	0.6	0.7	1.4	2.5	3.9	4.4	C94-C95
C76-C80	Other and unspecified	29.4	19.0	12.5	0.6	0.3	0.1	0.2	0.1	0.5	0.5	1.4	3.8	6.4	9.9	15.9	27.5	43.3	67.9	100.0	143.4	167.1	216.7	C76-C80
D05	Breast in situ	6.5	6.9	5.1							0.3			3.1	6.7	10.9	27.9	23.3	24.3	8.9	7.2	4.6	5.4	2.2	D05
D06	Cervix in situ	74.5	73.5	71.4					18.6	172.4	259.7	214.5	152.2	91.9	55.5	33.9	30.5	22.8	10.2	4.4	3.1	1.8	0.1	D06

Appendix B3(a) Percentage distribution for each age-group of new cases by site , England and Wales, 1994 - males

ICD10	Site	All ages	Under 1	1-4	5-9	10-14	15-19	20-24	25-29	30-34	35-39	40-44	45-49	50-54	55-59	60-64	65-69	70-74	75-79	80-84	85 and over	ICD10
C00-C97 xC44	All cancers excluding NMSC	100.0	100.0	100.0	100.0	100.0	100.0	100.0	100.0	100.0	100.0	100.0	100.0	100.0	100.0	100.0	100.0	100.0	100.0	100.0	100.0	C00-C97 xC44
C00-C14	Lip, mouth and pharynx	2.1	-	1.2	-	0.6	2.9	0.6	1.9	1.7	4.2	5.5	5.5	5.3	3.8	2.7	2.0	1.6	1.3	1.1	1.1	C00-C14
C00	Lip	0.2	-	-	-	-	-	0.2	-	-	0.2	0.2	0.3	0.3	0.2	0.2	0.2	0.2	0.1	0.1	0.1	C00
C01-C02	Tongue	0.5	-	-	-	-	0.4	0.2	0.7	0.4	0.9	1.3	1.4	1.1	0.9	0.6	0.5	0.3	0.3	0.1	0.1	C01-C02
C03-C06	Mouth	0.5	-	-	-	-	-	-	0.3	0.2	1.7	1.2	1.2	1.7	0.9	0.8	0.5	0.3	0.3	0.3	0.3	C03-C06
C07-C08	Salivary glands	0.2	-	-	-	-	-	-	0.4	0.5	0.4	0.6	0.6	0.3	0.3	0.2	0.2	0.2	0.1	0.1	0.3	C07-C08
C09-C10	Oropharynx	0.3	-	-	-	-	-	-	-	-	0.4	1.1	0.9	0.9	0.5	0.5	0.2	0.2	0.1	0.1	0.3	C09-C10
C11	Nasopharynx	0.1	-	1.2	-	0.6	2.5	-	0.4	0.4	0.4	0.4	0.3	0.3	0.2	0.1	0.2	0.1	0.1	0.0	0.0	C11
C12-C13	Hypopharynx	0.2	-	-	-	-	-	0.2	-	-	0.1	0.4	0.6	0.5	0.5	0.2	0.2	0.2	0.1	0.2	0.1	C12-C13
C14	Pharynx unspecified	0.1	-	-	-	-	-	-	-	-	0.2	0.3	0.3	0.1	0.2	0.2	0.2	0.1	0.0	0.1	0.0	C14
C15	Oesophagus	3.2	-	-	-	-	-	-	0.6	0.8	1.8	3.1	3.6	4.0	3.6	3.9	3.3	3.1	2.9	2.8	3.3	C15
C16	Stomach	5.5	-	-	-	-	0.4	0.4	1.1	1.6	2.3	4.1	4.3	4.8	4.9	5.6	5.8	5.8	5.7	6.2	5.7	C16
C17	Small intestine	0.2	-	-	-	-	-	-	0.1	0.3	0.4	0.4	0.5	0.4	0.2	0.3	0.2	0.2	0.2	0.2	0.1	C17
C18-C21	Colorectal	13.2	1.3	-	-	0.6	0.8	0.9	3.1	4.6	7.5	9.8	12.9	13.8	14.4	14.7	14.0	13.4	13.2	13.2	13.1	C18-C21
C18	Colon	7.8	-	-	-	0.6	0.4	0.6	2.3	2.6	3.6	5.0	7.0	7.0	7.4	8.1	8.2	8.0	8.2	8.4	8.6	C18
C19-C21	Rectum	5.4	1.3	-	-	-	0.4	0.2	0.9	2.0	3.8	4.7	6.0	6.7	7.1	6.6	5.8	5.4	5.0	4.8	4.5	C19-C21
C22	Liver	0.8	5.3	1.6	0.5	-	0.4	0.2	0.3	1.0	0.7	0.8	0.8	1.1	1.1	1.0	0.9	0.8	0.8	0.5	0.7	C22
C23-C24	Gallbladder	0.4	-	0.4	-	-	-	-	-	-	0.2	0.4	0.3	0.3	0.3	0.3	0.5	0.5	0.5	0.5	0.4	C23-C24
C25	Pancreas	2.5	-	-	-	-	-	-	0.4	0.6	1.2	2.0	2.9	2.7	3.0	2.8	2.4	2.4	2.7	2.6	2.8	C25
C26	Digestive organs etc. unspecified	0.2	-	-	-	-	-	-	0.1	0.1	-	0.2	0.2	0.2	0.2	0.2	0.2	0.1	0.2	0.3	0.5	C26
C30-C31	Nasal cavities	0.2	1.3	0.8	-	-	-	-	0.1	0.5	0.4	0.3	0.4	0.4	0.2	0.1	0.1	0.2	0.1	0.1	0.1	C30-C31
C32	Larynx	1.5	-	-	-	-	-	-	0.1	0.1	0.8	2.3	2.4	3.2	2.3	2.1	1.7	1.5	1.1	0.9	0.7	C32
C33-C34	Lung	20.8	1.3	-	-	-	0.4	0.9	1.4	2.6	4.3	10.3	15.3	19.2	21.2	22.3	24.3	23.7	22.2	19.5	17.3	C33-C34
C37-C39, C45	Respiratory system	1.0	-	-	-	0.6	1.2	0.6	0.6	0.4	1.0	1.4	1.2	1.6	1.6	1.5	1.3	0.9	0.8	0.5	0.6	C37-C39, C45
C40-C41	Bone	0.2	1.3	-	5.2	8.5	9.0	4.3	1.3	0.6	1.0	0.6	0.3	0.3	0.1	0.1	0.1	0.1	0.1	0.1	0.1	C40-C41
C43	Melanoma of skin	1.7	-	-	2.1	2.3	4.5	6.2	6.9	9.0	8.9	7.5	5.8	3.8	2.7	1.6	1.2	0.9	0.9	0.8	0.9	C43
C47, C49	Connective tissue	0.5	6.6	5.6	3.6	7.4	5.7	3.2	1.9	2.3	2.1	1.5	1.2	0.7	0.6	0.4	0.4	0.3	0.3	0.3	0.3	C47, C49
C48	Retroperitoneum and peritoneum	0.1	1.3	-	0.5	-	0.8	0.4	0.3	0.3	0.1	0.4	0.2	0.3	0.2	0.1	0.1	0.1	0.1	0.0	0.1	C48
C50	Breast	0.2	-	-	0.5	-	-	-	-	-	0.1	0.2	0.2	0.3	0.2	0.2	0.2	0.1	0.2	0.1	0.2	C50
C60, C63	Penis	0.3	-	-	-	-	-	-	0.6	0.5	0.6	1.2	0.5	0.4	0.5	0.3	0.2	0.3	0.3	0.3	0.4	C60, C63
C61	Prostate	17.3	1.3	1.6	-	-	-	-	0.1	0.1	0.4	0.4	1.3	4.5	8.3	12.5	16.2	20.0	22.9	25.2	27.6	C61
C62	Testis	1.2	5.3	3.2	0.5	2.3	11.9	29.9	35.3	31.8	18.4	8.7	3.5	1.1	0.6	0.1	0.1	0.1	0.1	0.1	0.1	C62
C64-C66, C68	Kidney	2.7	6.6	8.8	4.1	1.7	0.4	1.5	1.0	1.8	3.6	4.6	4.6	4.3	4.0	3.5	2.9	2.3	2.0	1.8	1.5	C64-C66, C68
C67	Bladder	7.6	1.3	0.8	-	1.1	1.6	1.9	1.7	1.4	3.5	4.6	5.3	6.8	7.5	7.8	7.9	7.8	7.9	8.5	8.6	C67
C69	Eye	0.2	17.1	4.8	0.5	0.6	-	0.4	-	0.3	0.5	0.3	0.4	0.2	0.2	0.2	0.2	0.1	0.1	0.1	0.1	C69
C70, C72	Other nervous system	0.1	1.3	2.4	0.5	1.1	0.4	0.4	0.1	0.8	0.4	0.3	0.3	0.1	0.1	0.1	0.1	0.0	0.0	0.0	0.3	C70, C72
C71	Brain	1.8	7.9	15.7	24.9	18.8	12.3	8.1	6.6	7.3	7.5	6.0	4.8	3.4	2.5	2.1	1.4	1.1	0.9	0.6	0.4	C71
C73	Thyroid	0.2	-	-	-	0.6	1.2	1.3	1.4	2.1	0.9	0.9	0.9	0.4	0.3	0.1	0.2	0.1	0.1	0.1	0.1	C73
C74-C75	Other endocrine glands	0.1	14.5	7.6	2.1	1.1	2.5	0.4	0.4	-	0.2	0.1	0.2	0.1	0.1	0.1	0.1	0.0	0.0	0.1	0.0	C74-C75
C81	Hodgkin's disease	0.6	1.3	0.8	8.8	8.5	13.1	17.9	13.0	6.9	4.6	3.1	1.3	0.8	0.6	0.3	0.2	0.1	0.1	0.1	0.0	C81
C82-C85, C91.4, C96	Non-Hodgkin's lymphoma	3.5	3.9	4.4	8.3	15.3	10.7	9.2	9.3	11.4	13.0	9.0	7.8	6.3	4.9	3.7	2.8	2.6	2.5	2.2	2.1	C82-C85, C91.4, C96
C88, C90	Multiple myeloma	1.3	-	-	-	0.6	-	0.2	0.7	0.1	0.1	0.1	1.7	1.6	1.6	1.3	1.3	1.3	1.2	1.3	1.4	C88, C90
C91-C95, xC91.4	Leukaemia	2.7	17.1	39.4	37.3	29.5	18.4	7.9	7.3	5.7	4.8	3.6	3.6	2.7	2.4	2.3	2.2	2.1	2.2	2.8	2.7	C91-C95, xC91.4
C91, xC91.4	Lymphoid leukaemia	1.3	5.3	33.3	31.1	21.6	9.8	2.8	2.1	1.3	1.5	1.2	1.4	1.1	1.1	1.1	1.0	1.0	1.0	1.3	1.3	C91, xC91.4
C92	Myeloid leukaemia	1.2	7.9	4.8	4.1	6.8	8.6	5.1	4.9	4.2	3.1	2.2	2.2	1.4	1.2	1.1	1.0	0.9	1.0	1.2	1.1	C92
C93	Monocytic leukaemia	0.0	2.6	-	0.5	0.6	-	-	-	0.1	0.1	-	-	0.0	0.0	0.0	0.0	0.0	0.0	0.0	0.0	C93
C94-C95	Other leukaemia	0.1	1.3	1.2	1.6	0.6	-	-	0.3	0.1	0.1	0.2	0.1	0.1	0.0	0.0	0.1	0.1	0.1	0.2	0.2	C94-C95
C76-C80	Other and unspecified	6.1	3.9	0.8	0.5	-	0.8	3.0	2.0	3.2	3.9	5.6	5.7	5.2	5.8	5.6	5.6	6.2	6.6	7.1	7.2	C76-C80

Appendix B3(b) Percentage distribution for each age-group of new cases by site, England and Wales, 1994 - females

ICD10	Site	All ages	Under 1	1-4	5-9	10-14	15-19	20-24	25-29	30-34	35-39	40-44	45-49	50-54	55-59	60-64	65-69	70-74	75-79	80-84	85 and over	ICD10
C00-C97 xC44	All cancers excluding NMSC	100.0	100.0	100.0	100.0	100.0	100.0	100.0	100.0	100.0	100.0	100.0	100.0	100.0	100.0	100.0	100.0	100.0	100.0	100.0	100.0	C00-C97 xC44
C00-C14	Lip, mouth and pharynx	1.2			1.2	2.6	1.6	1.3	1.1	0.8	1.0	1.5	1.2	0.9	1.3	1.3	1.3	1.0	1.1	1.2	1.2	C00-C14
C00	Lip	0.1			0.6	0.6			0.1		0.0	0.1	0.0		0.1	0.0	0.0	0.1	0.0	0.1	0.2	C00
C01-C02	Tongue	0.2				0.6		0.4	0.2	0.2	0.3	0.4	0.3	0.3	0.4	0.3	0.4	0.3	0.2	0.2	0.2	C01-C02
C03-C06	Mouth	0.3				0.6	0.5		0.3	0.1	0.3	0.5	0.4	0.1	0.4	0.3	0.4	0.3	0.3	0.4	0.4	C03-C06
C07-C08	Salivary glands	0.2					0.5	0.2	0.3	0.3	0.2	0.2	0.1	0.2	0.1	0.2	0.2	0.1	0.1	0.2	0.2	C07-C08
C09-C10	Oropharynx	0.1									0.1	0.1	0.1	0.2	0.1	0.1	0.2	0.1	0.1	0.1	0.1	C09-C10
C11	Nasopharynx	0.1				0.6	0.5	0.7	0.1	0.2	0.2	0.1	0.1	0.0	0.1	0.0	0.1	0.0	0.0	0.0	0.0	C11
C12-C13	Hypopharynx	0.1					0.5					0.1	0.1	0.1	0.1	0.1	0.1	0.2	0.0	0.1	0.1	C12-C13
C14	Pharynx unspecified	0.1										0.1	0.1	0.1	0.1	0.1	0.1	0.0	0.0	0.0	0.1	C14
C15	Oesophagus	2.2	1.7					0.2		0.1	0.2	0.6	0.6	1.0	1.4	1.6	2.0	2.5	3.0	3.4	3.6	C15
C16	Stomach	3.2							0.2	1.0	0.9	1.1	0.8	1.0	1.5	2.3	2.8	3.3	4.6	5.3	5.9	C16
C17	Small intestine	0.2							0.1	0.3	0.0	0.2	0.2	0.1	0.3	0.2	0.2	0.2	0.3	0.2	0.2	C17
C18-C21	Colorectal	12.6					0.5	2.2	1.9	1.9	3.7	3.7	5.0	7.1	9.7	11.1	13.3	14.2	15.2	17.8	18.9	C18-C21
C18	Colon	8.5					0.5	1.5	1.1	1.0	2.3	2.2	3.1	4.4	6.4	7.3	8.9	9.4	10.6	12.5	13.6	C18
C19-C21	Rectum	4.0						0.7	0.8	0.9	1.4	1.5	1.9	2.7	3.3	3.7	4.4	4.8	4.6	5.3	5.3	C19-C21
C22	Liver	0.6	8.3	0.9	1.2		1.6	0.4	0.2	0.2	0.4	0.3	0.2	0.3	0.3	0.5	0.7	0.6	0.7	0.8	0.8	C22
C23-C24	Gallbladder	0.7								0.1	0.0	0.1	0.2	0.3	0.5	0.5	0.7	0.7	0.9	1.0	1.1	C23-C24
C25	Pancreas	2.8					1.1		0.2	0.1	0.4	0.8	0.9	1.3	1.5	2.6	3.0	3.4	3.6	4.2	4.5	C25
C26	Digestive organs etc. unspecified	0.3									0.0	0.0		0.1	0.2	0.2	0.2	0.2	0.4	0.4	0.7	C26
C30-C31	Nasal cavities	0.1			0.6				0.2	0.2	0.2	0.1	0.0	0.1	0.2	0.2	0.1	0.2	0.1	0.1	0.1	C30-C31
C32	Larynx	0.3	1.7				0.5	0.4		0.1	0.2	0.1	0.2	0.3	0.5	0.4	0.6	0.4	0.2	0.2	0.2	C32
C33-C34	Lung	11.0	3.3	0.9			0.5	0.4	0.4	1.1	1.8	3.3	5.0	6.6	8.5	11.8	16.5	16.7	13.9	10.8	8.0	C33-C34
C37-C39, C45	Respiratory system	0.3	3.3	0.5		0.6	0.5	0.9	0.6	0.3	0.2	0.3	0.2	0.3	0.3	0.3	0.2	0.3	0.3	0.1	0.2	C37-C39, C45
C40-C41	Bone	0.1		0.5	6.7	9.6	8.7	1.3	1.0	0.5	0.1	0.2	0.2	0.1	0.1	0.1	0.0	0.1	0.1	0.1	0.1	C40-C41
C43	Melanoma of skin	2.5			3.0	1.9	7.1	17.2	16.6	10.3	6.6	5.2	4.9	3.2	2.5	2.0	1.9	1.5	1.6	1.5	1.4	C43
C47, C49	Connective tissue	0.4	16.7	2.3	3.0	5.1	2.7	2.4	2.0	1.2	1.0	0.7	0.4	0.4	0.3	0.3	0.4	0.3	0.4	0.3	0.4	C47, C49
C48	Retroperitoneum and peritoneum	0.1		1.4				0.2	0.1	0.3	0.1	0.1	0.2	0.1	0.2	0.2	0.4	0.3	0.1	0.1	0.1	C48
C50	Breast	28.2			0.6		2.7	4.8	18.3	32.8	44.9	51.2	52.4	48.7	39.9	32.2	23.6	21.5	19.4	18.9	19.2	C50
C51-C52	Other female genital	1.0	1.7	0.9			0.5	0.4	0.4	1.0	1.0	0.6	0.7	0.6	0.6	0.8	0.8	1.0	1.1	1.3	1.6	C51-C52
C53	Cervix	2.8				0.6	0.5	12.0	21.3	22.5	15.0	8.9	4.7	3.1	2.4	2.0	1.6	1.6	1.4	1.2	0.9	C53
C54	Uterus	3.6	11.7	5.9	1.2				0.7	0.3	1.5	1.9	2.8	4.7	6.0	5.9	4.6	3.4	3.3	2.7	2.3	C54
C55	Uterus unspecified	0.3	1.7					0.2		0.1	0.1	0.3	0.4	0.2	0.2	0.3	0.3	0.3	0.3	0.3	0.5	C55
C56-C57	Ovary	4.8		0.5	3.0	4.5	5.5	5.9	5.9	4.8	4.7	5.3	5.7	6.5	6.4	6.0	5.6	4.6	3.9	3.6	2.7	C56-C57
C58	Placenta	0.0				0.6	1.1		0.4	0.1	0.1											C58
C64-C66, C68	Kidney	1.6	5.0	16.0	6.1	3.2	0.5	1.1	0.4	1.1	0.9	0.9	1.3	1.3	1.6	1.9	1.8	1.7	1.6	1.6	1.3	C64-C66, C68
C67	Bladder	2.9		0.5				0.7	0.1	0.5	0.4	0.7	0.8	1.3	1.7	2.4	3.0	3.5	4.0	4.3	4.5	C67
C69	Eye	0.2	11.7	5.9	1.2		0.5	0.4	0.3	0.1	0.3	0.1	0.2	0.1	0.2	0.1	0.2	0.1	0.0	0.1	0.1	C69
C70, C72	Other nervous system	0.1	1.7	4.1	4.3	3.8	0.5	0.2	0.2	0.3	0.1	0.2	0.2	0.1	0.0	0.0	0.0	0.1	0.0	0.1	0.0	C70, C72
C71	Brain	1.4	10.0	15.1	28.7	19.2	13.1	5.2	4.4	3.5	2.3	1.6	1.7	1.3	1.5	1.5	1.3	1.2	1.1	0.7	0.4	C71
C73	Thyroid	0.7				1.9	8.2	10.5	4.7	4.4	2.6	2.0	1.0	0.6	0.5	0.4	0.4	0.4	0.4	0.4	0.3	C73
C74-C75	Other endocrine glands	0.1	6.7	5.5	3.0	2.6	1.1	1.1	0.6	0.2	0.2	0.2	0.1	0.1	0.1	0.1	0.0	0.0	0.1	0.0	0.0	C74-C75
C81	Hodgkin's disease	0.5		0.9	2.4	5.1	20.8	16.6	8.8	2.8	1.7	0.8	0.4	0.3	0.4	0.2	0.2	0.2	0.1	0.1	0.1	C81
C82-C85, C91.4, C96	Non-Hodgkin's lymphoma	3.0	3.3	3.2	7.3	7.1	6.0	5.2	4.5	3.0	2.9	2.5	2.6	2.9	3.0	3.1	3.0	3.4	3.3	2.8	2.2	C82-C85, C91.4, C96
C88, C90	Multiple myeloma	1.3			0.6				0.1	0.2	0.2	0.4	0.6	0.7	0.9	1.3	1.4	1.6	1.7	1.9	1.5	C88, C90
C91-C95 xC91.4	Leukaemia	2.2	20.0	38.8	26.2	26.3	13.1	6.5	2.7	2.4	1.5	1.7	1.3	1.3	1.3	1.3	1.9	1.9	2.4	2.6	3.2	C91-C95 xC91.4
C91 xC91.4	Lymphoid leukaemia	1.0	10.0	34.2	20.7	19.2	6.6	1.3	0.7	0.4	0.2	0.3	0.3	0.4	0.6	0.7	0.8	0.9	1.0	1.2	1.7	C91 xC91.4
C92	Myeloid leukaemia	1.1	5.0	3.7	5.5	5.8	5.5	4.8	1.9	1.9	1.3	1.3	1.0	0.8	0.7	0.7	1.0	0.9	1.2	1.2	1.2	C92
C93	Monocytic leukaemia	0.0							0.1					0.0			0.0	0.0	0.0	0.0	0.0	C93
C94-C95	Other leukaemia	0.1	5.0	0.9		1.3	1.1	0.4	0.1	0.1		0.1	0.0	0.1	0.0	0.1	0.1	0.1	0.2	0.2	0.2	C94-C95
C76-C80	Other and unspecified	6.9	3.3	1.8	0.6	1.9	1.1	2.0	1.1	1.7	2.8	2.9	2.9	3.2	4.3	5.0	6.5	7.6	9.5	9.6	11.7	C76-C80

Appendix B4(a) Percentage distribution for each site of new cases by age-group, England and Wales, 1994 - males

ICD10	Site	All ages	Under 1	1-4	5-9	10-14	15-19	20-24	25-29	30-34	35-39	40-44	45-49	50-54	55-59	60-64	65-69	70-74	75-79	80-84	85 and over	ICD10
C00-C97 xC44	All cancers excluding NMSC	100.0	0.1	0.2	0.2	0.2	0.2	0.4	0.6	0.9	1.0	1.5	2.8	4.2	6.8	10.4	15.3	20.0	14.9	12.6	7.8	C00-C97 xC44
C00-C14	Lip, mouth and pharynx	100.0	-	0.1			0.3	0.1	0.5	0.7	2.0	3.9	7.4	10.5	12.4	13.2	14.6	14.7	8.8	6.6	4.1	C00-C14
C00	Lip	100.0						0.5			1.0	2.0	4.0	7.9	9.4	10.9	18.3	18.8	11.4	9.4	6.4	C00
C01-C02	Tongue	100.0					0.2	0.2	1.0	0.8	2.0	4.3	8.8	10.6	13.5	13.1	15.5	13.9	9.0	3.5	3.5	C01-C02
C03-C06	Mouth	100.0							0.3	0.3	3.1	3.4	6.4	12.7	11.4	14.5	14.5	12.1	9.4	7.7	4.2	C03-C06
C07-C08	Salivary glands	100.0							1.3	2.1	1.7	4.2	7.6	6.4	11.0	10.2	11.0	15.7	9.7	8.5	10.6	C07-C08
C09-C10	Oropharynx	100.0								0.3	1.3	5.8	9.6	13.8	13.2	17.0	11.6	15.8	6.4	3.9	1.3	C09-C10
C11	Nasopharynx	100.0		2.5		0.8	4.9	0.8	2.5	3.3	4.1	5.7	7.4	10.7	10.7	9.0	12.3	13.9	7.4	3.3	0.8	C11
C12-C13	Hypopharynx	100.0									0.4	2.5	7.4	9.0	14.8	9.8	14.8	18.9	9.0	11.1	2.5	C12-C13
C14	Pharynx unspecified	100.0									1.6	4.0	6.4	5.6	14.4	17.6	20.8	12.8	6.4	8.0	2.4	C14
C15	Oesophagus	100.0							0.1	0.2	0.6	1.5	3.2	5.3	7.9	12.7	16.0	19.4	13.6	11.4	8.1	C15
C16	Stomach	100.0						0.0	0.1	0.2	0.4	1.1	2.2	3.7	6.1	10.7	16.1	21.3	15.5	14.4	8.1	C16
C17	Small intestine	100.0							0.4	1.2	2.1	2.9	6.6	7.1	5.4	14.9	14.9	15.8	15.4	9.5	3.7	C17
C18-C21	Colorectal	100.0	0.0				0.0	0.0	0.1	0.3	0.6	1.1	2.8	4.4	7.5	11.6	16.1	20.3	14.9	12.6	7.7	C18-C21
C18	Colon	100.0				0.0	0.0	0.0	0.2	0.3	0.5	1.0	2.5	3.8	6.5	10.8	16.0	20.5	15.7	13.6	8.6	C18
C19-C21	Rectum	100.0	0.0				0.0	0.1	0.1	0.3	0.7	1.3	3.1	5.2	8.9	12.7	16.3	20.0	13.7	11.2	6.5	C19-C21
C22	Liver	100.0	0.4	0.4	0.1			0.1	0.2	1.1	0.8	1.4	2.7	5.3	8.8	12.7	17.1	20.2	14.0	8.2	6.4	C22
C23-C24	Gallbladder	100.0		0.2						0.2	0.4	1.5	2.4	3.2	5.3	8.5	17.1	22.2	16.7	14.3	8.1	C23-C24
C25	Pancreas	100.0							0.1	0.2	0.5	1.2	3.2	4.5	8.2	11.4	14.4	19.2	15.6	13.0	8.5	C25
C26	Digestive organs etc. unspecified	100.0						0.4	0.4	0.4		1.3	2.1	4.6	5.4	8.8	13.8	13.8	12.6	17.6	19.2	C26
C30-C31	Nasal cavities	100.0	0.5	1.0			0.5	1.0	0.5	2.5	2.0	2.5	7.0	10.1	6.0	8.5	11.6	20.1	12.6	8.5	5.0	C30-C31
C32	Larynx	100.0							0.1	0.1	0.5	2.2	4.4	8.7	10.4	14.4	17.3	20.1	10.3	7.8	3.8	C32
C33-C34	Lung	100.0	0.0				0.0	0.0	0.0	0.1	0.2	0.7	2.1	3.9	7.0	11.2	17.8	22.8	15.9	11.8	6.5	C33-C34
C37-C39,C45	Respiratory system	100.0				0.1	0.3	0.2	0.4	0.4	1.0	2.0	3.3	6.6	10.9	15.5	19.7	17.6	11.6	6.3	4.3	C37-C39,C45
C40-C41	Bone	100.0	0.5		4.5	6.8	10.0	9.0	4.1	2.7	5.0	4.5	4.5	6.3	4.5	6.8	8.6	9.0	5.0	5.9	2.3	C40-C41
C43	Melanoma of skin	100.0			0.2	0.2	0.6	1.5	2.5	4.6	5.2	6.7	9.6	9.3	10.9	10.0	10.6	10.7	7.6	5.9	4.0	C43
C47,C49	Connective tissue	100.0	0.9	2.4	1.2	2.3	2.4	2.6	2.3	3.8	4.2	4.5	6.4	6.1	8.1	8.3	10.4	13.5	9.4	7.3	4.0	C47,C49
C48	Retroperitoneum and peritoneum	100.0	0.7		0.7		1.3	1.3	1.3	2.0	0.7	4.0	4.0	9.4	10.7	11.4	16.8	13.4	14.8	2.7	4.7	C48
C50	Breast	100.0			0.5					0.5		2.1	3.2	6.9	9.5	12.7	14.3	14.8	16.4	11.1	7.9	C50
C60,C63	Penis	100.0							1.1	1.4	1.9	5.6	4.7	4.7	10.0	9.2	10.3	17.3	13.4	11.7	8.6	C60,C63
C61	Prostate	100.0	0.0	0.0				0.0	0.0	0.0	0.0	0.0	0.2	1.1	3.3	7.5	14.3	23.1	19.7	18.3	12.4	C61
C62	Testis	100.0	0.3	0.6	0.1	0.3	2.1	10.1	17.9	22.2	15.0	10.6	8.0	3.6	3.5	1.1	1.4	1.1	0.7	0.9	0.4	C62
C64-C66,C68	Kidney	100.0	0.2	0.7	0.3	0.1	0.0	0.2	0.2	0.6	1.4	2.6	4.9	6.7	10.1	13.8	16.9	17.4	11.0	8.5	4.4	C64-C66,C68
C67	Bladder	100.0	0.0			0.0	0.0	0.1	0.1	0.2	0.5	0.9	2.0	3.8	6.7	10.7	15.9	20.6	15.5	14.2	8.8	C67
C69	Eye	100.0	6.4	5.9	0.5	0.5				1.5	2.9	2.5	5.9	4.4	8.8	9.8	13.7	15.2	10.8	7.8	3.4	C69
C70,C72	Other nervous system	100.0	1.2	7.1	1.2	2.4	1.2	2.4	1.2	9.4	4.7	5.9	10.6	7.1	5.9	9.4	11.8	8.2	5.9	4.7		C70,C72
C71	Brain	100.0	0.3	2.0	2.4	1.7	1.5	1.9	2.3	3.6	4.2	5.1	7.7	8.1	9.6	12.3	12.3	12.4	7.2	3.9	1.6	C71
C73	Thyroid	100.0	0.4	0.9	0.6	0.4	1.2	2.3	3.9	7.7	3.9	6.2	10.4	7.3	8.1	6.2	15.4	11.2	9.3	3.5	3.1	C73
C74-C75	Other endocrine glands	100.0	9.5	16.4	3.4	1.7	5.2	1.7	2.6		1.7	1.7	4.3	2.6	6.0	7.8	12.1	9.5	3.4	6.9	3.4	C74-C75
C81	Hodgkin's disease	100.0	0.1	0.3	2.5	2.2	4.7	12.5	13.5	9.9	7.7	7.7	5.9	5.6	6.4	5.9	5.8	4.7	2.4	1.3	0.6	C81
C82-C85, C91.4, C96	Non-Hodgkin's lymphoma	100.0	0.1	0.3	0.4	0.7	0.7	1.1	1.7	2.8	3.8	3.9	6.3	7.6	9.5	11.1	12.0	15.1	10.6	7.8	4.7	C82-C85,C91.4,C96
C88, C90	Multiple myeloma	100.0	0.0			0.1	0.0	0.1	0.3	0.1	0.6	1.1	3.8	5.1	8.3	10.5	15.0	20.3	13.7	12.3	8.7	C88,C90
C91-C95, xC91.4	Leukaemia	100.0	0.4	3.2	2.4	1.7	1.5	1.2	1.7	1.8	1.8	2.0	3.7	4.1	5.9	8.8	12.1	15.2	12.0	13.0	7.7	C91-C95,xC91.4
C91, xC91.4	Lymphoid leukaemia	100.0	0.3	5.6	4.0	2.6	1.6	0.9	1.0	0.9	1.1	1.3	3.0	3.6	5.7	9.0	11.7	15.5	11.5	12.8	7.8	C91,xC91.4
C92	Myeloid leukaemia	100.0	0.4	0.9	0.6	0.9	1.5	1.7	2.4	2.9	2.5	2.6	4.9	4.7	6.3	9.0	12.4	14.7	12.5	12.4	6.7	C92
C93	Monocytic leukaemia	100.0	6.9	3.4	3.4	3.4					3.4			6.9	10.3	3.4	3.4	20.7	10.3	13.8	13.8	C93
C94-C95	Other leukaemia	100.0	0.7	2.1	2.1	0.7			1.4	0.7	0.7	2.1	1.4	2.8	3.5	5.0	13.5	16.3	11.3	21.3	14.2	C94-C95
C76-C80	Other and unspecified	100.0	0.0	0.0	0.0	-	0.0	0.2	0.2	0.5	0.6	1.4	2.7	3.6	6.5	9.6	14.1	20.5	16.2	14.7	9.2	C76-C80

Appendix B4(b) Percentage distribution for each site of new cases by age-group, England and Wales, 1994 - females

ICD10	Site	All ages	Under 1	1-4	5-9	10-14	15-19	20-24	25-29	30-34	35-39	40-44	45-49	50-54	55-59	60-64	65-69	70-74	75-79	80-84	85 and over	ICD10
C00-C97 xC44	All cancers excluding NMSC	100.0	0.1	0.2	0.1	0.1	0.2	0.4	0.8	1.5	2.1	3.3	5.5	6.4	7.6	9.7	11.6	15.1	12.0	11.8	11.4	C00-C97 xC44
C00-C14	Lip, mouth and pharynx	100.0	-	-	0.2	0.3	0.2	0.5	0.8	1.0	1.8	4.2	5.8	4.8	8.3	10.6	13.3	13.5	11.5	11.9	11.4	C00-C14
C00	Lip	100.0	-	-	1.3	1.3	-	-	1.3	-	1.3	2.6	3.8	1.3	7.7	5.1	7.7	17.9	7.7	24.4	16.7	C00
C01-C02	Tongue	100.0	-	-	-	-	-	0.7	1.1	1.1	1.1	5.1	6.9	6.9	7.2	12.6	10.5	15.9	11.9	8.3	10.8	C01-C02
C03-C06	Mouth	100.0	-	-	0.3	-	-	-	0.5	0.3	2.1	4.8	6.1	2.1	9.5	9.3	14.3	14.0	11.6	12.7	12.4	C03-C06
C07-C08	Salivary glands	100.0	-	-	0.5	0.5	-	0.5	1.6	2.6	2.1	4.8	4.8	6.3	8.4	11.1	9.0	13.2	9.5	12.2	13.8	C07-C08
C09-C10	Oropharynx	100.0	-	-	-	-	-	-	-	-	2.1	2.8	5.6	9.9	8.5	12.0	14.8	12.7	12.0	10.6	9.2	C09-C10
C11	Nasopharynx	100.0	-	-	1.6	1.6	1.6	4.9	1.6	6.6	8.2	6.6	6.6	3.3	4.9	6.6	14.8	6.6	8.2	9.8	8.2	C11
C12-C13	Hypopharynx	100.0	-	-	0.8	0.8	-	-	-	-	-	1.6	4.9	5.7	7.3	10.6	22.0	8.1	17.9	13.8	6.5	C12-C13
C14	Pharynx unspecified	100.0	-	-	-	-	-	-	-	-	-	3.4	6.9	6.9	13.8	15.5	19.0	13.8	8.6	6.9	12.1	C14
C15	Oesophagus	100.0	0.0	-	-	-	-	0.0	-	0.1	0.2	0.9	1.6	3.0	4.8	7.2	10.8	17.4	16.8	18.4	18.4	C15
C16	Stomach	100.0	-	-	-	-	-	-	0.1	0.4	0.6	1.2	1.4	2.0	3.6	6.9	10.2	15.7	17.4	19.6	21.1	C16
C17	Small intestine	100.0	-	-	-	-	-	-	0.4	2.2	0.4	2.6	4.8	4.0	11.5	7.9	8.8	14.1	16.3	14.1	12.8	C17
C18-C21	Colorectal	100.0	-	0.0	-	0.0	0.0	0.1	0.1	0.2	0.6	1.0	2.2	3.6	5.9	8.6	12.3	17.0	14.5	16.7	17.2	C18-C21
C18	Colon	100.0	-	-	-	0.0	0.0	0.1	0.1	0.2	0.6	0.9	2.0	3.3	5.7	8.3	12.0	16.5	14.9	17.3	18.2	C18
C19-C21	Rectum	100.0	-	-	-	-	-	0.1	0.2	0.3	0.7	1.2	2.7	4.2	6.3	9.1	12.7	18.1	13.8	15.6	15.1	C19-C21
C22	Liver	100.0	0.8	0.3	0.3	-	0.5	0.3	0.3	0.5	1.4	1.6	3.0	3.2	3.6	7.1	11.4	17.1	14.7	17.4	16.6	C22
C23-C24	Gallbladder	100.0	-	0.3	-	-	-	-	0.1	0.3	0.6	0.5	1.5	3.4	5.4	8.1	12.3	15.6	15.2	17.9	18.9	C23-C24
C25	Pancreas	100.0	-	-	-	-	0.1	0.0	0.1	0.1	0.3	0.9	1.7	2.9	4.0	8.9	12.3	17.9	15.2	17.7	18.0	C25
C26	Digestive organs etc. unspecified	100.0	-	-	-	-	-	0.4	-	0.3	0.3	1.0	1.0	1.3	2.6	4.9	7.5	12.8	19.7	19.3	29.5	C26
C30-C31	Nasal cavities	100.0	-	0.7	0.7	1.4	-	-	1.4	2.8	2.8	2.1	1.4	5.6	9.1	11.9	7.7	15.4	13.3	10.5	13.3	C30-C31
C32	Larynx	100.0	-	-	-	-	-	0.6	-	0.3	1.1	0.6	4.1	5.0	11.0	10.8	23.2	19.9	9.1	6.9	7.5	C32
C33-C34	Lung	100.0	0.0	0.0	-	-	0.0	0.0	0.0	0.1	0.4	1.0	2.5	3.9	5.9	10.5	17.4	22.9	15.2	11.7	8.3	C33-C34
C37-C39,C45	Respiratory system	100.0	0.7	0.4	-	0.4	0.4	1.4	1.8	0.1	1.8	1.4	3.6	6.8	10.3	12.5	11.0	21.7	11.4	6.0	8.5	C37-C39,C45
C40-C41	Bone	100.0	-	0.7	7.3	9.9	10.6	4.0	6.0	5.3	2.0	4.6	4.0	4.0	7.5	7.6	6.0	9.3	7.3	7.3	7.9	C40-C41
C43	Melanoma of skin	100.0	-	0.7	0.2	0.1	0.5	2.8	5.3	6.0	5.5	6.9	10.7	8.2	7.5	7.6	8.7	9.1	7.5	7.2	6.3	C43
C47,C49	Connective tissue	100.0	2.0	1.0	1.0	1.6	1.0	2.2	3.6	4.0	4.6	5.0	4.6	5.8	5.2	6.9	11.1	11.1	10.7	7.3	11.1	C47,C49
C48	Retroperitoneum and peritoneum	100.0	-	2.2	-	0.7	-	0.7	0.7	3.7	2.2	3.7	4.4	5.2	12.6	12.6	7.4	13.3	8.9	12.6	8.9	C48
C50	Breast	100.0	0.0	-	0.0	-	0.0	0.1	0.5	1.7	3.4	6.0	10.3	11.0	10.8	10.3	9.7	11.5	7.9	7.8	7.8	C50
C51-C52	Other female genital	100.0	0.1	0.2	-	-	0.1	0.2	0.4	1.0	2.1	1.9	4.2	3.9	4.6	8.4	9.3	14.9	13.8	15.6	18.9	C51-C52
C53	Cervix	100.0	-	-	-	-	0.0	1.7	6.0	11.6	11.2	10.4	9.2	6.9	6.4	6.7	6.6	8.5	6.0	5.0	3.5	C53
C54	Uterus	100.0	-	0.0	-	-	1.0	1.0	-	0.1	0.9	1.7	4.3	8.4	12.6	15.8	14.7	14.3	11.0	8.8	7.1	C54
C55	Uterus unspecified	100.0	-	-	-	-	-	0.3	-	0.6	0.6	3.0	7.5	4.2	5.1	8.4	10.2	15.6	11.4	13.2	19.8	C55
C56-C57	Ovary	100.0	-	0.0	0.1	0.1	0.2	0.5	1.0	1.5	2.1	3.6	6.6	8.7	10.3	12.2	13.5	14.5	9.8	8.9	6.4	C56-C57
C58	Placenta	100.0	-	-	-	8.3	16.7	-	33.3	16.7	25.0	-	-	-	-	-	-	-	-	-	-	C58
C64-C66, C68	Kidney	100.0	0.2	2.0	0.6	0.3	0.1	0.3	0.2	1.0	1.2	1.9	4.7	5.3	7.8	11.8	13.5	16.5	12.1	11.6	9.3	C64-C66, C68
C67	Bladder	100.0	-	0.0	-	-	-	0.1	0.0	0.2	0.3	0.8	1.6	2.9	4.3	8.1	11.9	18.2	16.5	17.5	17.6	C67
C69	Eye	100.0	3.8	7.1	1.1	-	0.5	1.1	1.6	1.1	3.8	2.7	6.0	3.8	10.4	6.6	8.2	16.4	8.2	10.4	7.1	C69
C70-C72	Other nervous system	100.0	1.0	8.9	6.9	5.9	1.0	1.0	2.0	5.0	3.0	5.9	7.9	9.9	4.0	4.0	2.0	12.9	5.9	7.9	5.0	C70-C72
C71	Brain	100.0	0.4	2.2	3.1	2.0	1.6	1.6	2.6	3.8	3.6	3.9	6.8	6.0	8.4	10.9	11.2	12.9	9.7	6.3	3.0	C71
C73	Thyroid	100.0	-	-	-	0.4	1.9	6.1	5.3	9.2	8.0	9.4	8.3	5.7	5.3	6.0	7.0	9.3	6.6	6.5	4.8	C73
C74-C75	Other endocrine glands	100.0	3.5	10.4	4.3	3.5	1.7	4.3	4.3	3.5	4.3	7.0	5.2	5.2	7.0	5.2	7.0	7.0	8.7	4.3	3.5	C74-C75
C81	Hodgkin's disease	100.0	-	0.4	0.8	1.6	7.5	14.9	15.3	9.0	8.1	5.7	4.3	3.9	5.9	4.9	4.3	5.1	1.8	3.5	2.9	C81
C82-C85, C91.4, C96	Non-Hodgkin's lymphoma	100.0	0.1	0.2	0.4	0.3	0.3	0.7	1.2	1.5	2.1	2.7	4.8	6.2	7.6	10.1	11.6	17.2	13.3	11.2	8.4	C82-C85, C91.4, C96
C88, C90	Multiple myeloma	100.0	-	-	0.1	-	-	-	0.1	0.2	0.3	0.9	2.5	3.3	5.6	9.7	12.3	18.4	15.9	17.6	13.1	C88, C90
C91-C95, xC91.4	Leukaemia	100.0	0.5	3.4	1.7	1.6	1.0	1.2	1.0	1.6	1.4	2.5	3.5	3.7	4.6	6.7	9.7	13.0	12.8	13.9	16.3	C91-C95, xC91.4
C91, xC91.4	Lymphoid leukaemia	100.0	0.5	6.6	3.0	2.7	1.1	0.5	0.5	0.5	0.4	1.0	1.9	2.7	4.8	6.8	8.9	13.1	11.9	14.1	19.1	C91, xC91.4
C92	Myeloid leukaemia	100.0	0.2	0.7	0.7	0.7	0.8	1.8	1.4	2.6	2.6	4.1	5.3	4.5	4.7	6.7	10.9	12.9	13.4	12.9	13.0	C92
C93	Monocytic leukaemia	100.0	-	-	-	-	-	-	5.0	-	-	-	-	5.0	-	5.0	10.0	20.0	15.0	15.0	25.0	C93
C94-C95	Other leukaemia	100.0	2.1	1.4	-	1.4	1.4	1.4	-	0.7	-	1.4	2.1	5.5	2.7	5.0	6.2	12.3	15.1	20.5	20.5	C94-C95
C76-C80	Other and unspecified	100.0	0.0	0.1	0.0	0.0	0.0	0.1	0.1	0.4	0.9	1.4	2.3	2.9	4.7	7.1	11.0	16.6	16.5	16.5	19.3	C76-C80

Appendix B5(a) Percentage distribution of new cases by subsite and sex, England and Wales, 1994 – males

Males ICD9	Site	Number of cases	Subsite (%)									
			0	1	2	3	4	5	6	7	8	9
140	Lip	202	5.0	45.0	:	0.5	6.9	1.0	1.0	:	0.0	40.6
141	Tongue	510	24.9	1.0	16.5	3.3	1.8	0.4	1.2	:	2.0	49.0
142	Salivary glands	236	80.9	15.7	0.4	:	:	:	:	:	0.4	2.5
143	Gum	82	13.4	42.7	:	:	:	:	:	:	0.0	43.9
144	Floor of mouth	240	9.2	2.9	:	:	:	:	:	:	3.8	84.2
145	Mouth unspecified	292	18.5	1.7	8.2	18.8	1.4	11.0	15.4	:	2.1	22.9
146	Oropharynx	311	62.7	4.2	0.6	3.2	1.6	0.3	1.0	0.6	1.6	24.1
147	Nasopharynx	122	2.5	3.3	0.0	2.5	:	:	:	:	4.9	86.9
148	Hypopharynx	244	9.0	70.5	3.3	2.9	:	:	:	:	2.0	12.3
149	Pharynx unspecified	125	86.4	0.0	:	:	:	:	:	:	7.2	6.4
150	Oesophagus	3,537	0.2	0.1	0.1	2.0	5.5	27.7	:	:	0.9	63.5
151	Stomach	6,115	21.6	2.1	4.1	1.4	2.4	5.6	1.8	:	1.6	59.4
152	Small intestine	241	30.3	12.0	22.0	0.4	:	:	:	:	2.1	33.2
153	Colon	8,729	2.1	5.0	3.8	28.8	17.8	0.5	6.3	2.7	0.9	32.0
154	Rectum	6,082	18.9	77.2	1.4	1.6	:	:	:	:	1.0	:
155	Liver	942	76.3	17.3	6.4	:	:	:	:	:	:	:
156	Gallbladder	468	25.0	37.8	26.7	:	:	:	:	:	1.3	9.2
157	Pancreas	2,846	40.0	2.7	1.8	0.1	0.2	:	:	:	0.5	54.7
158	Retroperitoneum and peritoneum	149	51.7	:	:	:	:	:	:	:	16.1	32.2
159	Digestive organs, etc. unspecified	239	65.7	0.4	:	:	:	:	:	:	2.5	31.4
160	Nasal cavities	199	46.7	8.0	32.2	10.1	0.0	0.0	:	:	1.0	2.0
161	Larynx	1,714	47.1	16.1	1.0	0.7	:	:	:	:	1.2	33.8
162	Lung	23,314	0.2	:	7.1	22.7	2.3	10.4	:	:	0.9	56.4
163	Pleura	1,040	1.0	0.3	:	:	:	:	:	:	0.4	98.4
164	Thymus	85	16.5	4.7	8.2	1.2	:	:	:	:	0.0	69.4
165	Respiratory system - other	4	0.0	:	:	:	:	:	:	:	25.0	75.0
170	Bone	221	5.4	5.9	6.3	7.7	10.4	1.8	13.6	35.3	2.3	11.3
171	Connective tissue	577	9.9	:	13.7	28.1	6.1	7.3	7.1	9.7	1.4	16.8
172	Melanoma	1,909	0.3	0.7	2.2	11.1	5.8	35.2	15.2	17.5	0.5	11.5
173	Non-melanoma skin cancer	19,404	1.9	7.0	12.1	41.4	9.7	9.0	7.6	3.4	2.6	5.2
175	Breast	259	:	:	:	:	:	:	:	:	:	:
185	Prostate	19,399	:	:	:	:	:	:	:	:	:	:
186	Testis	1,381	0.9	:	:	:	:	:	:	:	:	99.1
187	Penis	359	:	10.0	19.8	1.4	56.5	0.0	0.6	10.3	0.8	0.6
188	Bladder	8,516	0.5	0.3	1.7	0.3	1.1	0.8	1.0	0.1	1.9	92.3
189	Kidney	2,985	87.9	4.7	4.1	1.5	0.0	:	:	:	0.7	1.1
190	Eye	204	7.8	3.9	1.0	6.9	1.5	13.7	45.6	1.5	1.0	17.2
191	Brain	1,986	4.1	16.6	11.1	15.9	3.6	1.1	3.3	1.7	8.5	34.2
192	Other nervous system	85	15.3	32.9	42.4	1.2	:	:	:	:	0.0	8.2
193	Thyroid	259	:	:	:	0.2	:	:	:	:	:	:
194	Other endocrine glands	116	62.1	4.3	0.0	15.5	11.2	1.7	3.4	:	0.0	1.7
200	Lymphosarcoma and reticulosarcoma	259	6.2	38.2	12.7	:	:	:	:	:	42.9	:
201	Hodgkin's disease	674	0.1	0.0	0.0	:	7.4	49.7	18.0	1.6	:	23.1
202	Lymphoid and histiocytic tissue	3,660	7.5	2.7	0.2	0.3	2.3	0.0	0.0	:	86.3	0.7
203	Multiple myeloma	1,442	98.1	0.6	:	:	:	:	:	:	1.2	:
204	Lymphoid leukaemia	1,482	22.7	74.6	0.0	:	:	:	:	:	0.6	2.1
205	Myeloid leukaemia	1,401	64.6	32.3	0.1	:	:	:	:	:	0.4	2.4
206	Monocytic leukaemia	29	69.0	20.7	0.0	:	:	:	:	:	0.0	10.3
207	Other specified leukaemia	11	54.5	9.1	27.3	:	:	:	:	:	9.1	:
208	Leukaemia unspecified	130	46.9	5.4	0.0	:	:	:	:	:	1.5	46.2

Appendix B5(b) Percentage distribution of new cases by subsite and sex, England and Wales, 1994 – females

Females

ICD9	Site	Number of cases	Subsite (%)									
			0	1	2	3	4	5	6	7	8	9
140	Lip	78	12.8	38.5	..	1.3	0.0	2.6	1.3	..	2.6	41.0
141	Tongue	277	14.8	1.8	20.9	4.0	3.2	0.0	1.4	..	0.7	53.1
142	Salivary glands	189	69.8	21.2	0.5	..	..	..	..	..	0.0	8.5
143	Gum	67	14.9	38.8	..	..	..	..	..	..	1.5	44.8
144	Floor of mouth	86	7.0	2.3	..	..	..	..	..	..	1.2	89.5
145	Mouth unspecified	225	29.8	0.9	8.4	12.0	1.8	15.1	9.8	..	2.2	20.0
146	Oropharynx	142	66.2	2.1	2.1	4.9	0.0	0.7	0.7	..	4.2	17.6
147	Nasopharynx	61	4.9	6.6	0.0	0.0	..	..	..	1.4	0.0	88.5
148	Hypopharynx	123	42.3	36.6	2.4	4.1	..	..	..	..	0.0	14.6
149	Pharynx unspecifed	58	93.1	0.0	..	..	..	..	..	..	3.4	3.4
150	Oesophagus	2,444	0.2	0.1	0.2	3.3	11.5	19.1	..	..	0.9	64.7
151	Stomach	3,599	10.8	2.6	6.8	1.2	2.6	4.9	1.6	..	1.5	68.0
152	Small intestine	227	36.6	7.0	24.7	1.3	..	..	..	..	0.9	29.5
153	Colon	9,575	2.0	5.6	3.5	24.6	20.7	0.7	7.3	2.3	0.9	32.4
154	Rectum	4,518	19.0	73.0	3.1	3.3	..	..	..	..	1.5	0.0
155	Liver	633	61.9	29.2	8.8	..	..	..	..	..	..	..
156	Gallbladder	737	46.0	28.9	17.6	..	..	..	..	..	0.5	6.9
157	Pancreas	3,170	39.7	2.1	1.7	0.3	0.3	..	..	..	0.7	55.1
158	Retroperitoneum and peritoneum	135	57.0	..	..	..	..	..	..	..	15.6	27.4
159	Digestive organs, etc. unspecified	305	68.9	1.3	..	..	..	..	..	..	5.9	23.9
160	Nasal cavities	143	46.9	7.7	29.4	8.4	0.0	2.8	..	..	2.1	2.8
161	Larynx	362	30.4	24.0	4.7	0.8	..	..	..	..	1.4	38.7
162	Lung	12,297	0.3	..	6.9	21.9	2.4	9.4	..	..	0.7	58.3
163	Pleura	196	1.0	0.0	..	..	..	..	..	..	..	99.0
164	Thymus	80	33.8	8.8	8.8	6.3	..	..	..	..	2.5	40.0
165	Respiratory system - other	5	0.0	..	..	..	..	..	..	..	60.0	40.0
170	Bone	151	5.3	2.6	9.9	5.3	11.3	1.3	17.2	31.8	1.3	13.9
171	Connective tissue	496	7.7	..	9.5	36.3	4.6	8.7	11.5	4.0	2.0	15.7
172	Melanoma	2,816	0.1	0.5	0.4	10.0	2.1	14.3	18.4	44.9	0.5	8.8
173	Non-melanoma skin cancer	17,383	2.7	7.9	2.0	48.1	7.1	6.8	5.4	13.2	2.2	4.7
174	Breast	31,671	1.3	3.9	5.0	2.3	19.0	3.5	0.5	..	5.6	58.8
179	Uterus unspecified	333	..	..	..	..	..	..	..	..	..	..
180	Cervix	3,173	7.3	9.0	..	..	..	..	..	..	2.9	80.8
181	Placenta	12	..	..	..	..	..	..	..	..	..	..
182	Uterus	4,044	99.7	0.0	..	..	0.0	..	..	..	0.2	..
183	Ovary	5,349	98.6	..	0.9	0.0	0.0	0.0	..	..	0.2	0.2
184	Other female genital	1,089	18.0	3.1	1.2	0.4	73.6	..	..	..	1.4	2.3
188	Bladder	3,286	0.4	0.2	2.1	0.2	0.9	0.6	0.8	0.1	1.8	92.9
189	Kidney	1,781	87.6	5.9	4.3	0.8	0.0	..	..	..	0.6	0.8
190	Eye	183	12.0	4.4	0.0	7.7	0.5	12.6	31.7	1.6	0.0	29.5
191	Brain	1,516	3.9	17.2	10.4	14.1	2.6	1.2	4.0	2.2	7.5	37.0
192	Other nervous system	101	15.8	33.7	41.6	5.0	..	..	..	..	0.0	4.0
193	Thyroid	786	..	..	..	..	..	..	..	..	0.9	0.9
194	Other endocrine glands	115	64.3	3.5	..	15.7	8.7	1.7	4.3	..	0.9	0.9
200	Lymphosarcoma and reticulosarcoma	199	11.6	35.7	4.5	..	..	..	..	..	48.2	0.0
201	Hodgkin's disease	509	0.0	0.0	0.2	..	5.7	58.3	9.8	1.0	..	25.0
202	Lymphoid and histiocytic tissue	3,153	9.9	1.9	0.1	0.2	1.1	0.1	0.0	..	85.9	0.7
203	Multiple myeloma	1,436	98.9	0.5	0.0	..	..	..	..	..	0.6	0.0
204	Lymphoid leukaemia	1,130	22.5	75.0	0.0	..	..	..	..	..	0.3	2.3
205	Myeloid leukaemia	1,210	68.9	28.4	0.1	0.1	..	..	..	..	0.2	2.2
206	Monocytic leukaemia	20	50.0	30.0	5.0	..	..	..	..	..	0.0	15.0
207	Other specified leukaemia	13	38.5	7.7	38.5	..	..	..	..	..	15.4	0.0
208	Leukaemia unspecified	133	56.4	3.8	0.0	..	..	..	..	..	1.5	38.3
233	Carcinoma in situ of breast and genitourinary system	22,049	7.8	88.9	0.3	1.7	..	..	..	1.2	..	0.1

Appendix B6 Number of new cases for selected cancers by sex, England and Wales, 1995 (provisional)

Males

ICD10	Site	All Ages	Under 1	1-4	5-9	10-14	15-19	20-24	25-29	30-34	35-39	40-44	45-49	50-54	55-59	60-64	65-69	70-74	75-79	80-84	85 and over	ICD10
C00-C97 x C44	All cancers excluding NMSC	109,700	60	220	220	200	230	510	720	1,060	1,200	1,720	3,120	4,740	7,580	11,100	17,200	21,100	16,400	13,500	8,920	C00-C97 x C44
C00-C14	Lip, mouth and pharynx	2,340	-	0	-	0	10	-	0	20	40	110	180	230	280	290	310	320	230	160	90	C00-C14
C15	Oesophagus	3,380	0	-	-	-	-	-	0	10	10	40	110	220	280	400	490	750	450	380	250	C15
C16	Stomach	6,010	-	-	-	-	0	0	10	10	30	80	160	220	420	580	1,010	1,260	950	780	510	C16
C18	Colon	8,620	0	-	-	0	0	0	10	30	50	100	200	390	570	980	1,370	1,760	1,330	1,130	710	C18
C19-C21	Rectum	5,680	0	-	-	-	-	-	0	10	40	60	160	280	490	670	1,010	1,110	780	670	400	C19-C21
C25	Pancreas	2,800	-	-	-	-	-	-	0	10	10	40	110	140	220	300	470	510	450	320	230	C25
C32	Larynx	1,590	-	0	-	-	-	-	0	0	10	40	100	140	170	220	260	280	180	130	60	C32
C33-C34	Lung	22,300	-	-	-	-	0	10	10	20	100	150	400	900	1,500	2,530	3,960	4,800	3,570	2,830	1,560	C33-C34
C43	Melanoma of skin	1,910	-	0	0	0	10	40	70	100	100	120	190	160	200	220	200	180	150	120	70	C43
C61	Prostate	18,700	-	-	-	-	0	-	-	0	0	10	40	190	610	1,490	2,860	3,950	3,680	3,340	2,550	C61
C62	Testis	1,520	0	10	0	0	30	160	240	350	250	130	120	70	40	20	20	20	10	20	0	C62
C64-C66,C68	Kidney	3,130	10	20	10	0	0	10	10	20	40	70	150	240	350	350	550	550	360	270	110	C64-C66,C68
C67	Bladder	8,620	-	0	-	0	0	0	10	20	20	80	190	320	590	840	1,450	1,750	1,420	1,110	800	C67
C71	Brain	1,960	10	30	60	30	30	30	40	80	90	110	150	150	210	210	220	270	130	90	30	C71
C82-C85,C96,C91.4	Non-Hodgkin's lymphoma	3,840	0	10	20	20	20	50	60	110	140	140	250	260	330	380	470	560	470	330	190	C82-C85,C96,C91.4
C88,C90	Multiple myeloma	1,620	-	0	0	0	-	0	10	0	10	20	70	70	130	190	240	290	240	220	140	C88,C90
C91-C95 xC91.4	Leukaemia	3,210	10	100	80	70	50	30	60	60	70	60	110	120	190	290	440	510	360	370	260	C91-C95 xC91.4

Females

ICD10	Site	All Ages	Under 1	1-4	5-9	10-14	15-19	20-24	25-29	30-34	35-39	40-44	45-49	50-54	55-59	60-64	65-69	70-74	75-79	80-84	85 and over	ICD10
C00-C97 x C44	All cancers excluding NMSC	110,000	60	190	170	160	210	480	970	1,620	2,320	3,550	5,900	7,490	8,490	9,920	13,000	16,100	13,600	13,100	12,600	C00-C97 x C44
C00-C14	Lip, mouth and pharynx	1,290	-	0	0	10	0	10	10	10	20	40	90	90	130	150	140	170	150	140	130	C00-C14
C15	Oesophagus	2,380	0	-	-	-	-	0	0	0	30	20	30	70	130	150	230	380	460	470	430	C15
C16	Stomach	3,400	-	-	-	-	-	0	0	20	30	30	40	80	120	270	300	540	590	680	700	C16
C18	Colon	9,410	0	-	-	0	-	10	10	10	40	90	170	340	540	670	1,150	1,470	1,410	1,750	1,770	C18
C19-C21	Rectum	4,120	0	0	10	-	-	0	0	10	20	50	130	160	280	360	480	690	660	650	630	C19-C21
C25	Pancreas	2,970	-	0	-	-	-	0	0	10	10	20	60	90	150	230	400	520	420	560	520	C25
C33-C34	Lung	12,600	-	-	-	-	-	0	10	20	60	140	290	560	750	1,180	2,170	2,820	2,060	1,550	1,010	C33-C34
C43	Melanoma of skin	2,770	0	0	0	0	20	60	40	160	160	230	250	210	200	190	280	280	190	240	170	C43
C50	Breast	31,000	10	-	-	-	0	20	140	550	1,050	1,760	3,020	3,770	3,320	3,150	2,980	3,530	2,840	2,350	2,520	C50
C53	Cervix	3,030	-	-	-	-	0	40	180	320	360	280	320	200	170	190	200	280	180	150	140	C53
C54	Uterus	3,910	-	-	-	-	-	0	0	10	30	80	200	300	540	550	600	560	420	330	290	C54
C56,C57.0-C57.4	Ovary	5,780	-	0	10	10	10	40	60	80	120	150	420	460	630	670	820	840	600	530	330	C56,C57.0-C57.4
C64-C66,C68	Kidney	1,780	0	20	10	0	10	10	0	20	20	30	80	90	130	190	270	270	260	180	210	C64-C66,C68
C67	Bladder	3,290	0	0	-	-	-	0	0	10	20	30	50	90	180	290	440	580	510	530	560	C67
C71	Brain	1,560	10	30	50	30	20	20	30	60	50	80	90	110	150	190	160	200	120	110	50	C71
C82-C85,C96,C91.4	Non-Hodgkin's lymphoma	3,310	0	10	10	10	10	30	40	70	90	90	150	210	240	300	390	490	430	430	310	C82-C85,C96,C91.4
C88,C90	Multiple myeloma	1,430	-	-	0	0	-	0	10	0	0	10	30	60	70	150	260	240	230	230	220	C88,C90
C91-C95 xC91.4	Leukaemia	2,420	10	80	60	50	30	30	10	50	60	60	80	80	120	150	260	310	300	320	370	C91-C95 xC91.4
D05	Breast in situ	1,750	-	-	-	-	-	0	0	0	60	90	200	430	340	270	100	100	60	40	20	D05
D06	Cervix in situ	20,500	-	-	0	-	340	3,240	5,630	4,510	2,790	1,540	1,090	500	390	310	120	50	20	20	10	D06

Appendix B7 Incidence rates per 100,000 population for selected cancers by sex, England and Wales, 1995 (provisional)

Males

ICD10	Site	CR	ESR	WSR	Under 1	1-4	5-9	10-14	15-19	20-24	25-29	30-34	35-39	40-44	45-49	50-54	55-59	60-64	65-69	70-74	75-79	80-84	85 and over
C00-C97 xC44	All cancers excluding NMSC	431.3	388.4	261.0	18.3	15.8	12.4	11.8	14.9	28.3	34.3	48.8	64.8	102.3	170.2	321.2	573.2	921.4	1,549.2	2,173.3	2,635.3	3,297.1	3,715.5
C00-C14	Lip, mouth and pharynx	9.2	8.9	6.3		0.2		0.2	0.4	0.2	0.3	0.9	2.1	6.8	10.1	15.5	25.2	24.3	27.6	33.0	37.3	38.6	38.3
C15	Oesophagus	13.3	12.1	8.1							0.1	0.3	0.7	2.6	5.7	14.6	21.5	33.6	44.0	76.9	71.5	93.0	103.7
C16	Stomach	23.6	21.1	13.8	0.3				0.1	0.1	0.5	0.6	1.5	4.5	8.5	14.9	31.6	48.0	91.4	129.7	153.3	189.2	212.4
C18	Colon	33.9	30.4	19.9	0.3			0.1	0.1	0.4	0.3	1.6	2.7	5.8	11.0	26.4	42.8	81.6	123.6	181.4	213.0	275.9	293.6
C19-C21	Rectum	22.3	20.3	13.5	0.3				0.1		0.3	0.5	1.9	3.8	8.5	18.7	37.2	55.6	91.6	114.0	125.7	163.1	165.8
C25	Pancreas	11.0	10.0	6.6							0.2	0.2	0.7	2.1	6.1	9.7	16.3	24.5	42.3	52.6	72.2	77.9	97.5
C32	Larynx	6.2	5.9	4.1							0.0	0.1	0.5	2.1	5.2	9.6	12.9	17.9	23.6	29.3	29.4	31.5	24.2
C33-C34	Lung	87.6	78.1	51.2		0.1		0.1		0.3	0.4	0.9	2.7	8.7	21.9	61.2	113.6	209.8	357.2	494.7	574.4	690.9	651.4
C43	Melanoma of skin	7.5	7.2	5.4				0.1	0.1	0.9	2.1	3.2	4.4	5.4	6.8	10.2	10.5	15.1	18.4	18.2	23.8	28.3	28.7
C61	Prostate	73.6	63.2	38.8									0.2	0.5	2.3	12.6	46.3	123.7	258.4	407.3	590.6	814.9	1,062.5
C62	Testis	6.0	5.6	5.2	0.6	0.4	0.2	0.2	1.9	9.2	11.5	16.3	13.7	7.9	6.7	4.8	3.1	2.0	2.2	2.3	1.9	3.7	1.7
C64-C66, C68	Kidney	12.3	11.5	8.1	1.5	1.4	0.4	0.1	0.3	0.3	0.6	1.1	2.1	4.1	8.4	16.5	26.7	29.1	49.9	57.0	58.5	66.4	47.1
C67	Bladder	33.9	30.2	19.7		0.1			0.3	0.4	0.6	1.0	2.0	4.6	10.5	21.4	44.3	69.4	131.2	180.0	227.9	270.5	334.0
C71	Brain	7.7	7.5	6.0	3.0	2.4	3.7	1.6	1.7	1.7	1.9	3.5	5.0	6.7	7.9	9.8	15.7	17.2	20.2	27.8	20.3	21.7	11.7
C82-C85, C91.4, C96	Non-Hodgkin's lymphoma	15.1	14.1	10.2	0.6	0.8	1.4	1.5	1.3	2.7	3.0	5.0	7.8	8.5	13.7	17.6	25.0	31.9	42.7	58.0	75.2	79.6	80.4
C88, C90	Multiple myeloma	6.4	5.8	3.8						0.1	0.2	0.0	0.5	1.1	3.6	4.5	10.1	15.4	21.7	29.7	39.1	53.7	58.7
C91-C95 xC91.4	Leukaemia	12.6	11.6	8.9	2.7	6.8	4.6	4.0	3.3	1.7	2.8	2.5	3.6	3.5	6.2	8.0	14.4	24.4	39.4	52.3	57.2	89.4	109.1

Females

ICD10	Site	CR	ESR	WSR	Under 1	1-4	5-9	10-14	15-19	20-24	25-29	30-34	35-39	40-44	45-49	50-54	55-59	60-64	65-69	70-74	75-79	80-84	85 and over
C00-C97 xC44	All cancers excluding NMSC	416.9	319.0	227.5	17.4	14.2	10.4	10.3	14.4	28.2	48.2	78.2	128.1	212.7	322.9	506.4	633.8	790.8	1,043.2	1,305.7	1,461.6	1,709.3	1,782.4
C00-C14	Lip, mouth and pharynx	4.9	3.9	2.8		0.1	0.1	0.3	0.1	0.5	0.7	0.6	1.3	2.5	5.0	5.9	8.6	12.0	11.2	13.5	16.3	18.7	18.5
C15	Oesophagus	9.0	5.6	3.6	0.3					0.2	0.0	0.1	0.4	1.4	1.9	4.9	9.9	11.7	18.1	30.5	49.3	60.5	60.4
C16	Stomach	12.9	7.9	5.1	0.3					0.2	0.0	0.1	0.9	2.0	2.4	5.5	8.7	21.1	24.4	43.9	62.8	88.2	99.3
C18	Colon	35.7	23.0	15.0				0.2		0.3	0.3	0.7	2.3	5.6	9.4	22.7	40.0	53.1	92.2	119.1	151.2	227.1	249.3
C19-C21	Rectum	15.6	10.7	7.1		0.1			0.3	0.6	1.0	2.1	3.2	6.5	9.2	21.2	28.7	38.9	55.6	64.5	70.4	83.9	89.3
C25	Pancreas	11.3	7.2	4.8				0.2		0.1	0.5	0.9	1.2	1.2	3.0	5.8	10.9	18.1	31.8	42.5	45.0	72.8	72.9
C33-C34	Lung	47.7	34.4	23.3	0.3	0.1		0.1		0.1	0.1	0.8	3.0	8.3	15.6	37.8	56.0	94.2	174.3	228.7	220.8	201.7	142.1
C43	Melanoma of skin	10.5	9.1	7.1					1.3	3.6	7.0	7.9	8.6	13.5	13.7	14.3	14.6	15.0	22.7	22.3	20.6	30.6	24.3
C50	Breast	117.7	101.8	74.2					0.1	1.4	9.1	20.0	57.8	105.1	165.3	248.2	255.1	251.1	238.9	286.7	304.4	305.8	355.3
C53	Cervix	11.5	10.3	8.3					0.1	1.4	10.3	15.4	20.0	17.0	17.2	12.9	13.2	14.9	16.0	23.0	19.2	19.1	19.8
C54	Uterus	14.8	12.3	8.7							0.2	0.4	1.7	4.5	10.8	20.2	40.2	44.2	48.5	45.8	44.8	43.2	41.0
C56, C57.0-C57.4	Ovary	21.9	18.3	13.3		0.1	0.3	0.6	1.0	2.1	3.2	3.8	6.5	9.2	23.1	31.4	46.9	53.5	65.6	68.0	64.5	68.7	46.2
C64-C66, C68	Kidney	6.7	5.1	3.7	0.9	1.6	0.5	0.2		0.1	0.1	0.8	0.9	1.6	4.3	5.8	9.9	14.8	21.8	22.2	27.9	23.3	29.2
C67	Bladder	12.5	8.2	5.4		0.1				0.1	0.1	0.4	1.0	2.0	2.5	5.9	13.3	23.4	34.9	47.4	54.4	69.2	78.5
C71	Brain	5.9	5.3	4.4	2.2	2.5	2.8	1.8	1.4	1.1	1.5	2.8	2.6	4.7	5.0	7.6	11.2	15.4	13.0	15.9	13.2	14.7	6.9
C82-C85, C91.4, C96	Non-Hodgkin's lymphoma	12.5	9.6	6.9	0.3	0.7	0.5	0.7	1.0	1.8	2.1	3.2	4.7	5.1	8.4	14.2	18.0	23.7	31.2	39.4	46.3	56.5	43.9
C88, C90	Multiple myeloma	5.4	3.7	2.4								0.2	0.1	0.8	1.7	3.9	5.0	12.0	15.7	19.1	24.5	29.3	30.6
C91-C95 xC91.4	Leukaemia	9.2	6.8	5.5	2.8	5.6	3.4	3.0	2.2	2.0	1.1	2.3	2.5	3.7	4.3	5.6	8.8	11.9	20.9	25.4	31.8	41.1	52.5
D05	Breast in situ	6.6	6.9	5.2							0.4	1.6	3.4	5.6	10.7	28.9	25.7	21.4	8.0	8.4	6.1	5.5	2.1
D06	Cervix in situ	77.8	77.1	75.4			0.1		23.1	190.1	281.0	217.6	153.9	92.1	59.6	33.7	28.8	24.8	9.2	4.1	2.2	2.0	0.7

Appendix B8 Number of new cases for selected cancers by sex, England and Wales, 1996 (provisional)

Males

ICD10	Site	All Ages	Under 1	1-4	5-9	10-14	15-19	20-24	25-29	30-34	35-39	40-44	45-49	50-54	55-59	60-64	65-69	70-74	75-79	80-84	85 and over	ICD10
C00-C97 x C44	All cancers excluding NMSC	110,000	60	260	180	160	260	480	720	960	1,110	1,720	3,140	4,780	7,370	11,400	16,500	20,600	17,300	13,600	9,310	C00-C97 x C44
C00-C14	Lip, mouth and pharynx	2,440	-	10	0	0	10	10	10	20	50	100	200	230	360	330	300	270	280	160	100	C00-C14
C15	Oesophagus	3,530	-	-	-	-	-	-	0	0	20	50	100	220	290	400	550	700	460	430	310	C15
C16	Stomach	5,860	-	-	-	-	0	0	10	20	30	60	140	210	370	570	930	1,190	980	810	560	C16
C18	Colon	8,900	0	-	-	0	0	0	20	30	60	110	220	350	590	970	1,410	1,700	1,490	1,150	810	C18
C19-C21	Rectum	6,240	0	-	-	-	0	0	10	10	30	80	190	340	530	760	1,110	1,140	890	700	450	C19-C21
C25	Pancreas	2,830	-	-	-	-	-	-	0	10	10	30	100	140	190	320	410	570	450	320	270	C25
C32	Larynx	1,570	-	-	-	-	-	-	0	0	10	30	80	130	170	240	240	250	220	120	80	C32
C33-C34	Lung	22,300	-	-	-	-	0	0	10	30	50	160	480	890	1,480	2,430	3,660	4,850	3,820	2,830	1,580	C33-C34
C43	Melanoma of skin	1,890	-	-	0	0	10	40	60	70	90	120	150	190	180	210	210	190	160	130	80	C43
C61	Prostate	19,200	-	-	0	0	-	-	0	-	0	10	40	220	670	1,660	2,940	3,950	3,860	3,240	2,620	C61
C62	Testis	1,540	10	10	-	0	20	170	260	330	260	160	140	70	30	30	10	20	10	20	10	C62
C64-C66,C68	Kidney	3,080	10	20	10	0	0	0	10	20	40	90	170	210	300	380	510	540	360	290	140	C64-C66,C68
C67	Bladder	8,580	0	0	-	-	0	10	10	20	30	90	170	320	520	890	1,310	1,740	1,520	1,140	820	C67
C71	Brain	1,900	10	30	40	40	20	30	30	70	90	100	130	180	210	220	220	240	130	90	30	C71
C82-C85,C91.4,C96	Non-Hodgkin's lymphoma	3,940	0	20	20	10	20	30	60	110	120	140	270	320	340	430	480	570	460	330	200	C96,C82-C85,C91.4
C88,C90	Multiple myeloma	1,550	-	-	-	0	-	0	10	0	10	20	70	80	130	160	230	280	230	200	130	C88,C90
C91-C95 x C91.4	Leukaemia	3,000	10	120	70	50	50	30	50	50	60	50	120	110	170	260	380	490	360	340	240	C91-C95 x C91.4

Females

ICD10	Site	All Ages	Under 1	1-4	5-9	10-14	15-19	20-24	25-29	30-34	35-39	40-44	45-49	50-54	55-59	60-64	65-69	70-74	75-79	80-84	85 and over	ICD10
C00-C97 xC44	All cancers excluding NMSC	111,900	50	230	160	150	190	400	860	1,640	2,350	3,650	6,010	8,250	8,370	9,720	12,400	16,200	15,100	12,800	13,400	C00-C97 x C44
C00-C14	Lip, mouth and pharynx	1,330	0	-	0	0	0	10	10	10	20	40	90	80	130	170	170	140	180	140	120	C00-C14
C15	Oesophagus	2,540	0	-	-	-	0	-	0	0	10	20	40	90	130	150	250	400	470	500	490	C15
C16	Stomach	3,230	-	-	-	0	0	-	0	20	20	30	50	70	140	220	310	470	580	600	740	C16
C18	Colon	9,940	0	-	-	0	0	10	10	20	60	100	180	390	570	730	1,060	1,580	1,650	1,610	1,980	C18
C19-C21	Rectum	4,610	-	-	-	-	0	10	10	10	30	60	160	220	320	400	550	760	720	720	670	C19-C21
C25	Pancreas	3,220	-	-	-	-	-	-	10	10	10	20	50	100	150	260	390	500	540	610	570	C25
C33-C34	Lung	12,900	0	-	-	0	0	0	0	20	50	160	320	630	800	1,240	2,130	2,820	2,320	1,390	1,020	C33-C34
C43	Melanoma of skin	2,600	0	-	-	0	10	50	110	180	170	190	250	250	160	190	240	230	180	200	170	C43
C67	Bladder	3,340	-	0	10	0	-	0	0	0	10	30	70	110	160	250	440	530	530	580	640	C67
C71	Brain	1,580	10	40	40	40	20	20	30	50	50	60	110	110	140	150	190	220	160	100	50	C71
C50	Breast	32,000	0	-	-	-	10	20	160	530	1,160	1,900	3,260	4,200	3,290	3,020	2,660	3,500	3,190	2,390	2,710	C50
C53	Cervix	2,850	-	-	-	0	0	40	180	310	320	310	280	210	150	170	180	260	200	120	130	C53
C54	Uterus	4,000	-	-	-	-	-	0	0	10	30	70	160	300	530	580	610	580	500	320	300	C54
C56,C57.0-C57.4	Ovary	5,810	-	-	0	10	20	30	60	80	120	180	370	510	590	710	720	870	660	520	350	C56,C57.0-C57.4
C64-C66,C68	Kidney	1,860	0	30	10	0	-	10	0	10	20	50	60	110	130	190	260	310	280	210	180	C64-C66,C68
C82-C85,C91.4,C96	Non-Hodgkin's lymphoma	3,380	0	10	10	10	10	20	50	60	70	100	190	210	270	330	370	470	480	370	360	C82-C85,C91.4,C96
C88,C90	Multiple myeloma	1,360	-	-	0	-	10	-	-	10	10	20	30	40	70	130	170	230	210	200	260	C88,C90
C91-C95 x C91.4	Leukaemia	2,530	10	90	40	40	30	30	20	50	50	60	60	100	120	170	280	320	270	350	440	C91-C95 x C91.4
D05	Breast in situ	1,800	-	-	-	-	-	-	10	30	70	80	200	450	330	270	120	120	70	30	20	D05
D06	Cervix in situ	19,700	-	-	0	-	300	2,920	5,550	4,590	2,640	1,370	1,000	480	320	330	110	50	30	20	-	D06

Appendix B9 Incidence rates per 100,000 population for selected cancers by sex, England and Wales, 1996 (provisional)

Males

ICD10	Site	CR	ESR	WSR	Under 1	1-4	5-9	10-14	15-19	20-24	25-29	30-34	35-39	40-44	45-49	50-54	55-59	60-64	65-69	70-74	75-79	80-84	85 and over	ICD10
C00-C97 xC44	All cancers excluding NMSC	430.4	385.6	258.8	18.7	18.7	10.0	9.7	16.3	28.3	35.0	43.7	58.1	102.1	170.4	310.6	562.5	948.4	1,494.7	2,159.7	2,634.6	3,338.5	3,740.6	C00-C97 xC44
C00-C14	Lip, mouth and pharynx	9.6	9.2	6.6	-	0.4	0.1	0.2	0.4	0.3	0.6	1.0	2.5	6.1	11.1	15.0	27.2	27.6	27.2	28.6	43.2	38.2	41.0	C00-C14
C15	Oesophagus	13.8	12.5	8.3	-	-	-	-	-	0.1	0.1	0.2	1.0	2.7	5.7	14.0	21.8	33.6	49.4	73.0	70.6	104.7	125.4	C15
C16	Stomach	22.9	20.3	13.1	-	-	-	-	0.1	0.1	0.4	0.8	1.6	3.7	7.4	13.6	28.1	47.1	84.4	124.8	148.3	196.9	223.0	C16
C18	Colon	34.8	31.0	20.3	0.3	-	-	0.2	0.1	0.2	0.9	1.5	3.1	6.2	12.1	22.5	44.9	80.4	127.3	177.6	226.9	282.3	323.9	C18
C19-C21	Rectum	24.4	22.2	14.9	0.6	-	-	-	-	-	0.4	0.6	1.4	4.7	10.5	21.8	40.3	63.4	100.6	119.2	134.6	172.0	181.6	C19-C21
C25	Pancreas	11.1	10.0	6.6	-	-	-	-	-	-	0.1	0.3	0.7	2.0	5.2	9.4	14.6	26.8	37.5	59.6	68.0	78.5	107.7	C25
C32	Larynx	6.1	5.8	4.0	-	-	-	0.1	-	-	0.1	0.1	0.5	1.6	4.3	8.5	13.0	20.1	21.7	26.2	33.8	28.6	30.1	C32
C33-C34	Lung	87.1	77.1	50.4	-	-	-	0.1	0.1	0.2	0.3	1.2	2.6	9.2	26.1	57.8	112.8	202.1	331.4	508.0	581.0	691.9	634.5	C33-C34
C43	Melanoma of skin	7.4	7.1	5.3	-	-	0.1	0.1	0.8	2.2	2.9	3.3	4.6	7.2	8.1	12.2	13.8	17.1	19.1	20.1	24.2	31.8	30.5	C43
C61	Prostate	75.2	64.4	39.8	-	0.2	0.1	0.1	-	-	0.0	-	0.1	0.5	2.3	14.6	51.0	137.9	266.0	413.7	586.2	793.6	1,050.8	C61
C62	Testis	6.0	5.7	5.3	1.5	0.5	0.1	0.2	1.4	10.0	12.3	14.9	13.5	9.5	7.8	4.2	2.5	2.4	1.3	2.0	1.4	4.2	2.4	C62
C64-C66,C68	Kidney	12.1	11.2	7.8	1.5	1.5	0.3	0.1	0.1	0.2	0.4	1.0	2.0	5.0	9.0	13.9	22.7	31.5	46.1	56.4	54.0	71.4	56.3	C64-C66,C68
C67	Bladder	33.6	29.7	19.3	0.3	0.1	-	-	0.3	0.4	0.7	1.0	1.7	5.2	9.1	20.6	39.9	73.7	118.5	181.8	230.8	278.4	327.5	C67
C71	Brain	7.4	7.2	5.7	2.4	2.2	2.4	2.1	1.5	1.8	1.5	3.2	4.4	6.0	7.2	11.6	16.2	18.0	20.2	24.7	19.5	21.3	10.4	C71
C82-C85,C91.4,C96	Non-Hodgkin's lymphoma	15.4	14.4	10.4	0.3	1.2	1.2	0.7	1.4	1.9	2.9	5.1	6.2	8.4	14.6	21.1	25.8	36.0	43.2	59.8	70.6	80.0	81.2	C82-C85,C91.4,C96
C88,C90	Multiple myeloma	6.1	5.5	3.6	-	-	-	0.1	0.1	0.1	0.3	0.1	0.4	1.1	3.5	5.4	9.7	13.7	20.8	29.5	35.3	48.4	53.4	C88,C90
C91-C95 xC91.4	Leukaemia	11.7	10.8	8.3	2.8	8.7	3.8	2.9	3.3	2.0	2.3	2.2	3.0	2.7	6.6	7.4	12.7	21.6	34.3	51.4	55.3	82.0	95.2	C91-C95 xC91.4

Females

ICD10	Site	CR	ESR	WSR	Under 1	1-4	5-9	10-14	15-19	20-24	25-29	30-34	35-39	40-44	45-49	50-54	55-59	60-64	65-69	70-74	75-79	80-84	85 and over	ICD10
C00-C97 xC44	All cancers excluding NMSC	423.0	322.1	228.8	15.2	17.2	9.2	9.3	12.8	24.9	43.4	77.8	125.8	217.8	327.3	534.1	630.0	780.5	998.8	1,356.4	1,553.1	1,680.9	1,865.8	C00-C97 xC44
C00-C14	Lip, mouth and pharynx	5.0	4.0	2.9	0.3	-	0.2	0.3	0.1	0.4	0.6	0.7	1.2	2.5	5.0	5.4	9.6	13.9	13.3	11.7	18.6	18.3	16.7	C00-C14
C15	Oesophagus	9.6	6.0	3.8	0.3	-	-	-	-	-	0.1	0.1	0.3	1.3	2.2	5.7	9.6	11.8	20.4	33.9	48.0	65.4	68.3	C15
C16	Stomach	12.2	7.4	4.7	-	-	-	0.1	-	-	0.1	1.0	1.2	1.7	2.7	4.3	10.2	17.3	24.6	39.4	59.1	78.4	103.0	C16
C18	Colon	37.6	24.4	16.0	0.3	-	-	0.1	0.3	0.3	0.5	1.0	3.3	5.8	9.9	24.9	42.9	58.9	85.3	132.1	169.5	211.6	274.5	C18
C19-C21	Rectum	17.4	12.1	8.1	-	0.2	-	-	0.1	0.1	0.3	0.4	1.7	3.2	8.4	14.0	23.9	32.2	44.7	63.5	74.2	94.5	92.3	C19-C21
C25	Pancreas	12.2	7.8	5.1	-	-	-	-	0.1	-	0.3	0.3	0.5	1.3	2.7	6.5	11.4	20.8	31.8	42.1	55.9	79.9	78.8	C25
C33-C34	Lung	48.8	35.6	24.2	0.3	-	-	0.1	0.2	0.1	0.2	1.0	2.4	9.3	17.4	40.9	60.4	99.7	172.0	236.4	237.9	182.7	142.1	C33-C34
C43	Melanoma of skin	9.8	8.5	6.7	-	0.2	0.4	0.1	0.9	3.3	5.8	8.3	9.2	11.5	13.8	15.9	12.2	15.0	19.4	19.5	18.8	25.9	23.3	C43
C50	Breast	120.9	104.0	75.7	1.0	-	0.1	-	0.3	1.0	7.9	25.1	62.1	113.2	177.4	272.0	247.7	242.2	214.9	293.8	327.3	313.8	376.6	C50
C53	Cervix	10.8	9.7	7.8	-	-	-	-	0.1	2.5	9.2	14.5	17.3	18.6	15.1	13.4	11.5	13.4	14.3	21.6	20.2	15.5	18.2	C53
C54	Uterus	15.1	12.5	8.7	-	-	-	-	-	0.1	0.1	0.4	1.5	3.3	8.9	19.6	40.2	46.8	49.2	49.0	51.3	41.6	41.6	C54
C56,C57.0-C57.4	Ovary	21.9	18.2	13.2	-	0.2	0.2	0.5	1.0	1.7	3.0	3.7	6.3	10.9	19.9	33.2	44.3	57.0	58.5	73.3	67.4	68.8	49.1	C56,C57.0-C57.4
C64-C66,C68	Kidney	7.0	5.3	3.8	1.0	2.2	0.4	0.1	-	0.5	0.2	0.7	1.1	2.8	3.4	6.8	9.4	14.8	21.3	25.7	28.9	27.9	25.4	C64-C66,C68
C67	Bladder	12.6	8.1	5.3	-	0.1	-	-	-	0.1	0.1	0.2	0.5	1.8	3.5	6.9	11.9	20.4	35.5	44.4	54.0	76.2	88.4	C67
C71	Brain	6.0	5.3	4.3	1.6	2.9	2.5	2.2	1.3	1.4	1.4	2.5	2.6	3.8	5.8	6.9	10.7	12.0	15.4	18.5	16.5	13.6	6.2	C71
C82-C85,C91.4,C96	Non-Hodgkin's lymphoma	12.8	9.9	7.1	0.3	0.6	0.3	0.5	0.9	1.4	2.4	2.6	4.0	6.1	10.4	13.8	20.0	26.6	29.9	39.5	49.3	48.0	49.4	C82-C85,C91.4,C96
C88,C90	Multiple myeloma	5.2	3.4	2.3	-	-	0.1	-	-	-	-	0.3	0.3	1.1	1.5	2.5	5.0	10.8	14.0	19.4	21.6	25.7	35.4	C88,C90
C91-C95 xC91.4	Leukaemia	9.6	7.0	5.6	3.2	7.0	2.6	2.5	1.9	2.1	1.2	2.1	2.6	3.3	3.4	6.6	8.9	13.2	22.9	27.0	27.5	45.9	60.6	C91-C95 xC91.4
D05	Breast in situ	6.8	7.0	5.2	-	-	-	-	-	-	0.4	1.4	3.8	4.8	10.7	29.2	25.0	21.8	9.7	9.6	7.4	4.5	2.2	D05
D06	Cervix in situ	74.5	73.9	72.4	-	-	-	-	19.8	180.3	281.3	218.1	141.2	81.9	54.4	31.1	24.3	26.1	8.6	4.0	3.0	2.0	-	D06

Appendix B10 Number of new cases for selected cancers by sex, England and Wales, 1997 (provisional)

Males

ICD10	Site	All Ages	Under 1	1-4	5-9	10-14	15-19	20-24	25-29	30-34	35-39	40-44	45-49	50-54	55-59	60-64	65-69	70-74	75-79	80-84	85 and over	ICD10
C00-C97 xC44	All cancers excluding NMSC	108,600	70	240	230	190	300	450	730	1,020	1,150	1,840	2,960	5,140	7,630	10,800	16,300	20,200	17,800	12,600	8,950	C00-C97 xC44
C00-C14	Lip, mouth and pharynx	2,380	0	-	-	-	10	10	10	20	60	100	170	240	310	280	340	310	280	140	100	C00-C14
C15	Oesophagus	3,560	-	-	-	-	-	-	-	10	20	40	120	250	280	400	570	650	530	410	280	C15
C16	Stomach	5,840	-	-	-	-	0	0	10	20	30	70	130	230	420	530	920	1,190	930	750	610	C16
C18	Colon	8,890	-	-	-	0	0	10	40	40	50	110	190	380	580	950	1,300	1,750	1,630	1,120	760	C18
C19-C21	Rectum	6,020	0	-	-	0	0	10	20	20	50	90	160	320	490	690	1,070	1,160	870	680	410	C19-C21
C25	Pancreas	2,740	-	-	-	0	-	0	10	10	20	30	100	140	210	310	460	510	410	330	240	C25
C32	Larynx	1,530	-	-	-	-	-	-	-	10	10	30	70	120	180	250	260	230	200	100	60	C32
C33-C34	Lung	21,000	-	-	0	0	-	0	0	20	60	170	430	900	1,530	2,170	3,530	4,340	3,840	2,490	1,460	C33-C34
C43	Melanoma of skin	2,020	-	-	0	0	20	30	50	80	100	120	180	190	190	240	220	210	190	100	90	C43
C61	Prostate	18,300	-	-	-	-	-	-	-	-	-	10	50	250	700	1,480	2,790	3,920	3,840	2,830	2,420	C61
C62	Testis	1,440	0	10	0	0	30	170	240	290	220	180	120	60	40	20	20	10	10	10	10	C62
C64-C66,C68	Kidney	3,040	0	30	10	0	10	10	10	20	40	70	150	260	300	400	500	470	370	310	120	C64-C66,C68
C67	Bladder	8,530	-	-	10	0	0	10	20	20	40	100	160	340	490	850	1,300	1,720	1,540	1,140	820	C67
C71	Brain	2,050	10	30	60	30	20	30	40	80	80	100	130	200	230	210	260	250	170	80	40	C71
C82-C85,C91.4,C96	Non-Hodgkin's lymphoma	3,960	0	10	30	20	30	30	60	140	110	170	200	350	370	430	460	540	480	330	180	C82-C85,C91.4,C96
C88,C90	Multiple myeloma	1,480	-	-	-	-	-	10	0	0	10	20	50	80	120	150	210	290	210	200	130	C88,C90
C91-C95 xC91.4	Leukaemia	2,920	10	90	70	60	60	30	70	50	70	60	100	110	220	230	320	460	330	310	270	C91-C95 xC91.4

Females

ICD10	Site	All Ages	Under 1	1-4	5-9	10-14	15-19	20-24	25-29	30-34	35-39	40-44	45-49	50-54	55-59	60-64	65-69	70-74	75-79	80-84	85 and over	ICD10
C00-C97 xC44	All cancers excluding NMSC	112,900	60	190	150	180	260	410	960	1,670	2,530	3,620	5,900	8,820	8,750	10,300	12,600	15,500	15,700	12,500	13,000	C00-C97 xC44
C00-C14	Lip, mouth and pharynx	1,440	-	-	-	-	0	10	10	20	30	50	80	100	140	150	160	200	150	150	140	C00-C14
C15	Oesophagus	2,500	-	-	-	-	0	0	0	0	10	20	30	90	100	140	280	390	460	440	530	C15
C16	Stomach	3,280	-	-	-	-	0	0	20	20	30	40	40	90	120	220	350	510	620	590	670	C16
C18	Colon	9,420	-	-	0	-	10	0	20	20	50	100	150	410	510	770	960	1,480	1,620	1,570	1,780	C18
C19-C21	Rectum	4,530	-	-	-	0	0	0	10	20	30	70	130	210	320	400	540	640	790	660	730	C19-C21
C25	Pancreas	2,990	10	-	-	0	0	0	0	10	10	20	50	120	130	260	350	470	500	520	540	C25
C33-C34	Lung	12,300	-	0	-	-	10	50	110	140	150	200	270	240	840	1,110	1,940	2,660	2,280	1,420	930	C33-C34
C43	Melanoma of skin	2,670	-	-	0	0	10	50	110	180	190	250	270	240	190	250	270	230	210	190	140	C43
C50	Breast	33,100	-	-	-	-	0	20	180	580	1,170	1,860	3,190	4,440	3,470	3,390	2,950	3,540	3,260	2,460	2,550	C50
C53	Cervix	2,850	-	-	-	0	0	40	210	310	340	300	260	200	170	140	160	250	210	130	130	C53
C54	Uterus	4,000	-	-	-	0	0	0	0	10	30	80	160	370	570	600	590	540	480	270	300	C54
C56,C57.0-C57.4	Ovary	6,130	10	0	10	10	30	30	70	90	130	190	400	530	630	680	800	860	750	560	390	C56,C57.0-C57.4
C64-C66,C68	Kidney	1,800	10	30	10	0	-	0	0	20	30	30	70	120	120	210	260	250	300	180	160	C64-C66,C68
C67	Bladder	3,550	-	-	10	0	-	0	10	10	30	30	60	120	180	240	480	590	560	590	670	C67
C71	Brain	1,480	10	30	30	20	10	20	40	60	50	60	80	110	130	180	190	170	150	90	40	C71
C82-C85,C91.4,C96	Non-Hodgkin's lymphoma	3,400	0	10	10	10	10	50	40	40	90	90	150	230	240	290	410	530	510	390	330	C82-C85,C91.4,C96
C88,C90	Multiple myeloma	1,390	-	-	-	0	0	20	60	40	10	60	40	120	80	120	180	220	240	220	220	C88,C90
C91-C95 xC91.4	Leukaemia	2,340	10	70	50	40	30	20	20	60	50	60	70	120	110	160	210	280	300	300	390	C91-C95 xC91.4
D05	Breast in situ	2,010	-	-	-	-	-	-	0	30	100	100	210	550	360	350	130	110	60	40	20	D05
D06	Cervix in situ	17,900	0	-	0	-	260	2,630	4,840	4,230	2,490	1,290	830	520	310	280	120	40	20	10	10	D06

Appendix B11 Incidence rates per 100,000 population for selected cancers by sex, England and Wales, 1997 (provisional)

Males

ICD10	Site	CR	ESR	WSR	Under 1	1-4	5-9	10-14	15-19	20-24	25-29	30-34	35-39	40-44	45-49	50-54	55-59	60-64	65-69	70-74	75-79	80-84	85 and over	ICD10
C00-C97 xC44	All cancers excluding NMSC	422.8	377.4	254.3	20.7	17.4	12.6	11.1	18.8	27.8	35.6	46.0	58.0	107.2	169.1	307.2	578.9	899.1	1,479.3	2,136.5	2,568.2	3,136.4	3,472.5	C00-C97 xC44
C00-C14	Lip, mouth and pharynx	9.3	8.8	6.3	0.3	0.2	-	-	0.6	0.3	0.5	1.0	2.8	5.5	9.7	14.2	23.8	23.0	30.9	32.5	41.1	35.1	38.0	C00-C14
C15	Oesophagus	13.8	12.5	8.3	-	-	-	-	-	-	0.1	0.3	0.9	2.4	6.9	15.1	21.1	33.2	51.5	69.1	76.8	100.8	108.6	C15
C16	Stomach	22.7	20.1	13.1	-	-	-	-	0.1	0.1	0.4	1.0	1.4	4.3	7.3	13.5	31.6	44.2	83.2	126.5	134.4	187.0	236.2	C16
C18	Colon	34.6	30.5	19.9	-	-	-	0.1	0.2	0.2	0.7	1.8	2.3	6.3	11.0	22.9	44.2	78.9	118.1	185.2	236.2	279.1	294.0	C18
C19-C21	Rectum	23.4	21.1	14.1	0.3	-	-	-	0.1	0.1	0.3	0.8	2.4	5.3	9.4	18.9	37.5	57.5	96.9	122.9	126.0	168.5	158.3	C19-C21
C25	Pancreas	10.7	9.6	6.4	-	-	-	0.1	-	-	0.1	0.3	0.9	1.7	5.7	8.1	15.6	25.4	41.2	53.6	59.8	81.7	91.9	C25
C32	Larynx	5.9	5.6	3.9	-	-	-	-	-	-	0.1	0.1	0.6	1.9	4.2	7.4	13.6	21.1	23.8	24.6	28.5	25.4	21.7	C32
C33-C34	Lung	81.6	72.0	47.2	-	-	0.1	0.1	0.1	0.2	0.2	1.0	2.9	9.7	24.4	54.0	116.0	180.1	319.5	460.0	555.5	620.1	566.7	C33-C34
C43	Melanoma of skin	7.8	7.5	5.6	-	-	0.1	0.1	0.9	2.0	2.5	3.8	5.2	6.8	10.3	11.5	14.1	20.2	19.8	22.5	27.3	24.9	34.9	C43
C61	Prostate	71.2	60.7	37.7	-	0.1	-	-	0.1	0.1	-	-	0.1	0.4	3.0	15.1	52.7	122.4	252.8	415.9	554.6	704.5	937.5	C61
C62	Testis	5.6	5.4	5.0	0.6	0.4	-	0.2	2.1	10.1	11.9	13.0	11.1	10.3	7.0	3.5	2.7	1.8	1.9	1.6	1.4	3.5	1.9	C62
C64-C66,C68	Kidney	11.8	11.1	7.7	1.2	2.0	0.3	0.2	-	0.3	0.5	1.0	1.9	3.9	8.3	15.5	22.9	33.2	45.3	49.4	52.8	75.9	46.5	C64-C66,C68
C67	Bladder	33.2	29.1	18.9	-	-	0.1	-	0.2	0.4	0.5	0.9	1.9	5.6	9.1	20.0	37.0	70.8	117.7	182.3	222.8	283.0	318.9	C67
C71	Brain	8.0	7.7	6.1	3.3	2.3	3.2	2.0	1.5	1.8	2.0	3.6	3.9	6.0	7.3	12.1	17.5	17.3	23.8	26.1	24.7	18.9	14.4	C71
C82-C85,C91.4,C96	Non-Hodgkin's lymphoma	15.4	14.9	10.8	1.2	1.0	1.6	1.1	1.9	2.1	3.1	6.3	5.7	10.0	11.4	20.6	28.2	35.3	41.7	57.5	70.0	83.1	70.2	C82-C85,C91.4,C96
C88,C90	Multiple myeloma	5.8	5.2	3.4	-	-	-	0.1	-	0.1	0.3	0.1	0.4	1.2	3.1	5.0	9.4	12.0	18.7	30.7	30.9	49.3	50.0	C88,C90
C91-C95 xC91.4	Leukaemia	11.4	10.5	8.1	3.0	6.7	4.2	3.3	3.9	1.6	3.2	2.2	3.6	3.6	5.7	6.6	16.7	19.0	29.1	48.4	47.8	78.2	105.9	C91-C95 xC91.4

Females

ICD10	Site	CR	ESR	WSR	Under 1	1-4	5-9	10-14	15-19	20-24	25-29	30-34	35-39	40-44	45-49	50-54	55-59	60-64	65-69	70-74	75-79	80-84	85 and over	ICD10
C00-C97 xC44	All cancers excluding NMSC	425.6	325.8	232.0	19.2	14.9	7.8	9.4	11.7	26.6	49.6	79.2	131.3	212.7	337.4	525.7	655.3	820.8	1,028.5	1,334.6	1,549.2	1,684.9	1,775.8	C00-C97 xC44
C00-C14	Lip, mouth and pharynx	5.4	4.3	3.1	-	-	-	0.3	0.1	0.6	0.7	1.0	1.5	2.9	4.3	6.2	10.3	12.3	16.0	14.0	19.6	20.5	18.5	C00-C14
C15	Oesophagus	9.4	5.8	3.7	0.3	-	-	-	-	-	-	0.1	0.3	1.2	1.8	5.1	7.8	11.4	23.1	33.7	45.4	59.2	72.8	C15
C16	Stomach	12.4	7.6	4.9	-	-	-	-	-	-	0.1	0.7	1.5	2.4	2.4	5.2	8.7	17.3	28.4	43.4	61.7	79.8	91.1	C16
C18	Colon	35.5	23.0	15.1	0.3	-	-	0.1	-	0.2	0.5	0.9	2.6	5.9	8.6	24.2	38.0	61.3	77.8	126.6	160.0	212.6	243.8	C18
C19-C21	Rectum	17.1	11.8	7.9	0.3	-	-	-	0.7	0.1	0.3	0.8	1.5	3.9	7.5	12.5	24.0	32.2	43.6	55.1	77.7	88.5	100.3	C19-C21
C25	Pancreas	11.3	7.3	4.8	-	-	-	-	-	0.1	0.1	0.6	0.8	1.3	3.0	7.1	10.0	20.8	28.2	40.3	49.3	70.3	74.3	C25
C33-C34	Lung	46.4	33.7	22.8	0.3	0.2	-	0.1	-	0.1	0.2	0.9	2.8	7.8	15.6	38.5	62.7	88.9	158.4	228.4	225.8	191.5	127.1	C33-C34
C43	Melanoma of skin	10.0	8.8	6.9	-	-	0.2	0.1	0.7	3.0	5.7	6.7	7.8	11.8	15.5	14.3	14.5	20.1	22.2	20.1	20.3	25.7	19.2	C43
C50	Breast	124.6	107.0	78.1	1.3	-	-	-	0.2	1.4	9.3	27.5	60.9	109.5	182.1	264.4	259.5	271.5	240.6	303.7	322.2	332.5	348.4	C50
C53	Cervix	10.7	9.6	7.8	-	-	-	0.1	0.2	2.8	10.7	14.9	17.6	14.8	14.8	12.2	12.7	10.8	12.9	21.5	20.7	17.2	17.1	C53
C54	Uterus	15.1	12.6	8.9	-	-	-	-	-	0.1	0.2	0.6	1.8	4.6	9.3	21.7	42.5	48.1	47.8	46.1	47.1	36.1	41.5	C54
C56,C57.0-C57.4	Ovary	23.1	19.0	13.7	0.3	0.1	0.3	0.6	0.7	1.7	3.7	4.2	6.9	10.9	22.7	31.7	47.1	54.7	65.3	73.8	74.1	75.3	53.6	C56,C57.0-C57.4
C64-C66,C68	Kidney	6.8	5.1	3.8	1.9	2.4	0.4	0.1	0.1	0.3	0.2	0.8	1.4	2.0	4.1	6.9	9.0	16.6	21.3	21.8	29.6	23.8	22.2	C64-C66,C68
C67	Bladder	13.4	8.6	5.7	-	0.2	-	-	-	0.1	0.3	0.3	0.6	1.7	3.7	6.9	13.3	18.9	39.4	50.7	55.7	80.0	91.8	C67
C71	Brain	5.6	5.0	4.1	2.8	2.5	2.0	1.7	1.4	1.4	1.9	2.9	2.3	3.7	4.5	6.6	9.7	14.2	15.3	14.3	14.4	12.4	5.7	C71
C82-C85,C91.4,C96	Non-Hodgkin's lymphoma	12.8	10.1	7.3	0.9	0.8	0.5	0.8	0.7	-	2.4	2.0	4.8	5.1	8.5	13.5	18.2	23.2	33.0	45.3	50.9	52.4	45.0	C82-C85,C91.4,C96
C88,C90	Multiple myeloma	5.2	3.5	2.3	-	-	-	-	-	-	-	0.2	0.2	0.2	2.1	2.9	6.1	9.5	14.8	19.1	24.1	29.7	29.8	C88,C90
C91-C95 xC91.4	Leukaemia	8.8	6.5	5.2	1.9	5.8	3.0	2.8	1.9	1.5	1.2	2.7	2.4	3.4	3.8	6.9	8.5	12.4	16.9	23.6	29.5	40.6	53.9	C91-C95 xC91.4
D05	Breast in situ	7.6	7.8	5.8	-	-	-	-	-	-	0.2	1.6	2.9	5.9	11.9	32.8	27.1	27.8	10.8	9.1	5.7	5.0	2.5	D05
D06	Cervix in situ	67.3	67.5	66.0	0.3	-	0.2	-	17.0	170.0	250.1	200.2	129.2	75.6	47.2	30.7	23.5	22.2	9.5	3.2	2.2	1.6	0.7	D06

Appendix B12(a) Directly age-standardised incidence rates* per 100,000 population for selected cancers, England and Wales, 1971-1997 - males

ICD10	Site	1971	1972	1973	1974	1975	1976	1977	1978	1979	1980	1981	1982	1983	1984	1985	1986	1987	1988	1989	1990	1991	1992	1993	1994	1995**	1996**	1997**	ICD10
	All registrations	378.2	384.2	396.4	404.5	406.5	407.9	414.7	411.8	417.8	424.7	435.5	434.3	437.1	436.8	460.2	449.5	455.2	474.5	465.3	465.0	472.0	488.6	480.9	492.7	:	:	:	C00-C97
C00-C97	All cancers	370.8	373.5	381.5	389.9	391.9	391.6	401.3	399.8	403.5	409.4	419.2	418.3	420.8	420.4	444.7	434.3	440.3	456.4	448.3	446.9	452.6	468.4	460.4	470.6	:	:	:	C00-C97 xC44
C00-C97 xC44	All cancers excluding NMSC	332.1	335.1	343.2	350.2	350.9	348.4	356.2	353.4	359.0	363.1	370.8	370.7	371.8	370.5	387.8	374.3	381.0	391.4	383.8	384.5	386.7	398.3	394.2	400.8	388.4	385.6	377.4	C00-C14
C00-C14	Lip, mouth and pharynx	8.7	9.1	8.3	8.5	8.6	8.1	7.8	7.3	7.8	7.8	7.7	8.1	8.0	7.9	8.0	7.8	7.9	8.3	7.8	8.0	8.5	8.9	8.6	9.1	8.9	9.2	8.8	C00
C00	Lip	1.8	1.9	1.8	1.6	1.5	1.4	1.2	0.8	1.2	1.2	0.9	1.1	0.9	0.8	0.8	0.8	0.8	0.8	0.7	0.6	0.5	0.7	0.6	0.7	:	:	:	C01-C02
C01-C02	Tongue	1.4	1.5	1.3	1.5	1.4	1.4	1.4	1.3	1.4	1.4	1.5	1.7	1.4	1.5	1.5	1.5	1.6	1.7	1.7	1.6	1.8	1.8	1.8	2.0	:	:	:	C03-C06
C03-C06	Mouth	1.9	2.0	1.8	2.0	2.0	1.9	1.9	1.9	1.8	1.7	2.0	1.9	2.0	1.5	2.0	2.2	2.1	2.3	2.1	2.1	2.4	2.4	2.1	2.4	:	:	:	C07-C08
C07-C08	Salivary glands	1.0	0.9	1.1	1.0	1.0	1.2	1.1	1.0	0.8	0.7	0.7	0.8	0.8	0.7	0.8	0.6	0.8	0.8	0.7	0.7	0.8	0.9	0.8	0.9	:	:	:	C09-C10
C09-C10	Oropharynx	1.0	1.0	0.9	0.9	0.9	0.7	0.9	1.0	0.9	1.0	0.9	1.1	1.0	0.9	0.9	0.9	1.0	0.9	1.0	1.2	1.1	1.2	1.3	1.2	:	:	:	C11
C11	Nasopharynx	0.6	0.5	0.5	0.6	0.7	0.5	0.4	0.5	0.5	0.6	0.6	0.5	0.6	0.6	0.5	0.5	0.6	0.6	0.5	0.4	0.5	0.6	0.5	0.5	:	:	:	C12-C13
C12-C13	Hypopharynx	0.9	1.1	1.0	0.8	1.0	0.9	0.8	0.7	0.8	0.8	0.9	0.8	1.0	1.0	1.1	0.9	0.9	0.9	0.9	0.9	1.0	1.0	1.0	0.9	:	:	:	C14
C14	Pharynx unspecified	0.2	0.4	0.4	0.3	0.3	0.3	0.4	0.3	0.4	0.4	0.4	0.4	0.4	0.4	0.4	0.4	0.4	0.4	0.4	0.5	0.4	0.4	0.4	0.5	:	:	:	C15
C15	Oesophagus	7.7	7.9	8.0	8.2	8.6	8.9	8.6	9.0	8.1	8.9	9.1	9.4	9.2	9.6	10.1	10.5	11.0	11.5	11.2	11.6	11.7	12.4	12.5	12.8	12.1	12.5	12.5	C16
C16	Stomach	32.0	31.9	32.1	31.0	31.5	30.3	30.3	29.8	28.9	30.1	29.3	28.4	28.9	27.3	27.7	27.0	25.5	25.8	24.7	23.5	22.6	22.9	21.3	21.6	21.1	20.3	20.1	C17
C17	Small intestine	0.8	0.7	0.7	0.7	0.8	0.8	0.7	0.7	0.6	0.6	0.7	0.6	0.7	0.7	0.7	0.7	0.7	0.7	0.7	0.9	0.8	1.0	0.8	0.8	:	:	:	C18-C21
C18-C21	Colorectal	43.1	43.0	45.0	45.4	46.0	45.4	46.4	45.6	46.5	47.2	48.7	48.5	49.3	49.2	50.7	48.5	50.3	51.3	51.7	51.9	51.6	54.6	53.9	53.0	50.6	53.2	51.6	C18
C18	Colon	23.4	23.5	24.5	25.2	25.3	25.4	25.6	25.4	25.5	25.9	26.7	27.1	27.5	27.3	28.0	27.2	28.8	29.2	29.8	29.8	29.8	31.2	31.6	31.0	30.4	31.0	30.5	C19-C21
C19-C21	Rectum	19.9	20.0	21.1	20.6	21.1	20.7	21.1	20.5	21.1	21.3	22.0	21.4	21.8	21.9	22.7	21.4	21.5	22.2	21.9	22.1	21.9	23.5	22.4	22.0	20.3	22.2	21.1	C22
C22	Liver	2.0	2.0	1.9	2.0	2.1	2.1	2.1	2.0	2.5	2.4	2.6	2.4	2.5	2.4	2.7	2.6	2.9	2.9	2.8	3.0	3.1	3.4	3.3	3.4	:	:	:	C23-C24
C23-C24	Gallbladder	2.0	2.1	1.9	1.8	1.7	2.1	2.0	2.1	1.9	2.0	2.0	1.9	1.9	1.9	1.9	1.9	1.7	2.1	1.8	1.8	2.1	2.1	1.7	1.6	:	:	:	C25
C25	Pancreas	11.7	11.7	11.7	12.2	12.3	11.9	12.2	12.5	13.0	12.2	12.3	12.1	12.1	11.9	12.0	11.6	11.3	11.5	11.0	11.3	11.2	11.0	10.7	10.2	10.0	10.0	9.6	C26
C26	Digestive organs etc. unspecified	0.2	0.2	0.3	0.2	0.1	0.3	0.3	0.3	0.6	0.7	0.7	0.6	0.7	0.6	0.8	0.7	0.7	0.9	0.9	0.9	0.8	0.9	0.8	0.9	:	:	:	C30-C31
C30-C31	Nasal cavities	1.2	1.1	1.3	1.2	1.1	1.1	1.1	1.0	1.0	0.8	0.8	0.9	0.8	0.8	0.8	0.9	0.9	0.9	0.9	0.9	0.8	0.9	0.8	0.7	:	:	:	C32
C32	Larynx	5.5	5.7	5.8	6.1	5.8	5.9	5.8	5.9	5.8	5.8	6.0	6.1	5.8	6.2	6.3	6.2	6.2	6.7	6.3	6.3	6.1	6.3	6.0	6.4	5.9	5.8	5.6	C33-C34
C33-C34	Lung	108.3	107.2	110.4	112.7	111.0	110.5	112.7	110.0	111.0	109.7	110.0	109.0	106.8	103.3	107.5	100.1	98.0	98.0	93.8	91.5	90.1	89.7	83.7	82.6	78.1	77.1	72.0	C37-C39, C45
C37-C39, C45	Respiratory system	0.7	0.8	0.7	0.8	1.0	1.0	1.2	1.2	1.3	1.5	1.5	1.7	2.0	2.1	2.3	2.7	2.6	2.7	3.3	3.4	3.7	4.0	4.0	4.2	:	:	:	C40-C41
C40-C41	Bone	1.2	1.0	1.0	1.1	1.1	1.0	1.0	1.0	1.0	0.9	1.0	0.9	1.0	1.0	0.9	0.9	0.9	0.9	0.9	1.0	1.0	0.9	0.9	0.9	:	:	:	C43
C43	Melanoma of skin	1.7	2.0	2.1	2.3	2.4	2.5	2.5	2.6	2.8	3.0	3.2	3.3	3.6	3.5	4.6	4.7	5.3	6.3	5.7	6.0	5.9	6.4	7.4	7.3	7.2	7.1	7.5	C44
C44	Non-melanoma of skin	38.9	38.8	38.9	40.2	41.4	43.7	45.3	46.6	44.7	46.5	48.6	47.8	49.4	50.2	57.1	60.3	59.5	65.4	64.7	62.7	66.3	70.3	66.2	69.7	:	:	:	C47, C49
C47, C49	Connective tissue	1.5	1.4	1.7	1.8	1.8	2.0	1.7	1.9	1.9	1.9	2.0	1.8	1.9	1.8	1.9	1.8	1.9	2.3	2.2	2.2	2.4	2.7	2.2	2.2	:	:	:	C48
C48	Retroperitoneum and peritoneum	0.5	0.7	0.6	0.6	0.5	0.5	0.5	0.6	0.6	0.5	0.5	0.5	0.6	0.6	0.5	0.5	0.5	0.6	0.5	0.5	0.4	0.4	0.5	0.7	:	:	:	C50
C50	Breast	0.7	0.8	0.8	0.8	0.9	0.9	1.0	1.0	0.7	0.7	0.8	0.8	0.7	0.8	0.8	0.7	0.8	0.7	0.8	0.8	0.8	0.7	0.9	0.7	:	:	:	C60, C63
C60, C63	Penis	1.3	1.2	1.3	1.2	1.2	1.2	1.3	1.2	1.4	1.5	1.4	1.5	1.4	1.3	1.5	1.3	1.4	1.4	1.4	1.4	1.4	1.4	1.4	1.3	:	:	:	C61
C61	Prostate	29.2	29.6	31.7	32.8	32.8	33.8	33.9	32.8	35.3	36.5	37.4	38.4	39.2	39.8	42.4	42.2	43.6	45.8	46.0	47.4	49.8	54.5	58.9	66.1	63.2	64.4	60.7	C62
C62	Testis	2.9	2.9	3.0	3.3	3.2	3.2	3.3	3.3	3.3	3.7	3.9	3.7	3.7	3.9	4.3	4.5	4.7	5.0	5.1	4.8	5.2	5.3	5.2	5.2	5.6	5.7	5.4	C64-C66, C68
C64-C66, C68	Kidney	5.8	6.0	6.2	6.5	7.1	6.6	7.2	6.9	7.0	7.1	7.7	8.1	8.2	8.4	8.6	8.3	9.2	9.2	10.1	9.8	9.8	10.5	10.7	11.1	11.5	11.2	11.1	C67
C67	Bladder	23.5	24.5	24.3	23.5	24.0	23.8	25.1	25.6	26.3	27.1	28.0	27.7	28.3	28.8	30.1	29.0	29.9	31.4	30.4	30.8	30.4	30.8	30.7	30.2	30.2	29.7	29.1	C69
C69	Eye	0.8	0.7	0.9	0.9	0.8	0.9	0.8	0.8	0.7	0.7	0.8	0.7	0.8	0.8	0.8	0.9	0.8	0.9	0.8	1.0	1.0	1.0	0.9	0.8	:	:	:	C70, C72
C70, C72	Other nervous system	0.7	0.8	0.7	0.8	0.7	0.7	0.8	0.7	0.7	0.2	0.3	0.2	0.2	0.3	0.3	0.2	0.3	0.3	0.3	0.4	0.3	0.3	0.3	0.3	:	:	:	C71
C71	Brain	5.8	5.6	5.2	5.2	5.3	5.2	5.4	5.4	6.1	6.2	6.4	6.2	6.5	6.7	7.0	6.8	7.2	7.2	7.4	7.4	7.4	8.0	7.6	7.3	7.5	7.2	7.7	C73
C73	Thyroid	0.8	0.8	0.9	0.8	0.8	0.9	0.9	0.8	0.8	0.9	0.9	0.9	0.9	0.9	1.0	0.9	0.9	0.9	1.0	0.9	0.9	0.9	1.0	1.0	:	:	:	C74-C75
C74-C75	Other endocrine glands	0.6	0.8	0.4	0.5	0.5	0.4	0.5	0.5	0.4	0.5	0.4	0.3	0.3	0.4	0.4	0.4	0.5	0.6	0.5	0.7	0.7	0.6	0.6	0.5	:	:	:	C81
C81	Hodgkin's disease	3.5	4.0	3.8	3.8	3.7	3.5	3.8	3.7	3.2	3.3	3.4	3.1	3.0	2.8	3.1	3.0	2.7	2.6	2.7	2.7	2.5	2.8	2.7	2.6	:	:	:	C82-C85, C91,4, C96
C82-C85, C91,4, C96	Non-Hodgkin's lymphoma	5.2	5.6	5.9	6.0	6.4	6.6	6.9	7.1	7.7	7.9	8.4	8.7	9.0	9.6	10.4	10.8	11.8	12.5	12.5	13.2	13.6	14.1	14.7	14.7	14.1	14.4	14.9	C88, C90
C88, C90	Multiple myeloma	2.9	3.0	3.1	3.4	3.5	3.6	3.9	4.1	4.1	4.2	4.3	4.2	4.8	4.6	5.3	5.0	5.3	5.5	5.1	5.1	5.4	5.6	5.3	5.2	5.8	5.5	5.2	C91-C95 x C91.4
C91-C95 x C91.4	Leukaemia	8.3	8.7	8.9	9.7	9.5	9.3	9.3	9.7	9.9	9.6	10.1	10.2	10.3	10.2	11.1	10.0	10.5	11.3	10.9	10.9	11.4	10.9	11.0	11.1	11.6	10.8	10.5	C91 x C91.4
C91 x C91.4	Lymphoid leukaemia	3.8	4.3	4.1	4.6	4.3	4.5	4.3	4.5	4.5	4.5	4.7	4.6	5.2	5.0	5.2	4.7	5.0	5.6	5.1	5.3	5.3	5.0	5.5	5.5	5.8	5.5	5.1	C92
C92	Myeloid leukaemia	3.3	3.8	3.8	4.0	4.3	4.5	4.0	4.1	4.3	4.0	4.4	4.4	4.1	4.8	4.6	4.5	4.7	4.7	4.9	4.8	5.2	5.1	5.1	5.5	5.8	5.7	5.1	C93
C93	Monocytic leukaemia	0.3	0.3	0.3	0.3	0.3	0.3	0.3	0.3	0.3	0.2	0.3	0.3	0.3	0.2	0.2	0.2	0.2	0.2	0.1	0.1	0.2	0.2	0.1	0.1	:	:	:	C94-C95
C94-C95	Other leukaemia	1.9	1.9	2.1	2.3	2.0	1.9	2.3	2.2	0.9	0.9	0.9	1.1	0.9	0.8	0.9	0.7	0.2	0.8	0.7	0.6	0.7	0.6	0.6	0.5	:	:	:	C76-C80
C76-C80	Other and unspecified	12.3	13.0	14.3	15.4	15.1	15.5	16.2	16.7	16.9	17.6	18.8	20.1	19.1	20.6	21.4	21.2	23.1	22.7	22.7	23.0	23.4	22.4	23.5	24.1	:	:	:	

* Directly age-standardised to the European standard population
** Provisional

Appendix B12(b) Directly age-standardised incidence rates* per 100,000 population for selected cancers, England and Wales, 1971-1997 - females

ICD10	Site	1971	1972	1973	1974	1975	1976	1977	1978	1979	1980	1981	1982	1983	1984	1985	1986	1987	1988	1989	1990	1991	1992	1993	1994	1995**	1996**	1997**	ICD10
C00-C97	All registrations	281.7	291.8	300.1	314.2	312.8	316.8	321.5	321.9	331.5	342.4	357.1	359.2	366.3	376.8	412.8	415.1	436.0	453.9	452.4	465.0	468.9	480.5	463.7	477.3	:	:	:	C00-C97
C00-C97	All cancers	266.6	274.2	278.6	291.1	289.4	289.4	294.7	296.3	297.0	304.9	313.5	314.9	316.3	318.5	339.2	335.3	347.9	357.0	359.0	357.7	365.3	378.9	364.5	372.6	:	:	:	C00-C97
C00-C97 xC44	All cancers excluding NMSC	243.3	249.5	254.1	266.0	263.8	262.1	266.8	267.5	268.9	276.3	282.6	284.7	284.4	286.6	304.6	297.5	308.8	314.4	316.1	316.5	322.7	331.7	322.0	327.0	319.0	322.1	325.8	C00-C97 xC44
C00-C14	Lip, mouth and pharynx	4.0	3.9	3.8	4.2	4.2	3.8	3.7	3.6	3.6	3.4	3.7	3.5	3.6	3.7	3.6	3.3	3.6	3.6	3.5	3.7	3.5	4.0	3.7	3.8	3.9	4.0	4.3	C00-C14
C00	Lip	0.2	0.2	0.2	0.2	0.1	0.1	0.1	0.1	0.1	0.1	0.2	0.2	0.1	0.2	0.2	0.1	0.2	0.2	0.2	0.1	0.1	0.2	0.2	0.2	:	:	:	C00
C01-C02	Tongue	0.6	0.7	0.7	0.8	0.8	0.8	0.7	0.6	0.8	0.7	0.7	0.7	0.7	0.8	0.8	0.7	0.7	0.8	0.7	0.8	0.8	0.9	0.9	0.8	:	:	:	C01-C02
C03-C06	Mouth	0.8	0.8	0.7	0.8	0.8	0.8	0.8	0.8	0.8	0.7	0.9	0.9	0.9	0.9	0.8	0.8	0.9	0.9	1.0	1.1	0.9	1.1	1.0	1.1	:	:	:	C03-C06
C07-C08	Salivary glands	1.0	0.9	0.9	0.9	0.9	0.8	0.8	0.7	0.6	0.5	0.6	0.5	0.5	0.6	0.5	0.6	0.5	0.5	0.5	0.5	0.5	0.7	0.5	0.6	:	:	:	C07-C08
C09-C10	Oropharynx	0.4	0.3	0.3	0.3	0.3	0.3	0.2	0.3	0.3	0.3	0.3	0.4	0.3	0.4	0.3	0.3	0.3	0.3	0.4	0.4	0.4	0.4	0.4	0.4	:	:	:	C09-C10
C11	Nasopharynx	0.2	0.2	0.2	0.3	0.2	0.2	0.2	0.2	0.3	0.3	0.2	0.3	0.2	0.3	0.2	0.2	0.3	0.2	0.2	0.2	0.2	0.2	0.2	0.2	:	:	:	C11
C12-C13	Hypopharynx	0.8	0.7	0.7	0.7	0.7	0.7	0.6	0.7	0.6	0.5	0.6	0.5	0.5	0.5	0.5	0.4	0.4	0.4	0.3	0.4	0.4	0.5	0.4	0.4	:	:	:	C12-C13
C14	Pharynx unspecified	0.1	0.1	0.1	0.2	0.2	0.1	0.2	0.2	0.2	0.2	0.2	0.2	0.2	0.2	0.2	0.2	0.2	0.2	0.3	0.4	0.4	0.5	0.4	0.4	:	:	:	C14
C15	Oesophagus	4.2	4.2	4.4	4.5	4.7	4.6	4.8	4.8	4.6	4.6	5.0	5.0	4.9	5.0	5.0	4.9	5.2	5.3	5.5	5.4	5.4	5.7	5.5	5.9	5.6	6.0	5.8	C15
C16	Stomach	15.1	14.9	14.7	14.6	14.9	13.9	13.8	13.7	13.2	12.6	12.8	12.3	11.9	11.7	11.5	10.6	11.2	10.2	10.0	9.3	9.2	8.9	8.4	8.3	7.9	7.4	7.6	C16
C17	Small intestine	0.5	0.5	0.5	0.4	0.5	0.5	0.5	0.5	0.6	0.4	0.5	0.4	0.5	0.5	0.5	0.6	0.5	0.5	0.5	0.5	0.5	0.6	0.6	0.6	0.6	0.6	0.6	C17
C18-C21	Colorectal	33.0	33.5	34.5	34.8	35.7	35.1	34.9	35.6	34.5	34.8	36.5	36.0	36.1	35.2	36.3	35.7	36.0	36.3	36.0	35.5	35.6	37.4	35.5	35.6	33.7	36.4	34.8	C18-C21
C18	Colon	22.0	22.2	23.0	23.3	23.7	23.4	23.1	23.2	22.5	22.4	24.0	23.4	23.3	22.8	23.8	23.5	23.8	24.1	23.6	23.6	23.8	25.0	23.5	23.8	23.0	24.4	23.0	C18
C19-C21	Rectum	11.2	11.7	12.0	11.8	12.3	12.1	11.9	12.5	12.0	12.4	12.5	12.6	12.7	12.3	12.5	12.2	12.2	12.3	12.4	11.9	11.8	12.4	12.1	11.8	10.7	12.1	11.8	C19-C21
C22	Liver	0.9	0.9	0.9	1.0	0.9	0.8	0.9	0.9	1.0	1.2	1.2	1.2	1.2	1.2	1.3	1.3	1.4	1.4	1.3	1.3	1.5	1.6	1.6	1.6	:	:	:	C22
C23-C24	Gallbladder	2.2	2.1	2.1	2.2	2.0	2.1	2.1	2.1	2.0	2.1	2.1	1.9	2.0	2.1	2.1	1.9	2.0	1.8	2.0	1.9	1.8	1.9	1.8	1.8	:	:	:	C23-C24
C25	Pancreas	6.9	7.0	7.0	7.5	7.4	7.7	7.5	7.5	7.8	7.6	7.7	7.9	7.8	7.6	8.2	7.8	8.2	8.0	8.3	8.0	8.0	7.9	7.8	7.7	7.2	7.8	7.3	C25
C26	Digestive organs etc. unspecified	0.1	0.1	0.1	0.2	0.1	0.2	0.2	0.2	0.4	0.5	0.5	0.6	0.5	0.5	0.6	0.5	0.6	0.7	0.7	0.7	0.5	0.6	0.6	0.6	:	:	:	C26
C30-C31	Nasal cavities	0.7	0.6	0.7	0.7	0.6	0.6	0.6	0.5	0.5	0.4	0.4	0.6	0.5	0.5	0.5	0.5	0.4	0.4	0.5	0.5	0.4	0.5	0.5	0.4	:	:	:	C30-C31
C32	Larynx	0.8	0.9	0.9	1.0	1.1	1.0	1.0	1.0	1.0	1.0	1.0	1.0	0.9	1.0	1.0	1.1	1.3	1.2	1.1	1.2	1.1	1.1	1.2	1.1	1.2	1.1	0.0	C32
C33-C34	Lung	19.1	19.3	20.3	22.2	22.5	23.1	23.4	24.9	26.0	27.1	28.6	29.1	29.3	30.0	32.0	31.2	32.9	33.4	32.8	32.7	32.8	33.9	33.2	33.7	34.4	35.6	33.7	C33-C34
C37-C39, C45	Respiratory system	0.2	0.2	0.2	0.2	0.2	0.3	0.3	0.4	0.6	0.5	0.5	0.5	0.6	0.5	0.5	0.8	0.6	0.6	0.7	0.8	0.8	1.3	0.9	0.9	0.5	0.5	0.5	C37-C39, C45
C40-C41	Bone	0.7	0.8	0.7	0.7	0.7	0.7	0.6	0.7	0.7	0.6	0.7	0.6	0.6	0.7	0.7	0.8	0.7	0.7	0.7	0.7	0.6	0.6	0.6	0.5	:	:	:	C40-C41
C43	Melanoma of skin	3.1	3.4	3.7	3.9	3.9	4.2	4.2	4.4	4.8	5.0	5.4	5.7	6.1	6.1	7.6	7.4	8.4	9.0	8.4	7.5	7.7	8.4	9.5	9.4	9.1	8.5	8.8	C43
C44	Non-melanoma of skin	23.5	24.9	24.7	25.4	25.9	27.5	28.1	29.0	28.2	28.7	31.0	30.3	32.0	32.0	34.8	38.0	39.2	43.0	43.0	41.3	42.9	47.4	42.5	45.6	:	:	:	C44
C47, C49	Connective tissue	1.6	1.6	1.6	1.3	1.3	1.4	1.2	1.3	1.4	1.5	1.5	1.3	1.4	1.4	1.5	1.5	1.6	1.7	1.6	1.7	1.7	2.0	1.6	1.6	:	:	:	C47, C49
C48	Retroperitoneum and peritoneum	0.6	0.7	0.6	0.5	0.5	0.6	0.5	0.5	0.5	0.5	0.6	0.5	0.5	0.5	0.5	0.5	0.4	0.4	0.4	0.4	0.3	0.4	0.4	0.4	:	:	:	C48
C50	Breast	66.9	71.9	73.0	77.6	74.6	73.0	74.7	76.2	74.4	77.8	77.5	79.6	78.0	78.6	85.7	85.1	88.3	89.9	95.2	98.4	105.1	107.6	101.4	104.7	101.8	104.0	107.0	C50
C51-C52	Other female genital	3.0	3.1	3.0	3.2	3.2	3.2	3.2	3.0	2.9	3.0	2.7	3.0	2.8	3.0	2.9	2.8	2.8	2.9	2.8	2.8	2.5	2.9	2.8	2.8	:	:	:	C51-C52
C53	Cervix	15.9	15.1	15.3	15.0	15.0	14.6	15.0	14.4	14.7	15.3	15.4	14.7	14.8	15.4	16.4	16.1	15.8	16.5	15.1	15.4	12.8	12.0	11.6	11.0	10.3	9.7	9.6	C53
C54	Uterus	10.9	11.0	11.4	11.5	11.7	12.2	11.9	11.6	11.7	11.8	11.9	11.9	11.9	12.0	12.4	12.0	11.7	12.3	12.2	12.3	12.6	12.5	12.7	12.8	12.3	12.5	12.6	C54
C55	Uterus unspecified	:	1.0	:	:	:	:	:	:	1.4	1.5	1.4	1.5	1.5	1.3	1.2	1.2	1.2	1.3	1.2	1.1	1.0	1.3	0.9	0.9	5.3	:	5.0	C55
C56-C57	Ovary	14.5	14.6	14.0	14.9	14.6	14.8	15.4	14.9	15.1	15.8	15.8	15.8	16.3	16.2	16.9	16.6	17.4	17.2	17.1	16.7	17.4	17.5	17.1	17.0	18.3	18.2	19.0	C56-C57
C58	Placenta	0.1	0.1	0.1	0.1	0.1	0.1	0.1	0.1	0.1	0.1	0.1	0.1	0.1	0.1	0.1	0.0	0.0	0.0	0.0	0.0	0.0	0.0	0.0	0.0	:	:	:	C58
C64-C66, C68	Kidney	2.9	2.8	3.0	2.9	3.0	3.1	3.2	3.3	3.3	3.4	3.5	3.6	3.8	3.9	4.0	4.0	4.2	4.3	4.5	4.7	4.9	4.9	4.8	5.2	5.1	5.3	5.1	C64-C66, C68
C67	Bladder	5.7	5.8	6.0	6.2	5.9	6.1	6.5	6.3	6.9	6.9	7.6	7.4	7.6	7.6	8.2	8.3	8.1	8.7	8.8	8.1	8.1	8.7	8.7	8.0	8.2	8.1	8.6	C67
C69	Eye	0.6	0.7	0.8	0.7	0.7	0.7	0.6	0.6	0.7	0.6	0.7	0.7	0.6	0.7	0.6	0.6	0.8	0.7	0.7	1.0	0.8	0.8	0.7	0.6	:	:	:	C69
C70, C72	Other nervous system	0.8	0.7	0.6	0.7	0.6	0.7	0.6	0.6	0.2	0.3	0.3	0.2	0.3	0.3	0.3	0.4	0.2	0.2	0.4	0.4	0.4	0.3	0.3	0.4	:	:	:	C70, C72
C71	Brain	3.8	3.6	3.4	3.7	3.4	3.5	3.7	3.5	4.1	4.1	4.2	4.2	4.4	4.6	4.6	4.4	4.6	5.0	5.0	4.8	4.8	5.4	5.3	5.1	5.3	5.3	5.0	C71
C73	Thyroid	1.6	1.6	1.7	1.9	1.9	2.1	2.0	1.7	1.9	2.1	2.1	2.2	2.1	2.0	1.8	2.0	2.0	2.1	2.1	2.3	2.3	2.4	2.3	2.7	:	:	:	C73
C74-C75	Other endocrine glands	0.7	0.6	0.4	0.5	0.4	0.4	0.4	0.4	0.5	0.4	0.4	0.3	0.6	0.5	0.5	0.4	0.5	0.6	0.5	0.5	0.6	0.4	0.4	0.4	:	:	:	C74-C75
C81	Hodgkin's disease	2.0	2.2	2.1	2.2	2.1	2.1	2.0	1.9	1.9	2.1	1.9	1.8	1.8	1.9	2.0	1.8	1.8	1.9	1.7	1.8	1.6	1.9	1.8	1.9	:	:	:	C81
C82-C85, C91.4, C96	Non-Hodgkin's lymphoma	3.5	3.7	4.0	4.2	4.7	4.5	4.7	4.6	5.0	5.3	5.7	5.9	6.3	6.5	7.3	7.3	7.9	8.7	8.6	8.4	9.1	9.5	9.5	9.9	9.6	9.9	10.1	C82-C85, C91.4, C96
C88, C90	Multiple myeloma	2.1	2.2	2.2	2.7	2.6	2.5	2.8	2.7	2.7	2.9	2.9	3.1	3.2	3.2	3.6	3.5	3.6	3.5	3.5	3.5	3.4	3.6	3.6	3.7	3.4	3.4	3.5	C88, C90
C91-C95 x C91.4	Leukaemia	4.9	5.3	5.7	5.8	5.9	6.0	6.2	5.8	5.8	5.8	6.1	6.1	6.6	6.4	6.8	6.3	6.8	6.9	6.8	6.6	6.5	6.6	6.8	7.0	6.8	7.0	6.5	C91-C95 x C91.4
C91 x C91.4	Lymphoid leukaemia	1.9	2.0	2.2	2.2	2.4	2.3	2.4	2.3	2.2	2.5	2.6	2.6	2.8	2.8	2.9	2.6	2.8	3.0	2.8	2.8	2.8	2.7	2.9	3.1	:	:	:	C91 x C91.4
C92	Myeloid leukaemia	2.4	2.7	2.8	2.9	3.0	3.1	2.9	2.8	2.9	2.6	2.9	2.8	2.8	3.1	3.3	3.2	3.4	3.4	3.4	3.4	3.3	3.5	3.4	3.4	:	:	:	C92
C93	Monocytic leukaemia	0.2	0.2	0.1	0.2	0.2	0.2	0.2	0.2	0.2	0.1	0.1	0.1	0.2	0.1	0.1	0.2	0.1	0.1	0.1	0.1	0.0	0.1	0.1	0.0	:	:	:	C93
C94-C95	Other leukaemia	1.0	1.1	1.1	1.3	1.1	1.1	1.4	1.3	0.5	0.6	0.6	0.6	0.6	0.5	0.5	0.4	0.5	0.5	0.4	0.4	0.3	0.4	0.4	0.4	:	:	:	C94-C95
C76-C80	Other and unspecified	8.9	9.5	10.3	11.3	11.4	11.4	11.8	12.1	12.9	13.5	14.0	14.8	14.8	14.9	16.0	15.6	16.5	16.4	16.0	16.6	17.3	16.9	18.0	19.0	:	:	:	C76-C80
D05	Breast in situ	:	:	:	:	:	:	:	:	3.2	3.3	2.4	2.3	2.8	3.6	3.6	3.8	3.6	3.8	3.8	4.6	6.2	6.9	6.7	6.9	6.9	7.0	7.8	D05
D06	Cervix in situ	10.1	10.5	11.7	13.2	14.1	16.1	17.2	17.2	18.8	20.6	25.0	24.8	25.9	38.3	53.1	58.5	66.4	71.3	67.6	80.1	73.3	69.7	67.9	73.5	77.1	73.9	67.5	D06

* Directly age-standardised to the European standard population
** Provisional

Appendix C1(a) Number of deaths, England and Wales, 1999 - males

ICD10	Site	All ages	Under 1	1-4	5-9	10-14	15-19	20-24	25-29	30-34	35-39	40-44	45-49	50-54	55-59	60-64	65-69	70-74	75-79	80-84	85 and over	ICD10
C00-C97	All cancers	68,968	5	56	62	45	76	96	134	232	370	640	1,353	2,908	4,212	6,399	9,282	12,202	13,834	8,852	8,210	C00-C97
C00-C97 xC44	All cancers excluding NMSC	68,756	5	56	62	45	76	96	132	231	369	638	1,349	2,904	4,204	6,390	9,265	12,170	13,800	8,803	8,161	C00-C97 xC44
C00-C14	Lip, mouth and pharynx	1,075	-	-	-	1	2	1	3	8	13	30	80	118	130	132	139	137	127	71	83	C00-C14
C00	Lip	11	-	-	-	-	-	-	-	-	-	-	-	-	-	-	-	2	2	1	3	C00
C01-C02	Tongue	234	-	-	-	-	-	-	2	1	3	9	18	29	36	26	27	31	23	15	14	C01-C02
C03-C06	Mouth	238	-	-	-	-	2	-	-	1	4	9	18	28	13	26	40	29	34	12	22	C03-C06
C07-C08	Salivary glands	89	-	-	-	-	-	-	-	1	-	-	3	2	12	4	7	17	18	12	13	C07-C08
C09-C10	Oropharynx	206	-	-	-	-	-	-	-	1	3	2	17	36	35	24	28	20	21	11	8	C09-C10
C11	Nasopharynx	87	-	-	-	1	-	-	-	1	1	3	6	11	11	14	9	14	9	3	3	C11
C12-C13	Hypopharynx	88	-	-	-	-	-	1	-	1	1	2	7	5	15	15	13	11	6	5	7	C12-C13
C14	Pharynx unspecified	122	-	-	-	-	-	-	1	2	1	5	10	5	8	23	15	13	14	12	13	C14
C15	Oesophagus	3,716	-	-	-	-	-	-	3	8	15	41	107	229	298	407	522	657	708	401	320	C15
C16	Stomach	3,814	-	-	-	-	1	1	5	9	11	34	59	119	215	317	553	695	782	513	500	C16
C17	Small intestine	124	-	-	-	-	-	-	-	-	2	1	5	10	12	10	21	18	22	12	11	C17
C18-C21	Colorectal	7,450	-	-	-	1	1	3	9	13	22	54	127	298	475	696	1,040	1,343	1,467	962	937	C18-C21
C18	Colon	4,781	-	-	-	-	1	3	7	13	16	31	81	181	268	447	654	871	963	613	632	C18
C19-C21	Rectum	2,669	-	-	-	-	-	-	2	3	6	23	46	117	207	249	386	472	504	349	305	C19-C21
C22	Liver	1,108	-	-	-	2	2	-	3	5	7	15	37	56	84	105	174	180	232	115	91	C22
C23-C24	Gallbladder	197	-	1	-	-	-	-	-	-	3	2	3	5	11	16	27	37	39	29	25	C23-C24
C25	Pancreas	2,791	-	-	-	-	-	1	-	3	18	30	61	157	211	320	449	462	513	330	236	C25
C26	Digestive organs etc. unspecified	810	-	-	-	1	-	-	1	-	4	3	19	35	38	68	114	163	137	125	101	C26
C30-C31	Nasal cavities	60	-	-	-	-	-	-	-	-	-	3	5	6	9	9	8	6	7	4	2	C30-C31
C32	Larynx	578	-	-	-	-	-	-	-	2	2	6	12	35	73	63	73	106	90	63	53	C32
C33-C34	Lung	18,255	-	-	-	-	1	-	2	7	43	102	267	682	1,143	1,861	2,767	3,795	3,931	2,129	1,525	C33-C34
C37-C39,C45	Respiratory system	465	-	-	-	-	-	2	2	3	2	2	8	21	44	64	75	90	94	33	25	C37-C39,C45
C40-C41	Bone	125	-	-	2	9	12	-	-	8	2	2	7	8	6	3	5	22	12	3	7	C40-C41
C43	Melanoma of skin	761	-	-	-	2	2	4	10	21	27	33	50	77	65	86	94	85	92	67	48	C43
C44	Non-melanoma of skin	212	-	-	-	-	-	-	2	1	1	2	4	4	8	9	17	32	34	49	49	C44
C47,C49	Connective tissue	320	1	3	2	3	5	4	6	12	7	11	15	23	21	26	35	50	35	31	30	C47,C49
C48	Retroperitoneum and peritoneum	88	-	-	-	-	-	-	-	1	1	1	1	7	10	19	16	11	10	5	4	C48
C50	Breast	65	-	-	-	-	-	-	-	-	-	3	3	5	5	5	5	8	15	13	11	C50
C60,C63	Penis	91	-	-	-	-	-	1	-	-	1	-	3	3	6	12	5	5	8	18	15	C60,C63
C61	Prostate	8,496	-	-	-	-	-	-	-	-	-	-	9	67	166	394	763	1,298	2,032	1,690	2,074	C61
C62	Testis	70	-	1	-	-	-	3	7	7	7	9	5	8	4	2	4	2	2	3	-	C62
C64-C66,C68	Kidney	1,684	-	1	1	-	-	-	5	5	3	6	27	64	121	138	224	277	283	264	147	C64-C66,C68
C67	Bladder	2,845	-	-	-	1	-	1	-	-	1	8	16	48	103	194	320	507	652	440	552	C67
C69	Eye	38	-	-	-	1	-	-	-	-	-	-	-	1	-	3	10	6	7	1	3	C69
C70,C72	Other nervous system	20	-	3	-	-	-	-	-	-	1	-	2	-	1	3	-	2	6	3	3	C70,C72
C71	Brain	1,621	1	17	26	8	8	18	13	37	51	63	120	193	176	215	215	216	148	64	29	C71
C73	Thyroid	106	-	-	-	-	-	-	-	2	2	2	4	7	7	11	9	20	18	12	13	C73
C74-C75	Other endocrine glands	79	1	13	8	-	-	1	-	3	3	5	2	5	5	7	4	6	8	3	4	C74-C75
C81	Hodgkin's disease	139	-	-	-	7	7	4	7	-	5	4	4	8	11	-	15	14	20	9	5	C81
C82-C85,C91.4,C96	Non-Hodgkin's lymphoma	2,087	-	-	-	4	9	12	15	24	34	48	75	134	163	225	264	323	351	216	188	C82-C85,C91.4,C96
C88,C90	Multiple myeloma	1,098	-	-	-	-	-	-	1	5	10	23	42	64	97	157	202	232	150	115	115	C88,C90
C91-C95 xC91.4	Leukaemia	1,925	2	17	18	26	22	22	33	27	35	35	93	99	145	247	296	331	223	235	235	C91-C95 xC91.4
C91 xC91.4	Lymphoid leukaemia	652	1	11	14	16	14	7	11	6	8	24	28	51	80	96	101	91	75	75	75	C91 xC91.4
C92	Myeloid leukaemia	1,114	1	6	3	8	7	14	18	20	24	27	65	65	86	150	179	194	104	134	134	C92
C93	Monocytic leukaemia	7	-	-	-	-	-	-	-	-	-	-	-	2	-	-	-	4	-	-	-	C93
C94-C95	Other leukaemias	152	-	1	1	2	1	1	4	1	3	4	6	6	17	21	32	28	25	25	25	C94-C95
C76-C80	Other and unspecified	6,826	1	1	1	2	7	7	9	12	34	53	127	292	412	657	872	1,150	1,431	946	819	C76-C80

Appendix C1(b) Number of deaths, England and Wales, 1999 - females

ICD10	Site	All ages	Under 1	1-4	5-9	10-14	15-19	20-24	25-29	30-34	35-39	40-44	45-49	50-54	55-59	60-64	65-69	70-74	75-79	80-84	85 and over	ICD10
C00-C97	All cancers	64,097	8	32	47	41	43	62	135	308	611	1,003	1,716	3,008	3,678	5,220	6,851	9,442	11,560	8,444	11,888	C00-C97
C00-C97 xC44	All cancers excluding NMSC	63,940	8	32	47	41	43	62	135	308	611	1,003	1,713	3,003	3,674	5,217	6,847	9,434	11,532	8,421	11,809	C00-C97 xC44
C00-C14	Lip, mouth and pharynx	591	-	1	-	-	2	-	2	7	11	10	18	37	51	51	47	72	83	65	134	C00-C14
C00	Lip	11																1	2	1	7	C00
C01-C02	Tongue	135							1	2	3	1	2	5	9	11	18	13	16	21	34	C01-C02
C03-C06	Mouth	165				1				1	3	5	4	11	12	12	9	19	26	22	40	C03-C06
C07-C08	Salivary glands	71							1		2		4	4	5	4	4	9	15	7	19	C07-C08
C09-C10	Oropharynx	72								1	2	1	6	9	12	15	5	9	5	3	6	C09-C10
C11	Nasopharynx	36		1			1			2	2		4	2	2	2	5	4	4	3	3	C11
C12-C13	Hypopharynx	53									1	2		6	2	4	4	9	8	3	14	C12-C13
C14	Pharynx unspecified	48							1		1		1		9	3	2	8	7	5	11	C14
C15	Oesophagus	2,304					1		1	2	4	7	38	64	86	128	240	320	473	367	573	C15
C16	Stomach	2,306						1	2	9	6	16	32	46	81	129	173	345	438	415	613	C16
C17	Small intestine	120									3	3	3	6	8	3	12	21	17	23	21	C17
C18-C21	Colorectal	7,090						2	7	11	32	64	100	222	281	453	711	953	1,279	1,130	1,843	C18-C21
C18	Colon	5,078						1	5	7	18	44	65	141	204	323	516	698	920	822	1,314	C18
C19-C21	Rectum	2,012				1	2	1	2	4	14	20	35	81	77	130	195	255	359	308	529	C19-C21
C22	Liver	817	1							2	4	11	13	35	42	66	89	117	151	140	142	C22
C23-C24	Gallbladder	333									1	3	3	11	12	20	34	47	74	55	73	C23-C24
C25	Pancreas	3,115								3	12	21	47	102	137	239	343	473	630	485	623	C25
C26	Digestive organs etc. unspecified	864				1		2	2	2	2	7	7	20	45	56	72	129	162	126	231	C26
C30-C31	Nasal cavities	35									1		1	2	2	5		6	8	4	4	C30-C31
C32	Larynx	156									1		3	7	12	18	21	29	30	20	15	C32
C33-C34	Lung	11,092						1	2	7	25	100	233	438	637	1,024	1,450	2,287	2,396	1,355	1,137	C33-C34
C37-C39,C45	Respiratory system	117							1	1	1	1	3	5	10	12	18	24	21	15	6	C37-C39,C45
C40-C41	Bone	86				2	6	4	4	3	6	3	3	5	1		7	4	13	8	6	C40-C41
C43	Melanoma of skin	711			4	2	2	2	7	15	22	27	30	52	51	65	69	75	89	90	115	C43
C44	Non-melanoma of skin	157				1							3	5	4	3	4	8	28	23	79	C44
C47,C49	Connective tissue	310	1			1	2	4	5	2	8	6	14	30	30	20	32	43	34	36	38	C47,C49
C48	Retroperitoneum and peritoneum	85			4	1				1			4	4	2	4	12	16	17	9	5	C48
C50	Breast	11,532					1	1	20	101	232	369	590	939	931	1,063	1,087	1,378	1,618	1,125	2,078	C50
C51-C52	Other female genital	413								3	2	2	5	8	20	14	23	46	69	76	145	C51-C52
C53	Cervix	1,104						6	18	33	66	77	86	92	65	87	87	112	164	105	106	C53
C54	Uterus	800							1		3	4	13	17	45	80	113	114	156	92	162	C54
C55	Uterus unspecified	432					2				2	3	7	15	28	42	47	51	69	70	98	C55
C56-C57	Ovary	3,942			1	2	2	2	9	10	30	57	131	292	340	497	518	613	642	396	403	C56-C57
C58	Placenta	2									1	1										C58
C64-C66,C68	Kidney	1,028				1	2		1	3	9	15	24	37	44	98	136	152	195	147	163	C64-C66,C68
C67	Bladder	1,459								3	2	6	11	18	36	67	130	195	299	241	451	C67
C69	Eye	45				1					1			2	2	11	2	5	10	4	8	C69
C70,C72	Other nervous system	25			1	1						1		2	2	2	6	2	3	1	3	C70,C72
C71	Brain	1,168	2	9	19	7	7	9	6	31	31	45	56	101	111	133	189	144	149	67	52	C71
C73	Thyroid	195	1			2		1	2	1	1	1	5	8	13	13	13	33	40	34	32	C73
C74-C75	Other endocrine glands	66		4	3	2	1	1	2	4	2	5	6	5	2	3	4	12	5	3	3	C74-C75
C81	Hodgkin's disease	100					2	1	3	10	3	4	3	8	14	6	8	10	13	6	10	C81
C82-C85,C91.4,C96	Non-Hodgkin's lymphoma	1,926		1		4	2	8	13	8	25	25	55	74	107	177	220	270	369	251	317	C82-C85,C91.4,C96
C88,C90	Multiple myeloma	1,049									2	12	13	28	48	77	112	172	223	176	186	C88,C90
C91-C95 xC91.4	Leukaemia	1,715	3	17	16	17	11	16	20	21	30	36	47	61	81	114	148	228	256	200	393	C91-C95 xC91.4
C91 xC91.4	Lymphoid leukaemia	496	1	9	10	10	9	8	3	3	7	5	9	13	21	23	36	60	70	49	150	C91 xC91.4
C92	Myeloid leukaemia	1,067	1	6	4	6	2	8	14	17	21	29	35	42	55	87	107	149	157	129	198	C92
C93	Monocytic leukaemia	14				1				1				1	1			2	2		8	C93
C94-C95	Other leukaemia	138	1	2	2			1	3	1	2		3	5	4	5	5	17	27	22	37	C94-C95
C76-C80	Other and unspecified	7,004				4	1	1	8	15	32	62	114	217	307	438	687	969	1,377	1,120	1,652	C76-C80

Appendix C2(a) Mortality per 100,000 population, England and Wales, 1999 (provisional) - males

ICD10	Site	CR	ESR	WSR	Under 1	1-4	5-9	10-14	15-19	20-24	25-29	30-34	35-39	40-44	45-49	50-54	55-59	60-64	65-69	70-74	75-79	80-84	85 and over	ICD10
C00-C97	All cancers	267.1	232.8	149.9	1.5	4.2	3.5	2.6	4.6	6.1	6.7	10.6	18.0	36.6	79.4	166.5	313.1	522.8	843.8	1,298.7	1,899.4	2,299.8	3,047.5	C00-C97
C00-C97 xC44	All cancers excluding NMSC	266.3	232.1	149.5	1.5	4.2	3.5	2.6	4.6	6.1	6.6	10.5	18.0	36.5	79.2	166.3	312.5	522.1	842.3	1,295.2	1,894.7	2,287.1	3,029.3	C00-C97 xC44
C00-C14	Lip, mouth and pharynx	4.2	3.9	2.7	-	-	-	-	0.1	0.1	0.2	0.4	0.6	1.7	4.7	6.8	9.7	10.8	12.6	14.6	17.4	18.4	30.8	C00-C14
C00	Lip	0.0	0.0	0.0	-	-	-	-	-	-	-	-	-	-	0.1	0.1	-	-	-	0.2	0.3	0.3	1.1	C00
C01-C02	Tongue	0.9	0.9	0.6	-	-	-	-	-	-	0.1	0.0	-	0.5	1.1	1.7	2.7	2.1	2.5	3.3	3.2	3.9	5.2	C01-C02
C03-C06	Mouth	0.9	0.9	0.6	-	-	-	0.1	-	-	-	0.0	0.2	0.5	1.1	1.6	1.0	2.1	3.6	3.1	4.7	3.1	8.2	C03-C06
C07-C08	Salivary glands	0.3	0.3	0.2	-	-	-	-	-	-	-	0.0	-	0.1	0.2	0.1	0.9	0.3	0.6	1.8	2.5	3.1	4.8	C07-C08
C09-C10	Oropharynx	0.8	0.8	0.5	-	-	-	-	-	-	-	0.0	0.1	0.1	1.0	2.1	2.6	2.0	2.5	2.1	2.9	2.9	3.0	C09-C10
C11	Nasopharynx	0.3	0.3	0.2	-	-	-	-	0.1	-	-	0.0	0.0	0.2	0.4	0.6	0.8	0.8	1.1	0.8	1.5	1.2	1.1	C11
C12-C13	Hypopharynx	0.3	0.3	0.2	-	-	-	-	-	-	-	0.0	0.0	0.1	0.4	0.4	1.1	1.2	1.2	1.2	0.8	1.3	2.6	C12-C13
C14	Pharynx unspecified	0.5	0.4	0.3	-	-	-	-	-	0.1	0.1	-	0.0	0.3	0.6	0.3	0.6	1.2	1.4	1.4	1.9	3.1	4.8	4.8	C14
C15	Oesophagus	14.4	12.8	8.4	-	-	-	-	-	0.2	-	0.4	0.7	2.3	6.3	13.1	22.2	33.3	47.5	69.9	97.2	104.2	118.8	C15
C16	Stomach	14.8	12.8	8.1	-	-	-	-	0.1	0.1	0.3	0.4	0.5	1.9	3.5	6.8	16.0	25.9	50.3	74.0	107.4	133.3	185.6	C16
C17	Small intestine	0.5	0.4	0.3	-	-	-	-	-	-	-	-	0.1	0.1	0.3	0.6	0.9	0.8	1.9	1.9	3.0	3.1	4.1	C17
C18-C21	Colorectal	28.9	25.1	16.0	-	-	-	-	0.1	0.2	0.5	0.7	1.1	3.1	7.5	17.1	35.3	56.9	94.5	142.9	201.4	249.9	347.8	C18-C21
C18	Colon	18.5	16.1	10.2	-	-	-	-	0.1	0.2	0.4	0.6	0.8	1.8	4.8	10.4	19.9	36.5	59.5	92.7	132.2	159.3	234.6	C18
C19-C21	Rectum	10.3	9.1	5.8	-	-	-	-	-	-	0.1	0.1	0.3	1.3	2.7	6.7	15.4	20.3	35.1	50.2	69.2	90.7	113.2	C19-C21
C22	Liver	4.3	3.8	2.5	-	0.1	-	0.1	0.1	-	0.2	0.2	0.1	0.9	2.2	3.2	6.2	8.6	15.8	19.2	31.9	29.9	33.8	C22
C23-C24	Gallbladder	0.8	0.7	0.4	-	-	-	-	-	-	-	-	0.1	0.1	0.3	0.3	0.3	1.3	2.5	3.9	5.4	7.5	9.3	C23-C24
C25	Pancreas	10.8	9.6	6.3	-	-	-	-	-	0.1	-	0.1	0.9	1.7	3.6	9.0	15.7	26.1	40.8	49.2	70.4	85.7	87.6	C25
C26	Digestive organs etc. unspecified	3.1	2.7	1.7	-	-	-	-	0.1	-	0.1	0.0	0.2	0.2	1.1	2.0	2.8	5.6	10.4	17.3	18.8	32.5	37.5	C26
C30-C31	Nasal cavities	0.2	0.2	0.2	-	-	-	-	-	-	-	0.0	0.0	0.3	0.3	0.3	0.7	0.7	0.5	0.6	1.0	1.0	0.7	C30-C31
C32	Larynx	2.2	2.0	1.3	-	-	-	-	-	-	-	-	0.1	0.3	0.7	2.0	5.4	5.1	6.6	11.3	12.4	16.4	19.7	C32
C33-C34	Lung	70.7	61.3	39.5	-	-	-	-	0.1	-	-	-	0.3	2.1	5.8	15.7	39.1	85.0	152.0	251.5	403.9	539.7	553.1	566.1	C33-C34
C37-C39,C45	Respiratory system	1.8	1.6	1.1	-	-	-	-	0.1	0.1	0.1	0.1	0.1	0.1	0.5	1.2	3.3	5.2	6.8	9.6	12.9	8.6	9.3	C37-C39,C45
C40-C41	Bone	0.5	0.5	0.4	-	-	0.1	0.3	0.5	0.8	0.5	0.2	0.4	0.1	0.4	0.2	0.4	0.2	0.5	2.3	1.6	0.8	2.6	C40-C41
C43	Melanoma of skin	2.9	2.7	1.9	-	-	-	-	0.1	0.3	0.5	1.0	1.3	1.9	2.9	4.4	4.8	7.0	8.5	9.0	12.6	17.4	17.8	C43
C44	Non-melanoma of skin	0.8	0.7	0.4	-	-	-	-	0.1	0.1	0.1	0.0	0.0	0.1	0.2	0.2	0.6	0.7	1.5	3.4	4.7	12.7	18.2	C44
C47,C49	Connective tissue	1.2	1.1	0.8	0.3	0.2	0.1	0.2	0.3	0.3	0.3	0.5	0.3	0.6	0.9	1.3	1.6	2.1	3.2	5.3	4.8	8.1	11.1	C47,C49
C48	Retroperitoneum and peritoneum	0.3	0.3	0.2	-	-	-	-	-	-	-	0.0	0.1	0.1	0.1	0.4	0.7	1.6	1.5	1.2	1.4	1.3	1.5	C48
C50	Breast	0.3	0.2	0.1	-	-	-	-	-	-	-	0.1	0.0	0.3	0.1	0.3	0.4	0.6	0.5	0.9	2.1	3.4	4.1	C50
C60,C63	Penis	0.4	0.3	0.2	-	-	-	-	-	-	-	0.0	-	0.2	0.2	0.3	0.9	0.4	0.5	0.9	2.5	3.9	5.6	C60,C63
C61	Prostate	32.9	27.3	15.7	-	-	-	-	-	-	-	-	-	0.2	0.5	3.8	12.3	32.2	69.4	138.1	279.0	439.1	769.9	C61
C62	Testis	0.3	0.3	0.2	-	-	0.1	0.1	0.5	0.2	0.4	0.3	0.3	0.5	0.3	0.5	0.3	0.2	0.4	0.2	0.3	0.8	-	C62
C64-C66,C68	Kidney	6.5	5.9	4.0	-	0.1	-	-	-	0.3	-	0.1	0.3	1.5	3.8	6.9	10.3	18.3	25.2	30.1	36.2	38.2	46.0	C64-C66,C68
C67	Bladder	11.0	9.3	5.6	-	-	-	-	-	0.1	-	-	-	0.5	0.9	2.4	7.7	15.8	29.1	54.0	89.5	114.3	204.9	C67
C69	Eye	0.1	0.1	0.1	-	-	-	0.1	-	-	-	-	-	-	0.1	0.1	0.4	0.2	0.9	0.6	1.0	0.3	1.1	C69
C70,C72	Other nervous system	6.3	6.0	4.6	0.3	1.3	1.5	0.6	0.5	1.1	0.7	1.7	2.5	3.6	7.0	11.1	13.1	17.6	19.5	23.0	20.3	16.6	10.8	C70,C72
C71	Brain	0.4	0.4	0.2	-	-	-	-	-	-	-	-	-	-	0.1	0.1	0.5	0.9	0.8	2.1	2.5	3.1	4.8	C71
C73	Thyroid	0.4	0.4	0.3	-	-	-	-	-	-	-	-	-	0.1	0.5	0.5	0.6	0.8	1.0	2.5	3.1	1.5	C73
C74-C75	Other endocrine glands	0.3	0.3	0.3	0.3	1.0	0.5	0.1	0.4	0.1	-	0.1	0.1	0.3	0.1	0.3	0.4	0.6	0.4	0.6	1.1	0.8	1.5	C74-C75
C81	Hodgkin's disease	0.5	0.5	0.4	-	-	-	-	0.4	0.3	0.4	0.3	0.2	0.2	0.2	0.5	0.8	1.6	1.4	1.5	2.7	2.3	1.9	C81
C82-C85,C91.4,C96	Non-Hodgkin's lymphoma	8.1	7.3	4.9	-	0.2	0.2	0.1	0.5	0.8	0.8	1.1	1.7	2.7	4.4	7.7	12.1	18.4	24.0	34.4	48.2	56.1	69.8	C82-C85,C91.4,C96
C88,C90	Multiple myeloma	4.3	3.7	2.3	-	-	-	-	-	-	-	0.0	0.2	0.6	1.4	2.4	4.8	7.9	14.3	21.5	31.9	39.0	42.7	C88,C90
C91-C95 xC91.4	Leukaemia	7.5	6.6	4.6	0.6	1.3	1.0	1.1	1.6	1.4	1.1	1.5	1.3	2.0	2.1	5.3	7.4	11.8	22.5	31.5	45.4	57.9	87.2	C91-C95 xC91.4
C91 xC91.4	Lymphoid leukaemia	2.5	2.2	1.6	0.3	0.8	0.8	0.6	1.0	0.9	0.4	0.5	0.3	0.5	0.5	1.4	2.1	4.2	7.3	10.2	13.9	23.6	27.8	C91 xC91.4
C92	Myeloid leukaemia	4.3	3.8	2.6	0.3	0.4	0.2	0.5	0.5	0.4	0.7	0.8	1.0	1.4	1.6	3.7	4.8	7.0	13.6	19.1	26.6	27.0	49.7	C92
C93	Monocytic leukaemia	0.0	0.0	0.0	-	-	-	-	-	-	-	-	-	-	-	0.2	-	0.2	-	-	0.5	-	0.4	C93
C94-C95	Other leukaemia	0.6	0.5	0.3	-	-	0.1	-	0.1	0.1	0.1	0.2	0.0	0.2	0.0	0.2	0.4	0.5	1.5	2.2	4.4	7.3	9.3	C94-C95
C76-C80	Other and unspecified	26.4	22.9	14.6	-	0.1	0.1	0.1	0.1	0.4	0.5	0.5	1.7	3.0	7.5	16.7	30.6	53.7	79.3	122.4	196.5	245.8	304.0	C76-C80

Appendix C2(b) Mortality per 100,000 population, England and Wales, 1999 (provisional) - females

ICD10	Site	CR	ESR	WSR	Under 1	1-4	5-9	10-14	15-19	20-24	25-29	30-34	35-39	40-44	45-49	50-54	55-59	60-64	65-69	70-74	75-79	80-84	85 and over
C00-C97	All cancers	240.9	161.8	108.4	2.6	2.5	2.8	2.5	2.7	4.1	7.1	14.7	30.8	58.0	100.9	171.5	269.6	412.4	564.0	823.8	1,098.0	1,198.2	1,590.9
C00-C97 xC44	All cancers excluding NMSC	240.3	161.5	108.2	2.6	2.5	2.8	2.5	2.7	4.1	7.1	14.7	30.8	58.0	100.7	171.2	269.3	412.1	563.7	823.1	1,095.3	1,195.0	1,580.3
C00-C14	Lip, mouth and pharynx	2.2	1.5	1.1		0.1			0.1		0.1	0.3	0.6	0.6	1.1	2.1	3.7	4.0	3.9	6.3	7.9	9.2	17.9
C00	Lip	0.0	0.0	0.0																0.1	0.2	0.1	0.9
C01-C02	Tongue	0.5	0.3	0.2					0.1				0.2	0.1	0.2	0.3	0.9	0.9	1.5	1.1	1.5	3.0	4.5
C03-C06	Mouth	0.6	0.4	0.3						0.1		0.1	0.2	0.3	0.2	0.6	0.9	0.9	0.7	1.7	2.5	3.1	5.4
C07-C08	Salivary glands	0.3	0.2	0.1										0.1	0.1	0.2	0.1	0.3	0.3	0.8	0.5	0.4	2.5
C09-C10	Oropharynx	0.3	0.2	0.2								0.0		0.1	0.4	0.3	0.3	1.2	0.4	0.8	0.5	0.4	0.8
C11	Nasopharynx	0.1	0.1	0.1		0.1				0.1		0.1	0.1	0.1	0.2	0.1	0.1	0.2	0.4	0.3	0.4	0.4	0.4
C12-C13	Hypopharynx	0.2	0.1	0.1								0.0	0.1			0.3	0.3	0.2	0.3	0.8	0.8	0.4	1.9
C14	Pharynx unspecified	0.2	0.1	0.1					0.1					0.1	0.1			0.2	0.2	0.7	0.7	0.7	1.5
C15	Oesophagus	8.7	5.2	3.3					0.1		0.1	0.1	0.2	0.4	2.2	3.6	6.3	10.1	19.8	27.9	44.9	52.1	76.7
C16	Stomach	8.7	5.0	3.2						0.1	0.1	0.4	0.3	0.9	1.9	2.6	5.9	10.2	14.2	30.1	41.6	58.9	82.0
C17	Small intestine	0.5	0.3	0.2									0.2	0.2	0.2	0.3	0.6	0.6	1.0	1.8	1.6	3.3	2.8
C18-C21	Colorectal	26.6	16.1	10.3					0.1	0.1	0.4	0.5	1.6	3.7	5.9	12.7	20.6	35.8	58.5	83.2	121.5	160.4	246.6
C18	Colon	19.1	11.5	7.3						0.1	0.3	0.3	0.9	2.5	3.8	8.0	15.0	25.5	42.5	60.9	87.4	116.6	175.8
C19-C21	Rectum	7.6	4.6	3.0					0.1		0.1	0.2	0.7	1.2	2.1	4.6	5.6	10.3	16.1	22.2	34.1	43.7	70.8
C22	Liver	3.1	2.0	1.3	0.3								0.1	0.6	0.8	2.0	3.1	5.2	7.3	10.2	14.3	19.9	19.0
C23-C24	Gallbladder	1.3	0.8	0.5									0.2	0.2	0.2	0.6	0.9	1.6	2.8	4.1	7.0	7.8	9.8
C25	Pancreas	11.7	7.4	4.8						0.1		0.1	0.6	1.2	2.8	5.8	10.0	18.9	28.2	41.3	59.8	68.8	83.4
C26	Digestive organs etc. unspecified	3.2	2.0	1.3										0.4	0.4	1.1	3.3	4.4	5.9	11.3	15.4	17.9	30.9
C30-C31	Nasal cavities	0.1	0.1	0.1									0.1	0.1	0.1	0.2	0.1	0.4		0.5	0.8	0.6	0.5
C32	Larynx	0.6	0.4	0.3										0.2	0.2	0.4	0.9	1.4	1.7	2.5	2.8	2.8	2.0
C33-C34	Lung	41.7	28.8	19.2							0.1	0.3	1.3	5.8	13.7	25.0	46.7	80.9	119.4	199.5	227.6	192.3	152.2
C37-C39,C45	Respiratory system	0.4	0.3	0.2						0.1		0.0	0.3	0.1	0.2	0.3	0.7	0.9	1.5	2.1	2.0	2.1	0.8
C40-C41	Bone	0.3	0.3	0.2			0.2	0.1	0.4	0.3	0.2	0.1	0.3	0.1	0.2	0.3	0.1	0.6	0.6	0.3	1.2	1.1	0.8
C43	Melanoma of skin	2.7	2.0	1.4					0.1	0.1	0.4	0.7	1.1	1.6	1.8	3.0	3.7	5.1	5.7	6.5	8.5	12.8	15.4
C44	Non-melanoma of skin	0.6	0.3	0.2									0.2	0.1	0.2	0.3	0.3	0.3	0.3	0.7	2.7	3.3	10.6
C47,C49	Connective tissue	1.2	0.9	0.7	0.3		0.2		0.1	0.3		0.1	0.5	0.3	0.8	1.7	2.2	1.6	2.6	3.8	3.2	5.1	5.1
C48	Retroperitoneum and peritoneum	0.3	0.2	0.2						0.1		0.0	0.1	0.3	0.2	0.2	0.4	0.9	1.0	1.4	1.6	1.3	0.7
C50	Breast	43.3	31.8	22.0							1.1	4.8	11.7	21.3	34.7	53.5	68.2	84.0	89.5	120.2	153.7	159.6	278.1
C51-C52	Other female genital	1.6	0.9	0.5									0.1	0.1	0.3	0.5	1.5	1.1	1.9	4.0	6.6	10.8	19.4
C53	Cervix	4.1	3.3	2.4						0.4	1.0	1.6	1.6	4.5	5.1	5.2	4.8	6.9	7.2	9.8	15.6	14.9	14.2
C54	Uterus	3.0	2.0	1.3									0.2	0.2	0.8	1.0	3.3	6.3	9.3	9.9	14.8	13.1	21.7
C55	Uterus unspecified	1.6	1.0	0.7									0.1	0.2	0.4	0.9	2.1	3.3	3.9	4.4	6.6	9.9	13.1
C56-C57	Ovary	14.8	11.2	7.7							0.5	0.5	1.5	3.3	7.7	16.6	24.9	39.3	42.6	53.5	61.0	56.2	53.9
C58	Placenta	0.0	0.0	0.0										0.1									
C64-C66,C68	Kidney	3.9	2.6	1.7					0.1	0.1	0.1	0.1	0.5	0.9	1.4	2.1	3.2	7.7	11.2	13.3	18.5	20.9	21.8
C67	Bladder	5.5	3.0	1.9								0.1	0.1	0.3	0.6	1.0	2.6	5.3	10.7	17.0	28.4	34.2	60.4
C69	Eye	0.2	0.1	0.1									0.1			0.1	0.1	0.2	0.2	0.4	0.9	0.6	1.1
C70,C72	Other nervous system	0.1	0.1	0.1			0.1							0.1		0.1	0.1	0.2	0.5	0.2	0.3	0.1	0.4
C71	Brain	4.4	3.7	2.9	0.6	0.7	1.1	0.4	0.4	0.6	0.3	1.0	1.6	2.6	3.3	5.8	8.1	10.5	15.6	12.6	14.2	9.5	7.0
C73	Thyroid	0.7	0.5	0.3					0.4	0.1	0.3		0.1	0.1	0.3	0.5	1.0	1.0	1.1	2.9	3.8	4.8	4.3
C74-C75	Other endocrine glands	0.2	0.2	0.2	0.3	0.1						0.2				0.3		0.2	0.5	1.0	0.5	0.1	0.4
C81	Hodgkin's disease	0.4	0.3	0.2					0.1	0.1	0.2		0.1	0.2	0.2	0.5	1.0	0.5	0.7	0.9	1.2	0.9	1.3
C82-C85,C91.4,C96	Non-Hodgkin's lymphoma	7.2	4.9	3.4		0.1			0.1	0.1	0.3	0.4	1.3	1.4	3.2	4.2	7.8	14.0	18.1	23.6	35.0	35.6	42.4
C88,C90	Multiple myeloma	3.9	2.5	1.6									0.4	0.7	0.8	1.6	3.5	6.1	9.2	15.0	21.2	25.0	24.9
C91-C95 xC91.4	Leukaemia	6.4	4.4	3.2	1.0	1.3	0.9	1.0	0.7	1.1	1.1	1.0	1.1	1.7	2.8	3.5	5.9	9.0	12.2	19.9	24.3	28.4	52.6
C91 xC91.4	Lymphoid leukaemia	1.9	1.2	0.9	0.3	0.7	0.6	0.6	0.6	0.5	0.2	0.1	0.4	0.3	0.5	0.7	1.5	1.8	3.0	5.2	6.6	7.0	20.1
C92	Myeloid leukaemia	4.0	2.8	2.0	0.3	0.5	0.2	0.4	0.1	0.5	0.7	0.1	1.1	1.1	2.1	2.4	4.0	6.9	8.8	13.0	14.9	18.3	26.5
C93	Monocytic leukaemia	0.1	0.0	0.0									0.1			0.1	0.3	0.3	0.4	0.2	0.2		1.1
C94-C95	Other leukaemias	0.5	0.3	0.2				0.1					0.1	0.1	0.2	0.3	0.3	0.3	0.4	1.5	2.6	3.1	5.0
C76-C80	Other and unspecified	26.3	16.1	10.3	0.2			0.2	0.2	0.2	0.4	0.7	1.6	3.6	6.7	12.4	22.5	34.6	56.6	84.5	130.8	158.9	221.1

Appendix C3(a) Percentage distribution for each age-group of deaths by site, England and Wales, 1999 - males

ICD10	Site	All ages	Under 1	1-4	5-9	10-14	15-19	20-24	25-29	30-34	35-39	40-44	45-49	50-54	55-59	60-64	65-69	70-74	75-79	80-84	85 and over	ICD10
C00-C97 xC44	All cancers excluding NMSC	100	100	100	100	100	100	100	100	100	100	100	100	100	100	100	100	100	100	100	100	C00-C97 xC44
C00-C14	Lip, mouth and pharynx	1.6				2.2	2.6	1.0	2.3	3.5	3.5	4.7	5.9	4.1	3.1	2.1	1.5	1.1	0.9	0.8	1.0	C00-C14
C00	Lip	0.0											0.1	0.1				0.0	0.0	0.0	0.0	C00
C01-C02	Tongue	0.3							1.5	0.4	0.8	1.4	1.3	1.0	0.9	0.4	0.3	0.3	0.2	0.2	0.2	C01-C02
C03-C06	Mouth	0.3					2.6			0.4	1.1	1.4	1.3	1.0	0.3	0.4	0.4	0.2	0.2	0.1	0.3	C03-C06
C07-C08	Salivary glands	0.1								0.4			0.2	0.1	0.3	0.1	0.1	0.1	0.1	0.1	0.2	C07-C08
C09-C10	Oropharynx	0.3								0.4	0.8	0.3	1.3	1.2	0.8	0.4	0.3	0.2	0.2	0.1	0.1	C09-C10
C11	Nasopharynx	0.1				2.2				0.4	0.3	0.5	0.4	0.4	0.3	0.2	0.1	0.1	0.0	0.1	0.0	C11
C12-C13	Hypopharynx	0.1						1.0		0.4	0.3	0.3	0.5	0.2	0.4	0.2	0.1	0.1	0.0	0.1	0.1	C12-C13
C14	Pharynx unspecified	0.2							0.8	0.9	0.3	0.8	0.7	0.2	0.2	0.4	0.2	0.1	0.1	0.1	0.2	C14
C15	Oesophagus	5.4							2.3	3.5	4.1	6.4	7.9	7.9	7.1	6.4	5.6	5.4	5.1	4.6	3.9	C15
C16	Stomach	5.5					1.3	1.0	3.8	3.9	3.0	5.3	4.4	4.1	5.1	5.0	6.0	5.7	5.7	5.8	6.1	C16
C17	Small intestine	0.2									0.5	0.2	0.4	0.3	0.3	0.2	0.2	0.1	0.2	0.1	0.1	C17
C18-C21	Colorectal	10.8					1.3	3.1	6.8	6.9	6.0	8.5	9.4	10.3	11.3	10.9	11.2	11.0	10.6	10.9	11.5	C18-C21
C18	Colon	7.0					1.3	3.1	5.3	5.6	4.3	4.9	6.0	6.2	6.4	7.0	7.1	7.2	7.0	7.0	7.7	C18
C19-C21	Rectum	3.9							1.5	1.3	1.6	3.6	3.4	4.0	4.9	3.9	4.2	3.9	3.7	4.0	3.7	C19-C21
C22	Liver	1.6		1.8		2.2	2.6		2.3	2.2	1.9	2.4	2.7	1.9	2.0	1.6	1.9	1.5	1.7	1.3	1.1	C22
C23-C24	Gallbladder	0.3									0.8	0.3	0.2	0.2	0.3	0.3	0.3	0.3	0.3	0.3	0.3	C23-C24
C25	Pancreas	4.1						1.0		1.3	4.9	4.7	4.5	5.4	5.0	5.0	4.8	3.8	3.7	3.7	2.9	C25
C26	Digestive organs etc. unspecified	1.2					1.3		0.8	0.4	1.1	0.5	1.4	1.2	0.9	1.1	1.2	1.3	1.0	1.4	1.2	C26
C30-C31	Nasal cavities	0.1								0.4		0.5	0.4	0.2	0.2	0.1	0.1	0.0	0.1	0.0	0.0	C30-C31
C32	Larynx	0.8								0.9	0.5	0.9	0.9	1.2	1.7	1.0	0.8	0.9	0.7	0.7	0.6	C32
C33-C34	Lung	26.6					1.3		1.5	3.0	11.7	16.0	19.8	23.5	27.2	29.1	29.9	31.2	28.5	24.2	18.7	C33-C34
C37-C39,C45	Respiratory system	0.7						2.1	1.5	1.3	0.5	0.3	0.6	0.7	1.0	1.0	0.8	0.7	0.7	0.4	0.3	C37-C39,C45
C40-C41	Bone	0.2			3.2	11.1	11.8	12.5	6.8	2.2	2.2	0.3	0.5	0.3	0.1	0.0	0.1	0.2	0.1	0.0	0.1	C40-C41
C43	Melanoma of skin	1.1			3.2		2.6	4.2	7.6	9.1	7.3	5.2	3.7	2.7	1.5	1.3	1.0	0.7	0.7	0.8	0.6	C43
C47,C49	Connective tissue	0.5		5.4	3.2	6.7	6.6	4.2	4.5	5.2	1.9	1.7	1.1	0.8	0.5	0.4	0.4	0.4	0.3	0.4	0.4	C47,C49
C48	Retroperitoneum and peritoneum	0.1								0.4	0.8	0.2	0.1	0.2	0.2	0.3	0.2	0.1	0.1	0.1	0.0	C48
C50	Breast	0.1											0.2	0.2	0.1		0.1	0.1	0.1	0.1	0.1	C50
C60,C63	Penis	0.1								0.4		0.5	0.2	0.2	0.3	0.1	0.1	0.1	0.1	0.2	0.2	C60,C63
C61	Prostate	12.4										0.5	0.7	2.3	3.9	6.2	8.2	10.7	14.7	19.2	25.4	C61
C62	Testis	0.1			1.6			3.1	5.3	3.0	3.5	1.4	0.4	0.3	0.1	0.0	0.0	0.0	0.0	0.0		C62
C64-C66,C68	Kidney	2.4		1.8				3.1	3.8	1.3	1.6	4.2	4.7	4.2	3.3	3.5	3.0	2.3	1.9	1.7	1.5	C64-C66,C68
C67	Bladder	4.1						1.0	0.8		0.8	1.3	1.2	1.7	2.5	3.0	3.5	4.2	4.7	5.0	6.8	C67
C69	Eye	0.1		5.4		2.2				0.4	0.3		0.1		0.1	0.0	0.1	0.0	0.1	0.0	0.0	C69
C70,C72	Other nervous system	0.0												0.0	0.0	0.0		0.0	0.0	0.0	-	C70,C72
C71	Brain	2.4	-	30.4	41.9	24.4	10.5	18.8	9.8	16.0	13.8	9.9	8.9	6.6	4.2	3.4	2.3	1.8	1.1	0.7	0.4	C71
C73	Thyroid	0.2								0.9	0.3	0.3	0.3	0.2	0.2	0.2	0.1	0.2	0.1	0.1	0.2	C73
C74-C75	Other endocrine glands	0.1	20.0	23.2	12.9	2.2		1.0		1.3	0.8	0.8	0.1	0.2	0.1	0.1	0.0	0.0	0.1	0.0	0.0	C74-C75
C81	Hodgkin's disease	0.2					9.2	4.2	5.3	3.0	1.4	0.6	0.3	0.3	0.3	0.3	0.2	0.1	0.1	0.1	0.1	C81
C82-C85,C91.4,C96	Non-Hodgkin's lymphoma	3.0			6.5	4.4	11.8	12.5	11.4	10.4	9.2	7.5	5.6	4.6	3.9	3.5	2.8	2.7	2.5	2.5	2.3	C82-C85,C91.4,C96
C88,C90	Multiple myeloma	1.6								0.4	1.4	1.6	1.7	1.4	1.5	1.5	1.7	1.7	1.7	1.7	1.4	C88,C90
C91-C95 xC91.4	Leukaemia	2.8	40.0	30.4	29.0	42.2	34.2	22.9	16.7	14.3	7.3	5.5	2.6	3.2	2.4	2.3	2.7	2.4	2.4	2.5	2.9	C91-C95 xC91.4
C91 xC91.4	Lymphoid leukaemia	0.9	20.0	19.6	22.6	21.1	21.1	14.6	5.3	4.8	1.6	1.3	0.6	0.8	0.7	0.8	0.9	0.8	0.7	1.0	0.9	C91 xC91.4
C92	Myeloid leukaemia	1.6	20.0	10.7	4.8	20.0	10.5	7.3	10.6	7.8	5.4	3.8	2.0	2.2	1.5	1.3	1.6	1.5	1.4	1.2	1.6	C92
C93	Monocytic leukaemia	0.0												0.0		0.0			0.0		0.0	C93
C94-C95	Other leukaemia	0.2			1.6		2.6	1.0	0.8	1.7	0.3	0.5		0.1	0.1	0.1	0.2	0.2	0.2	0.3	0.3	C94-C95
C76-C80	Other and unspecified	9.9	-	1.8	1.6	2.2	2.6	7.3	6.8	5.2	9.2	8.3	9.4	10.1	9.8	10.3	9.4	9.4	10.4	10.7	10.0	C76-C80

Appendix C3(b) Percentage distribution for each age-group of deaths by site, England and Wales, 1999 - females

ICD10	Site description	All ages	Under 1	1-4	5-9	10-14	15-19	20-24	25-29	30-34	35-39	40-44	45-49	50-54	55-59	60-64	65-69	70-74	75-79	80-84	85 and over	ICD10
C00-C97 xC44	All cancers excluding NMSC	100	100	100	100	100	100	100	100	100	100	100	100	100	100	100	100	100	100	100	100	C00-C97 xC44
C00-C14	Lip, mouth and pharynx	0.9		3.1			4.7		1.5	2.3	1.8	1.0	1.1	1.2	1.4	1.0	0.7	0.8	0.7	0.8	1.1	C00-C14
C00	Lip	0.0																		0.0	0.1	C00
C01-C02	Tongue	0.2								0.6	0.5	0.1	0.1	0.2	0.2	0.2	0.3	0.1	0.1	0.2	0.3	C01-C02
C03-C06	Mouth	0.3					2.3		0.7		0.5	0.5	0.2	0.4	0.3	0.2	0.1	0.2	0.2	0.3	0.3	C03-C06
C07-C08	Salivary glands	0.1								0.3				0.1	0.1	0.1	0.1	0.1	0.1	0.1	0.2	C07-C08
C09-C10	Oropharynx	0.1								0.3		0.1	0.4	0.3	0.3	0.3	0.1	0.0	0.0	0.0	0.1	C09-C10
C11	Nasopharynx	0.1					2.3		0.7	0.6	0.3	0.1	0.2	0.1	0.1	0.0	0.1	0.0	0.0	0.0	0.0	C11
C12-C13	Hypopharynx	0.1								0.3	0.2	0.1	0.2	0.2	0.1	0.1	0.1	0.1	0.1	0.0	0.1	C12-C13
C14	Pharynx unspecified	0.1										0.2	0.1		0.2	0.1	0.0	0.1	0.1	0.0	0.1	C14
C15	Oesophagus	3.6					2.3		0.7	0.6	0.7	0.7	2.2	2.1	2.3	2.5	3.5	3.4	4.1	4.4	4.9	C15
C16	Stomach	3.6						1.6	1.5	2.9	1.0	1.6	1.9	1.5	2.2	2.5	2.5	3.7	3.8	4.9	5.2	C16
C17	Small intestine	0.2									0.5	0.3	0.2	0.2	0.2	0.1	0.2	0.2	0.1	0.3	0.2	C17
C18-C21	Colorectal	11.1					4.7	3.2	5.2	3.6	5.2	6.4	5.8	7.4	7.6	8.7	10.4	10.1	11.1	13.4	15.6	C18-C21
C18	Colon	7.9						1.6	3.7	2.3	2.9	4.4	3.8	4.7	5.6	6.2	7.5	7.4	8.0	9.8	11.1	C18
C19-C21	Rectum	3.1					4.7	1.6	1.5	1.3	2.3	2.0	2.0	2.7	2.1	2.5	2.8	2.7	3.1	3.7	4.5	C19-C21
C22	Liver	1.3	12.5			2.4			2.2	0.6	0.7	1.1	0.8	1.2	1.1	1.3	1.3	1.2	1.3	1.7	1.2	C22
C23-C24	Gallbladder	0.5									0.2	0.3	0.2	0.4	0.3	0.4	0.5	0.5	0.6	0.7	0.6	C23-C24
C25	Pancreas	4.9								1.0	2.0	2.1	2.7	3.4	3.7	4.6	5.0	5.0	5.5	5.8	5.3	C25
C26	Digestive organs etc. unspecified	1.4					2.3	3.2	1.5	0.6	0.3	0.7	0.4	0.7	1.2	1.1	1.1	1.4	1.4	1.5	2.0	C26
C30-C31	Nasal cavities	0.1									0.2	0.1	0.1	0.1	0.1	0.1		0.1	0.1	0.0	0.0	C30-C31
C32	Larynx	0.2									0.2		0.2	0.2	0.3	0.3	0.3	0.3	0.3	0.2	0.1	C32
C33-C34	Lung	17.3						1.6	1.5	2.3	4.1	10.0	13.6	14.6	17.3	19.6	21.2	24.2	20.8	16.1	9.6	C33-C34
C37-C39,C45	Respiratory system	0.2								0.3	0.2	0.1	0.2	0.2	0.3	0.2	0.3	0.3	0.2	0.2	0.1	C37-C39,C45
C40-C41	Bone	0.1			8.5	4.9	14.0	6.5	3.0	1.0	1.0	0.3	0.2	0.2	0.0	0.1	0.1	0.0	0.1	0.1	0.1	C40-C41
C43	Melanoma of skin	1.1					4.7	3.2	5.2	4.9	3.6	2.7	1.8	1.7	1.4	1.2	1.0	0.8	0.8	1.1	1.0	C43
C47,C49	Connective tissue	0.5	12.5		8.5	2.4	4.7	6.5	3.7	0.6	1.3	0.6	0.8	1.0	0.8	0.4	0.5	0.5	0.3	0.4	0.3	C47,C49
C48	Retroperitoneum and peritoneum	0.1				2.4				0.3			0.2	0.1	0.1	0.2	0.2	0.2	0.1	0.1	0.0	C48
C50	Breast	18.0						1.6	14.8	32.8	38.0	36.8	34.4	31.3	25.3	20.4	15.9	14.6	14.0	13.4	17.6	C50
C51-C52	Other female genital	0.6							0.7	1.0	0.3	0.2	0.3	0.3	0.5	0.3	0.3	0.5	0.6	0.9	1.2	C51-C52
C53	Cervix	1.7						9.7	13.3	10.7	10.8	7.7	5.0	3.1	1.8	1.7	1.3	1.2	1.4	1.2	0.9	C53
C54	Uterus	1.3								0.3	0.5	0.4	0.8	0.6	1.2	1.5	1.7	1.2	1.4	1.1	1.4	C54
C55	Uterus unspecified	0.7									0.3	0.3	0.4	0.5	0.8	0.8	0.7	0.5	0.6	0.8	0.8	C55
C56-C57	Ovary	6.2					4.7	3.2	6.7	3.2	4.9	5.7	7.6	9.7	9.3	9.5	7.6	6.5	5.6	4.7	3.4	C56-C57
C58	Placenta	0.0									0.2	0.1										C58
C64-C66,C68	Kidney	1.6				2.4	4.7	1.6	0.7	1.0	1.5	1.5	1.4	1.2	1.2	1.9	2.0	1.6	1.7	1.7	1.4	C64-C66,C68
C67	Bladder	2.3								1.0	0.3	0.6	0.6	0.6	1.0	1.3	1.9	2.1	2.6	2.9	3.8	C67
C69	Eye	0.1									0.2			0.1	0.1	0.2	0.0	0.1	0.1	0.0	0.1	C69
C70,C72	Other nervous system	0.0			2.1	2.4						0.1		0.1	0.1	0.2	0.0	0.0	0.0	0.0	0.0	C70,C72
C71	Brain	1.8	25.0	28.1	40.4	17.1	16.3	14.5	4.4	10.1	5.1	4.5	3.3	3.4	3.0	2.5	2.8	1.5	1.3	0.8	0.4	C71
C73	Thyroid	0.3						1.6		0.3	0.2	0.1	0.3	0.4	0.4	0.2	0.2	0.3	0.3	0.4	0.3	C73
C74-C75	Other endocrine glands	0.1	12.5	12.5	6.4	4.9	2.3	1.6	1.5	1.3	0.3	0.5	0.4	0.3	0.1	0.1	0.1	0.1	0.0	0.0	0.1	C74-C75
C81	Hodgkin's disease	0.2					2.3	1.6	2.2	3.2	0.5	0.4	0.2	0.3	0.4	0.1	0.1	0.1	0.1	0.1	0.1	C81
C82-C85,C91.4,C96	Non-Hodgkin's lymphoma	3.0		3.1		9.8	4.7	12.9	9.6	2.6	4.1	2.5	3.2	2.5	2.9	3.4	3.2	2.9	3.2	3.0	2.7	C82-C85,C91.4,C96
C88,C90	Multiple myeloma	1.6									0.3	1.2	0.8	0.9	1.3	1.5	1.6	1.8	1.9	2.1	1.6	C88,C90
C91-C95 xC91.4	Leukaemia	2.7	37.5	53.1	34.0	41.5	25.6	25.8	14.8	6.8	4.9	3.6	2.7	2.0	2.2	2.2	2.2	2.4	2.2	2.4	3.3	C91-C95 xC91.4
C91 xC91.4	Lymphoid leukaemia	0.8	12.5	28.1	21.3	24.4	20.9	12.9	2.2	1.0	1.1	0.5	0.5	0.4	0.6	0.4	0.5	0.6	0.6	0.6	1.3	C91 xC91.4
C92	Myeloid leukaemia	1.7	12.5	18.8	8.5	14.6	4.7	12.9	10.4	5.5	3.4	2.9	2.0	1.4	1.5	1.7	1.6	1.6	1.4	1.5	1.7	C92
C93	Monocytic leukaemia	0.0												0.0	0.0			0.0	0.0			C93
C94-C95	Other leukaemia	0.2	12.5	6.3	4.3	2.4			2.2	0.3	0.3	0.2	0.2	0.2	0.1	0.1	0.1	0.2	0.0	0.3	0.3	C94-C95
C76-C80	Other and unspecified	11.0					2.3	1.6	5.9	4.9	5.2	6.2	6.7	7.2	8.4	8.4	10.0	10.3	11.9	13.3	14.0	C76-C80

Appendix C4(a) Percentage distribution for each site of deaths by age-group, England and Wales, 1999 - males

ICD10	Site	All ages	Under 1	1-4	5-9	10-14	15-19	20-24	25-29	30-34	35-39	40-44	45-49	50-54	55-59	60-64	65-69	70-74	75-79	80-84	85 and over	ICD10
C00-C97 xC44	All cancers excluding NMSC	100	0.0	0.1	0.1	0.1	0.1	0.1	0.2	0.3	0.5	0.9	2.0	4.2	6.1	9.3	13.5	17.7	20.1	12.8	11.9	C00-C97 xC44
C00-C14	Lip, mouth and pharynx	100	-	-	-	-	0.2	0.1	0.3	0.7	1.2	2.8	7.4	11.0	12.1	12.3	12.9	12.7	11.8	6.6	7.7	C00-C14
C00	Lip	100	-	-	-	-	-	-	-	-	-	-	9.1	18.2	-	-	-	18.2	18.2	9.1	27.3	C00
C01-C02	Tongue	100	-	-	-	-	-	0.9	-	0.4	1.3	3.8	7.7	12.4	15.4	11.1	11.5	13.2	9.8	6.4	6.0	C01-C02
C03-C06	Mouth	100	-	-	-	-	0.8	-	-	0.4	1.7	3.8	7.6	11.8	5.5	10.9	16.8	12.2	14.3	5.0	9.2	C03-C06
C07-C08	Salivary glands	100	-	-	-	-	-	-	-	1.1	-	-	3.4	2.2	13.5	4.5	7.9	19.1	20.2	13.5	14.6	C07-C08
C09-C10	Oropharynx	100	-	-	-	-	-	-	-	0.5	1.5	1.0	8.3	17.5	17.0	11.7	13.6	9.7	10.2	5.3	3.9	C09-C10
C11	Nasopharynx	100	-	-	1.1	-	-	-	-	1.1	1.1	3.4	6.9	12.6	12.6	16.1	10.3	16.1	10.3	3.4	3.4	C11
C12-C13	Hypopharynx	100	-	-	-	-	-	1.1	-	1.1	1.1	2.3	8.0	5.7	17.0	17.0	14.8	12.5	6.8	5.7	8.0	C12-C13
C14	Pharynx unspecified	100	-	-	-	-	-	-	0.8	1.6	0.8	4.1	8.2	4.1	6.6	18.9	12.3	10.7	11.5	9.8	10.7	C14
C15	Oesophagus	100	-	-	-	-	-	0.1	0.1	0.2	0.4	1.1	2.9	6.2	8.0	11.0	14.0	17.7	19.1	10.8	8.6	C15
C16	Stomach	100	-	-	-	-	0.0	0.0	0.1	0.2	0.3	0.9	1.5	3.1	5.6	8.3	14.5	18.2	20.5	13.5	13.1	C16
C17	Small intestine	100	-	-	-	-	-	-	-	-	1.6	0.8	4.0	8.1	9.7	8.1	16.9	14.5	17.7	9.7	8.9	C17
C18-C21	Colorectal	100	-	-	-	0.0	0.0	0.0	0.1	0.2	0.3	0.7	1.7	4.0	6.4	9.3	14.0	18.0	19.7	12.9	12.6	C18-C21
C18	Colon	100	-	-	-	-	0.0	0.1	0.1	0.3	0.3	0.6	1.7	3.8	5.6	9.3	13.7	18.2	20.1	12.8	13.2	C18
C19-C21	Rectum	100	-	-	-	-	-	-	0.1	0.1	0.2	0.9	1.7	4.4	7.8	9.3	14.5	17.7	18.9	13.1	11.4	C19-C21
C22	Liver	100	-	0.1	-	0.1	0.2	-	0.3	0.5	0.6	1.4	3.3	5.1	7.6	9.5	15.7	16.2	20.9	10.4	8.2	C22
C23-C24	Gallbladder	100	-	-	-	-	-	-	-	-	1.5	1.0	1.5	2.5	5.6	8.1	13.7	18.8	19.8	14.7	12.7	C23-C24
C25	Pancreas	100	-	-	-	-	-	0.0	-	0.1	0.6	1.1	2.2	5.6	7.6	11.5	16.1	16.6	18.4	11.8	8.5	C25
C26	Digestive organs etc. unspecified	100	-	-	-	-	0.1	-	0.1	0.1	0.5	0.4	2.3	4.3	4.7	8.4	14.1	20.1	16.9	15.4	12.5	C26
C30-C31	Nasal cavities	100	-	-	-	-	-	-	-	1.7	-	5.0	8.3	10.0	15.0	15.0	13.3	10.0	11.7	6.7	3.3	C30-C31
C32	Larynx	100	-	-	-	-	-	-	-	0.3	0.3	1.0	2.1	6.1	12.6	10.9	12.6	18.3	15.6	10.9	9.2	C32
C33-C34	Lung	100	-	-	-	-	-	0.0	0.0	0.0	0.2	0.6	1.5	3.7	6.3	10.2	15.2	20.8	21.5	11.7	8.4	C33-C34
C37-C39,C45	Respiratory system	100	-	-	-	-	-	0.4	0.4	0.6	0.4	0.9	1.7	4.5	9.5	13.8	16.1	19.4	20.2	7.1	5.4	C37-C39,C45
C40-C41	Bone	100	-	-	1.6	4.0	7.2	9.6	7.2	4.0	6.4	1.6	5.6	6.4	4.8	2.4	4.0	17.6	9.6	2.4	5.6	C40-C41
C43	Melanoma of skin	100	-	-	-	-	0.3	0.5	1.3	2.8	3.5	4.3	6.6	10.1	8.5	11.3	12.4	11.2	12.1	8.8	6.3	C43
C47,C49	Connective tissue	100	0.3	0.9	0.6	0.9	1.6	1.3	1.9	3.8	2.2	3.4	4.7	7.2	6.6	8.1	10.9	15.6	10.9	9.7	9.4	C47,C49
C48	Retroperitoneum and peritoneum	100	-	-	-	-	-	-	-	1.1	3.4	1.1	1.1	8.0	11.4	21.6	18.2	12.5	11.4	5.7	4.5	C48
C50	Breast	100	-	-	-	1.3	-	1.3	-	3.8	3.8	6.3	4.6	7.7	7.7	-	7.7	12.3	23.1	20.0	16.9	C50
C60,C63	Penis	100	-	-	-	-	-	-	-	1.1	-	3.3	3.3	6.6	13.2	5.5	5.5	8.8	19.8	16.5	16.5	C60,C63
C61	Prostate	100	-	-	-	-	-	-	-	-	-	0.0	0.1	0.8	2.0	4.6	9.0	15.3	23.9	19.9	24.4	C61
C62	Testis	100	-	1.4	-	-	-	4.3	10.0	10.0	18.6	12.9	7.1	11.4	5.7	2.9	5.7	2.9	2.9	4.3	-	C62
C64-C66,C68	Kidney	100	-	0.1	-	0.1	-	0.3	0.7	0.2	0.4	1.6	3.8	7.2	8.2	13.3	16.4	16.8	15.7	8.7	7.4	C64-C66,C68
C67	Bladder	100	1.3	-	-	-	-	0.0	0.0	0.6	0.1	0.3	0.6	1.7	3.6	6.8	11.2	17.8	22.9	15.5	19.4	C67
C69	Eye	100	-	-	-	2.6	-	-	-	-	-	-	2.6	2.6	13.2	7.9	26.3	15.8	18.4	2.6	7.9	C69
C70,C72	Other nervous system	100	-	15.0	-	-	-	-	-	5.0	5.0	-	10.0	-	5.0	5.0	-	10.0	30.0	15.0	-	C70,C72
C71	Brain	100	0.1	1.0	1.6	0.7	0.5	0.8	-	2.3	3.1	3.9	7.4	11.9	10.9	13.3	13.3	13.3	9.1	3.9	1.8	C71
C73	Thyroid	100	-	-	1.6	-	-	1.1	-	1.9	0.9	1.9	3.8	6.6	6.6	10.4	8.5	18.9	17.0	11.3	12.3	C73
C74-C75	Other endocrine glands	100	1.3	16.5	10.1	1.3	-	1.3	-	3.8	3.8	6.3	2.5	6.3	6.3	8.9	5.1	7.6	10.1	3.8	5.1	C74-C75
C81	Hodgkin's disease	100	-	1.0	-	-	5.0	2.9	5.0	5.0	3.6	2.9	2.9	5.8	7.9	13.7	10.8	10.1	14.4	6.5	3.6	C81
C82-C85,C91.4,C96	Non-Hodgkin's lymphoma	100	-	-	0.2	0.1	0.4	0.6	0.7	1.1	1.6	2.3	3.6	6.4	7.8	10.8	12.6	15.5	16.8	10.3	9.0	C82-C85,C91.4,C96
C88,C90	Multiple myeloma	100	-	-	-	-	-	-	-	0.1	0.5	0.9	2.1	3.8	5.8	8.8	14.3	18.4	21.1	13.7	10.5	C88,C90
C91-C95 xC91.4	Leukaemia	100	0.1	1.0	1.0	1.1	1.4	1.2	1.2	1.6	1.5	1.8	2.0	5.0	5.2	7.8	13.0	15.5	16.9	11.0	11.8	C91-C95 xC91.4
C91 xC91.4	Lymphoid leukaemia	100	0.2	1.7	2.1	1.5	2.5	2.1	1.1	1.7	0.9	1.2	1.2	3.7	4.3	7.8	12.3	14.7	15.5	14.0	11.5	C91 xC91.4
C92	Myeloid leukaemia	100	0.1	0.5	0.3	0.8	0.7	0.6	1.3	1.6	1.8	2.2	2.4	5.8	5.8	7.7	13.5	16.1	17.4	9.3	12.0	C92
C93	Monocytic leukaemia	100	-	-	-	-	-	-	-	-	-	-	-	-	-	28.6	-	-	57.1	-	14.3	C93
C94-C95	Other leukaemias	100	-	0.7	-	-	1.3	0.7	0.7	2.6	0.7	2.0	0.0	2.6	3.9	3.9	11.2	13.8	21.1	18.4	16.4	C94-C95
C76-C80	Other and unspecified	100	-	0.0	0.0	0.0	0.0	0.1	0.1	0.2	0.5	0.8	1.9	4.3	6.0	9.6	12.8	16.8	21.0	13.9	12.0	C76-C80

Appendix C4(b) Percentage distribution for each site of deaths by age-group, England and Wales, 1999 - females

ICD10	Site	All ages	Under 1	1-4	5-9	10-14	15-19	20-24	25-29	30-34	35-39	40-44	45-49	50-54	55-59	60-64	65-69	70-74	75-79	80-84	85 and over	ICD10
C00-C97 xC44	All cancers excluding NMSC	100	0.0	0.1	0.1	0.1	0.1	0.1	0.2	0.5	1.0	1.6	2.7	4.7	5.7	8.2	10.7	14.8	18.0	13.2	18.5	C00-C97 xC44
C00-C14	Lip, mouth and pharynx	100	-	0.2	-	-	0.3	-	0.3	1.2	1.9	1.7	3.0	6.3	8.6	8.6	8.0	12.2	14.0	11.0	22.7	C00-C14
C00	Lip	100	-	-	-	-	-	-	-	-	-	-	-	-	-	-	-	9.1	18.2	9.1	63.6	C00
C01-C02	Tongue	100	-	-	-	-	-	-	-	1.5	2.2	0.7	1.5	3.7	6.7	8.1	13.3	9.6	11.9	9.6	25.2	C01-C02
c03-C06	Mouth	100	-	-	-	-	0.6	-	0.6	-	1.8	3.0	2.4	6.7	7.3	7.3	5.5	11.5	15.8	13.3	24.2	C03-C06
C07-C08	Salivary glands	100	-	-	-	-	-	-	-	1.4	2.8	-	1.4	5.6	7.0	5.6	5.6	12.7	21.1	9.9	26.8	C07-C08
C09-C10	Oropharynx	100	-	-	-	-	-	-	-	1.4	-	1.4	8.3	12.5	16.7	20.8	6.9	12.5	6.9	4.2	8.3	C09-C10
C11	Nasopharynx	100	-	2.8	-	-	2.8	-	1.9	5.6	5.6	2.8	11.1	5.6	5.6	5.6	13.9	11.1	11.1	8.3	8.3	C11
C12-C13	Hypopharynx	100	-	-	-	-	-	-	-	1.9	1.9	-	-	11.3	3.8	7.5	7.5	17.0	15.1	5.7	26.4	C12-C13
C14	Pharynx unspecified	100	-	-	-	-	-	-	-	-	-	4.2	2.1	-	18.8	6.3	4.2	16.7	14.6	10.4	22.9	C14
C15	Oesophagus	100	-	-	-	-	0.0	-	0.0	0.1	0.2	0.3	1.6	2.8	3.7	5.6	10.4	13.9	20.5	15.9	24.9	C15
C16	Stomach	100	-	-	-	-	-	-	0.1	0.4	0.3	0.7	1.4	2.0	3.5	5.6	7.5	15.0	19.0	18.0	26.6	C16
C17	Small intestine	100	-	-	-	-	-	-	-	-	2.5	2.5	2.5	5.0	6.7	2.5	10.0	17.5	14.2	19.2	17.5	C17
C18-C21	Colorectal	100	-	-	-	-	0.0	0.0	0.1	0.2	0.5	0.9	1.4	3.1	4.0	6.4	10.0	13.4	18.0	15.9	26.0	C18-C21
C18	Colon	100	-	-	-	-	-	0.0	0.1	0.1	0.4	0.9	1.3	2.8	4.0	6.4	10.2	13.7	18.1	16.2	25.9	C18
C19-C21	Rectum	100	-	-	-	-	0.1	0.0	-	0.2	0.7	1.0	1.7	4.0	3.8	6.5	9.7	12.7	17.8	15.3	26.3	C19-C21
C22	Liver	100	0.1	-	-	0.1	-	-	0.4	0.2	0.5	1.3	1.6	4.3	5.1	8.1	10.9	14.3	18.5	17.1	17.4	C22
C23-C24	Gallbladder	100	-	-	-	-	-	-	-	-	0.3	0.9	0.9	3.3	3.6	6.0	10.2	14.1	22.2	16.5	21.9	C23-C24
C25	Pancreas	100	-	-	-	-	-	-	0.2	0.1	0.4	0.7	1.5	3.3	4.4	7.7	11.0	15.2	20.2	15.6	20.0	C25
C26	Digestive organs etc. unspecified	100	-	-	-	-	0.1	0.2	0.2	0.2	0.2	0.8	0.8	2.3	5.2	6.5	8.3	14.9	18.8	14.6	26.7	C26
C30-C31	Nasal cavities	100	-	-	-	-	-	-	-	-	2.9	2.9	2.9	8.6	5.7	14.3	13.5	17.1	22.9	11.4	11.4	C30-C31
C32	Larynx	100	-	-	-	-	-	-	-	-	0.6	-	1.9	4.5	7.7	11.5	13.5	18.6	19.2	12.8	9.6	C32
C33-C34	Lung	100	-	-	-	-	-	-	-	0.1	0.2	0.9	2.1	3.9	5.7	9.2	13.1	20.6	21.6	12.2	10.3	C33-C34
C37-C39,C45	Respiratory system	100	-	-	-	-	-	0.0	-	0.9	0.9	0.9	2.6	4.3	8.5	10.3	15.4	20.5	17.9	12.8	5.1	C37-C39,C45
C40-C41	Bone	100	-	-	4.7	2.3	7.0	4.7	4.7	3.5	7.0	3.5	3.5	5.8	1.2	8.1	8.1	4.7	15.1	9.3	7.0	C40-C41
C43	Melanoma of skin	100	0.3	-	-	-	0.3	0.3	1.0	2.1	3.1	3.8	4.2	7.3	7.2	9.1	9.7	10.5	12.5	12.7	16.2	C43
C47,C49	Connective tissue	100	-	-	1.3	0.3	0.6	1.3	1.6	0.6	2.6	1.9	4.5	9.7	9.7	6.5	10.3	13.9	11.0	11.6	12.3	C47,C49
C48	Retroperitoneum and peritoneum	100	-	-	-	1.2	-	-	-	1.2	-	-	4.7	4.7	5.9	12.9	14.1	18.8	20.0	10.6	5.9	C48
C50	Breast	100	-	-	-	-	-	0.0	-	-	2.0	3.2	5.1	8.1	8.1	9.2	9.4	11.9	14.0	9.8	18.0	C50
C51-C52	Other female genital	100	-	-	-	-	-	-	-	0.7	0.5	0.5	1.2	1.9	4.8	3.4	5.6	11.1	16.7	18.4	35.1	C51-C52
C53	Cervix	100	-	-	-	-	-	0.5	1.6	3.0	6.0	7.0	7.8	8.3	5.9	7.9	7.9	10.1	14.9	9.5	9.6	C53
C54	Uterus	100	-	-	-	-	-	-	-	0.1	0.4	0.5	1.6	2.1	5.6	10.0	14.1	14.3	19.5	11.5	20.3	C54
C55	Uterus unspecified	100	-	-	-	-	0.1	-	-	-	0.5	0.7	1.6	3.5	6.5	9.7	10.9	11.8	16.0	16.2	22.7	C55
C56-C57	Ovary	100	-	-	-	-	0.1	-	0.2	0.3	0.8	1.4	3.3	7.4	8.6	12.6	13.1	15.6	16.3	10.0	10.2	C56-C57
C58	Placenta	100	-	-	-	-	-	-	-	-	50.0	50.0	-	-	-	-	-	-	-	-	-	C58
C64-C66, C68	Kidney	100	0.1	-	-	-	0.2	0.1	0.1	0.3	0.9	1.5	2.3	3.6	4.3	9.5	13.2	14.8	19.0	14.3	15.9	C64-C66, C68
C67	Bladder	100	-	-	-	-	-	-	-	0.2	0.1	0.4	0.8	1.2	2.5	4.6	8.9	13.4	20.5	16.5	30.9	C67
C69	Eye	100	-	-	-	-	-	-	-	-	2.2	-	-	4.4	4.4	24.4	4.4	11.1	22.2	8.9	17.8	C69
C70,C72	Other nervous system	100	-	-	4.0	-	-	-	-	-	-	4.0	4.0	8.0	8.0	12.0	24.0	8.0	12.0	4.0	12.0	C70,C72
C71	Brain	100	0.2	0.8	1.6	0.6	0.6	0.8	0.5	2.7	2.7	3.9	4.8	8.6	9.5	11.4	16.2	12.3	12.8	5.7	4.5	C71
C73	Thyroid	100	-	-	-	-	-	-	0.5	0.5	0.5	0.5	2.6	4.1	6.7	6.7	6.7	16.9	20.5	17.4	16.4	C73
C74-C75	Other endocrine glands	100	1.5	6.1	4.5	3.0	1.5	1.5	3.0	6.1	3.0	7.6	9.1	7.6	3.0	4.5	6.1	18.2	7.6	1.5	4.5	C74-C75
C81	Hodgkin's disease	100	-	-	-	-	1.0	1.0	3.0	10.0	3.0	4.0	3.0	8.0	14.0	6.0	8.0	10.0	13.0	6.0	10.0	C81
C82-C85, C91.4, C96	Non-Hodgkin's lymphoma	100	-	0.1	-	0.2	0.1	0.4	0.7	0.4	1.3	1.3	2.9	3.8	5.6	9.2	11.4	14.0	19.2	13.0	16.5	C82-C85,C91.4,C96
C88, C90	Multiple myeloma	100	-	-	-	-	-	-	-	-	0.2	1.1	1.2	2.7	4.6	7.3	10.7	16.4	21.3	16.8	17.7	C88,C90
C91-C95 xC91.4	Leukaemia	100	0.1	1.0	0.9	1.0	0.7	1.0	1.1	1.3	1.8	2.2	2.8	3.6	4.9	7.0	9.1	13.4	14.5	11.3	22.6	C91-C95 xC91.4
C91 xC91.4	Lymphoid leukaemia	100	0.2	1.8	2.0	2.0	1.8	1.6	0.6	0.6	1.4	1.0	1.8	2.6	4.2	4.6	7.3	12.1	14.1	9.9	30.2	C91 xC91.4
C92	Myeloid leukaemia	100	0.1	0.6	0.4	0.6	0.2	0.7	1.3	1.6	2.0	2.7	3.3	3.9	5.2	8.2	10.0	14.0	14.7	12.1	18.6	C92
C93	Monocytic leukaemia	100	-	-	-	-	-	-	-	-	-	-	-	7.1	7.1	-	-	14.3	14.3	-	57.1	C93
C94-C95	Other leukaemias	100	0.7	1.4	1.4	0.7	0.0	-	2.2	0.7	1.4	1.4	2.2	3.6	2.9	2.9	3.6	12.3	19.6	15.9	26.8	C94
C76-C80	Other and unspecified	100	-	-	-	-	0.0	0.0	0.1	0.2	0.5	0.9	1.6	3.1	4.4	6.3	9.8	13.8	19.7	16.0	23.6	C76-C80

Appendix C5(a) Directly age-standardised mortality* per 100,000 population, England and Wales, 1971-1999 - males

ICD10	Site	1971	1972	1973	1974	1975	1976	1977	1978	1979	1980	1981	1982	1983	1984	1985	1986	1987	1988	1989	1990	1991	1992	1993	1994	1995	1996	1997	1998	1999**	ICD10
C00-C97	All cancers	276.3	279.0	278.6	280.3	277.0	281.1	277.3	279.1	279.3	277.0	275.8	272.9	277.8	284.3	282.7	277.4	277.4	278.9	275.7	274.1	272.5	270.7	261.5	256.6	251.7	247.6	239.2	211.5	232.8	C00-C97
C00-C97 xC44	All cancer excluding NMSC	275.7	279.4	278.9	280.9	277.6	281.6	278.0	279.9	280.0	277.7	276.5	273.6	278.3	277.5	275.8	270.6	270.5	272.1	269.0	267.2	265.6	263.8	262.3	257.6	253.0	246.8	239.5	238.7	232.1	C00-C97 xC44
C00-C14	Lip, mouth and pharynx	4.5	4.5	4.2	3.9	4.3	4.3	4.0	3.9	4.1	4.0	4.1	4.0	4.3	4.2	3.8	3.8	3.8	3.9	4.0	4.0	4.0	4.0	3.8	4.3	3.8	4.3	3.9	3.9	3.9	C00-C14
C00	Lip	0.3	0.2	0.2	0.2	0.2	0.2	0.2	0.2	0.2	0.1	0.1	0.1	0.1	0.1	0.1	0.1	0.1	0.1	0.0	0.0	0.1	0.1	0.1	0.1	0.1	0.1	0.1	0.0	0.0	C00
C01-C02	Tongue	1.0	1.0	1.0	0.8	0.8	0.9	0.9	0.9	0.9	1.0	0.9	0.8	0.9	0.9	0.9	0.8	0.9	0.9	1.0	0.9	0.8	0.9	0.8	1.0	0.9	0.8	0.8	0.9	0.9	C01-C02
C03-C06	Mouth	0.8	0.8	0.7	0.8	1.0	0.9	0.8	0.8	1.0	1.0	1.0	0.9	1.1	1.0	1.0	0.9	0.9	1.0	1.1	1.0	0.8	1.0	0.8	1.0	0.8	0.9	0.9	0.9	0.9	C03-C06
C07-C08	Salivary glands	0.5	0.4	0.5	0.5	0.5	0.5	0.5	0.4	0.4	0.4	0.4	0.3	0.3	0.4	0.3	0.5	0.4	0.4	0.4	0.4	0.4	0.4	0.4	0.4	0.4	0.3	0.4	0.4	0.3	C07-C08
C09-C10	Oropharynx	0.6	0.6	0.5	0.5	0.5	0.5	0.5	0.5	0.5	0.4	0.5	0.6	0.6	0.6	0.5	0.5	0.4	0.5	0.5	0.5	0.7	0.6	0.6	0.7	0.6	0.7	0.7	0.7	0.8	C09-C10
C11	Nasopharynx	0.3	0.3	0.3	0.4	0.3	0.3	0.3	0.3	0.4	0.4	0.4	0.4	0.4	0.4	0.3	0.4	0.4	0.3	0.4	0.3	0.3	0.4	0.4	0.3	0.3	0.3	0.3	0.3	0.3	C11
C12-C13	Hypopharynx	0.8	0.8	0.8	0.6	0.7	0.7	0.7	0.7	0.4	0.4	0.4	0.4	0.3	0.6	0.5	0.5	0.4	0.4	0.4	0.4	0.5	0.4	0.4	0.4	0.4	0.4	0.4	0.3	0.3	C12-C13
C14	Pharynx unspecified	0.2	0.3	0.3	0.2	0.2	0.2	0.2	0.2	0.4	0.3	0.4	0.4	0.3	0.3	0.4	0.4	0.4	0.4	0.4	0.5	0.5	0.4	0.4	0.5	0.4	0.5	0.4	0.4	0.4	C14
C15	Oesophagus	7.6	7.9	7.8	8.0	8.0	8.2	8.3	8.6	8.4	8.9	8.9	9.1	9.5	9.2	10.3	9.8	10.5	10.9	10.7	11.4	11.4	11.7	12.3	12.8	12.5	12.6	12.5	12.8	12.8	C15
C16	Stomach	31.7	31.3	30.7	29.8	29.5	28.9	27.4	27.5	26.4	25.4	24.7	23.4	24.6	23.2	22.0	21.7	20.9	20.2	19.6	18.5	17.9	17.3	16.2	16.1	14.8	14.5	13.7	13.1	12.8	C16
C17	Small intestine	0.5	0.5	0.4	0.4	0.5	0.5	0.4	0.5	0.5	0.5	0.4	0.4	0.5	0.4	0.4	0.4	0.4	0.4	0.4	0.4	0.4	0.5	0.5	0.4	0.4	0.4	0.3	0.4	0.4	C17
C18-C21	Colorectal	33.1	33.5	33.1	33.3	32.6	33.5	32.5	31.7	31.4	30.4	30.5	29.6	30.5	30.9	30.1	29.3	29.5	29.9	29.9	29.6	29.1	29.6	29.3	28.2	27.9	27.3	26.7	26.2	25.1	C18-C21
C18	Colon	18.7	19.3	18.9	19.9	18.5	19.2	18.6	18.2	17.5	17.1	17.5	17.9	17.4	18.6	18.0	17.9	18.1	18.4	18.5	18.4	18.1	18.7	18.7	18.1	18.0	17.4	17.4	16.8	16.1	C18
C19-C21	Rectum	14.4	14.3	14.2	13.3	14.1	14.3	14.0	13.5	14.0	13.4	13.2	11.8	13.2	12.3	12.1	11.5	11.4	11.4	11.4	11.2	11.2	10.9	10.6	10.1	9.9	9.9	9.3	9.4	9.1	C19-C21
C22	Liver	1.3	1.4	1.4	1.4	1.4	1.5	1.5	1.5	2.3	2.1	2.5	2.4	2.3	2.7	2.6	2.8	2.8	3.1	3.0	3.0	3.2	3.3	3.4	3.5	3.5	3.6	3.9	3.8	3.8	C22
C23-C24	Gallbladder	1.7	1.8	1.8	1.6	1.6	1.9	1.7	1.7	1.7	1.6	1.5	1.4	1.4	1.4	1.3	1.3	1.3	1.2	1.1	1.0	1.2	1.1	0.9	0.8	0.7	0.7	0.7	0.6	0.7	C23-C24
C25	Pancreas	12.2	12.1	12.0	11.9	12.6	12.0	12.1	12.5	12.5	12.4	11.7	11.2	11.5	11.5	11.4	10.9	11.0	10.6	10.6	10.4	10.4	10.3	10.3	9.9	9.8	9.9	9.6	9.6	9.6	C25
C26	Digestive organs etc. unspecified	0.2	0.2	0.2	0.2	0.2	0.3	0.3	0.3	0.8	0.7	1.0	1.0	0.8	0.8	1.2	1.5	1.5	1.6	1.6	1.8	1.8	1.9	2.1	2.3	2.5	2.7	2.7	2.7	2.7	C26
C30-C31	Nasal cavities	0.7	0.6	0.7	0.6	0.7	0.6	0.5	0.6	0.5	0.5	0.5	0.5	0.5	0.4	0.4	0.5	0.4	0.4	0.4	0.5	0.4	0.4	0.3	0.2	0.2	0.2	0.2	0.2	0.2	C30-C31
C32	Larynx	2.7	2.5	2.7	2.8	2.5	2.7	2.4	2.5	2.6	2.4	2.6	2.6	2.3	2.2	2.4	2.4	2.5	2.4	2.5	2.4	2.3	2.4	2.3	2.3	2.2	2.2	2.2	2.0	2.0	C32
C33-C34	Lung	106.6	108.3	108.7	110.0	107.2	108.2	107.8	107.5	105.8	104.2	101.4	99.5	100.7	97.5	96.4	92.7	90.1	89.4	85.4	84.4	82.2	79.1	76.6	74.3	71.2	68.3	64.8	64.0	61.3	C33-C34
C37-C39,C45	Respiratory system	..	..	..	..	1.0	1.1	1.0	1.0	0.8	1.0	0.9	0.9	1.3	1.4	1.4	1.6	1.7	2.0	1.9	1.9	2.0	2.2	1.6	1.6	1.6	1.7	1.7	1.7	1.6	C37-C39,C45
C40-C41	Bone	1.2	1.0	1.3	1.2	1.0	1.1	1.0	1.0	0.8	0.7	0.6	0.6	0.5	0.6	0.6	0.5	0.6	0.5	0.5	0.5	0.5	0.5	0.4	0.5	0.4	0.5	0.5	0.4	0.5	C40-C41
C43	Melanoma of skin	1.0	1.2	1.2	1.2	1.2	1.4	1.4	1.4	1.6	1.6	1.8	1.8	1.8	2.0	2.0	1.9	1.9	2.1	2.1	2.0	2.4	2.3	2.4	2.4	2.6	2.5	2.5	2.3	2.7	C43
C44	Non-melanoma of skin	1.2	1.2	1.3	1.0	1.1	1.2	1.1	1.2	1.0	0.9	0.9	0.9	1.0	0.9	1.1	1.1	0.9	0.9	1.1	1.0	1.2	1.0	0.9	0.9	0.9	0.8	0.7	0.8	0.7	C44
C47,C49	Connective tissue	0.5	0.6	0.6	0.6	0.7	0.7	0.6	0.8	0.9	0.9	1.0	0.9	1.0	1.1	1.0	1.1	1.0	1.0	1.1	1.0	1.2	1.1	1.2	1.2	1.1	1.2	1.1	1.2	1.1	C47,C49
C48	Retroperitoneum and peritoneum	0.6	0.7	0.6	0.6	0.5	0.5	0.5	0.6	0.4	0.4	0.4	0.4	0.4	0.4	0.4	0.4	0.4	0.3	0.4	0.4	0.4	0.4	0.4	0.3	0.3	0.3	0.3	0.3	0.3	C48
C50	Breast	0.4	0.4	0.3	0.3	0.3	0.4	0.4	0.4	0.3	0.3	0.3	0.4	0.5	0.5	0.3	0.5	0.3	0.3	0.4	0.4	0.4	0.4	0.4	0.4	0.3	0.4	0.2	0.3	0.2	C50
C60,C63	Penis	0.4	0.4	0.5	0.5	0.5	0.4	0.4	0.4	0.4	0.5	0.5	0.5	0.5	0.5	0.5	0.5	0.5	0.4	0.4	0.5	0.4	0.4	0.4	0.4	0.3	0.4	0.3	0.3	0.3	C60,C63
C61	Prostate	19.8	20.2	20.3	20.4	20.3	20.8	20.5	20.6	20.7	21.1	21.3	21.5	22.3	23.4	24.2	24.6	25.1	25.7	26.8	27.0	28.2	28.3	29.5	29.3	29.5	28.8	27.6	27.4	27.3	C61
C62	Testis	1.1	1.1	1.2	1.0	1.0	1.2	1.1	0.9	0.7	0.8	0.6	0.5	0.6	0.6	0.4	0.5	0.5	0.5	0.5	0.5	0.4	0.4	0.4	0.3	0.3	0.4	0.2	0.3	0.3	C62
C64-C66,C68	Kidney	3.3	4.1	3.9	4.1	4.4	4.0	4.1	4.5	4.6	4.3	4.6	4.9	5.0	4.0	5.0	5.1	5.2	5.1	5.4	5.5	5.4	5.7	5.8	5.8	5.8	5.7	5.7	5.8	5.9	C64-C66,C68
C67	Bladder	12.4	12.7	12.5	12.6	12.3	12.6	12.5	12.3	12.1	12.1	12.0	11.9	12.1	11.8	11.9	11.3	11.1	11.1	11.0	11.4	11.3	11.6	11.3	10.9	11.0	10.5	10.0	9.8	9.3	C67
C69	Eye	0.3	0.3	0.3	0.4	0.3	0.4	0.3	0.3	0.3	0.3	0.3	0.2	0.3	0.3	0.3	0.4	0.2	0.1	0.2	0.1	0.3	0.2	0.1	0.1	0.2	0.1	0.1	0.1	0.1	C69
C70,C72	Other nervous system	4.5	:	:	:	:	:	:	5.4	4.9	5.0	4.8	4.3	5.0	5.4	5.4	5.0	5.5	5.5	5.7	5.5	5.5	5.9	5.7	5.7	5.7	5.8	6.0	5.9	6.0	C70,C72
C71	Brain	:	4.7	4.3	4.9	4.5	4.7	4.8	:	:	:	:	:	:	:	:	:	:	:	:	:	:	:	:	:	:	:	:	:	:	C71
C73	Thyroid	0.5	0.5	0.5	0.5	0.5	0.5	0.5	0.5	0.5	0.4	0.4	0.4	0.4	0.4	0.4	0.4	0.4	0.4	0.4	0.4	0.4	0.3	0.4	0.3	0.4	0.3	0.3	0.4	0.4	C73
C74-C75	Other endocrine glands	0.2	0.2	0.2	0.2	0.2	0.1	0.2	0.2	0.2	0.3	0.2	0.2	0.3	0.3	0.3	0.2	0.2	0.3	0.3	0.2	0.3	0.3	0.4	0.4	0.3	0.4	0.3	0.4	0.3	C74-C75
C81	Hodgkin's disease	2.2	2.1	1.9	1.9	1.9	1.7	1.7	1.7	1.4	1.4	1.3	1.3	1.2	1.2	1.0	1.1	1.1	1.0	0.8	0.8	0.8	0.9	0.8	0.6	0.6	0.6	0.5	0.5	0.5	C81
C82-C85,C91.4,C96	Non-Hodgkin's lymphoma	3.8	3.9	3.9	4.1	4.0	4.0	4.1	4.8	4.6	4.9	4.4	5.0	5.4	5.6	5.5	5.9	6.3	6.7	6.7	6.9	6.7	6.9	7.4	7.2	7.6	7.4	7.3	7.4	7.3	C82-C85,C91.4,C96
C88,C90	Multiple myeloma	2.3	2.5	2.5	2.7	2.8	2.8	2.8	3.1	3.2	3.1	3.2	3.1	3.4	3.4	3.7	3.7	4.0	3.9	3.9	3.6	3.6	3.6	3.9	3.7	3.8	3.7	3.6	3.7	3.7	C88,C90
C91-C95 xC91.4	Leukaemia	7.2	7.1	7.4	7.1	7.4	7.2	7.1	7.6	7.3	7.5	7.3	7.3	7.4	7.2	7.2	7.1	7.1	7.1	7.0	6.7	7.1	6.8	6.8	6.7	6.9	6.4	6.5	6.5	6.6	C91-C95 xC91.4
C91 xC91.4	Lymphoid leukaemia	2.9	2.8	2.7	2.5	2.8	2.7	2.7	2.8	2.8	2.8	2.4	2.5	2.7	2.5	2.8	2.7	2.5	2.8	2.6	2.6	2.6	2.6	2.3	2.3	2.4	2.3	2.3	2.3	2.2	C91 xC91.4
C92	Myeloid leukaemia	3.3	3.3	3.7	3.7	3.9	3.7	3.7	3.9	3.9	3.9	3.9	3.7	4.0	4.0	4.0	4.0	4.2	4.0	4.0	3.8	4.2	4.0	4.0	3.8	3.8	3.6	3.7	3.7	3.8	C92
C93	Monocytic leukaemia	0.3	0.3	0.3	0.2	0.2	0.3	0.3	0.3	0.3	0.2	0.2	0.2	0.3	0.1	0.2	0.2	0.1	0.1	0.1	0.1	0.1	0.1	0.5	0.0	0.1	0.1	0.1	0.1	0.0	C93
C94-C95	Other leukaemias	0.3	0.4	0.3	0.3	0.2	0.3	0.3	0.3	0.6	0.7	0.8	0.4	0.5	0.6	0.6	0.5	0.5	0.6	0.1	0.4	0.4	0.4	0.5	0.0	0.6	0.4	0.4	0.5	0.5	C94-C95
C76-C80	Other and unspecified	8.5	8.6	9.0	9.9	9.8	11.0	11.2	11.4	12.5	13.3	16.0	18.1	15.4	18.0	18.4	18.4	19.3	19.9	20.7	20.7	20.5	20.5	24.2	23.6	23.4	23.7	22.9	24.2	22.9	C76-C80

* Directly age-standardised to the European standard population
** Provisional

Appendix C5(b) Directly age-standardised mortality* per 100,000 population, England and Wales, 1971-1999 - females

ICD10	Site	1971	1972	1973	1974	1975	1976	1977	1978	1979	1980	1981	1982	1983	1984	1985	1986	1987	1988	1989	1990	1991	1992	1993	1994	1995	1996	1997	1998	1999**	ICD10
C00-C97	All cancers	171.4	171.3	172.3	173.2	173.9	176.7	174.9	175.9	177.7	178.0	178.0	177.9	178.3	185.2	185.0	183.6	185.0	185.4	185.6	182.0	181.0	179.8	175.4	173.0	170.6	169.3	165.9	162.7	161.8	C00-C97
C00-C97 xC44	All cancers excluding NMSC	171.2	171.6	172.6	173.7	174.1	177.1	175.4	176.5	178.1	178.4	178.6	178.5	178.9	181.4	181.5	180.0	181.1	181.6	181.7	178.2	177.2	176.0	176.0	174.0	171.6	168.9	166.2	163.9	161.5	C00-C97 xC44
C00-C14	Lip, mouth and pharynx	2.0	1.9	1.8	1.9	1.9	1.9	1.8	1.7	2.0	1.8	1.9	1.8	1.7	1.8	1.9	1.7	1.8	1.6	1.7	1.6	1.6	1.6	1.5	1.7	1.5	1.5	1.6	1.6	1.5	C00-C14
C00	Lip	0.0	0.0	0.0	0.0	0.0	0.0	0.0	0.0	0.0	0.0	0.0	0.0	0.0	0.0	0.0	0.0	0.0	0.0	0.0	0.0	0.0	0.0	0.0	0.0	0.0	0.0	0.0	0.0	0.0	C00
C01-C02	Tongue	0.4	0.4	0.4	0.4	0.5	0.5	0.4	0.3	0.4	0.4	0.4	0.4	0.4	0.4	0.5	0.4	0.4	0.4	0.4	0.4	0.4	0.4	0.4	0.4	0.4	0.4	0.4	0.4	0.3	C01-C02
c03-C06	Mouth	0.3	0.3	0.3	0.3	0.3	0.3	0.3	0.3	0.4	0.4	0.4	0.3	0.3	0.4	0.5	0.4	0.4	0.4	0.3	0.4	0.4	0.4	0.3	0.5	0.3	0.3	0.4	0.4	0.4	c03-C06
C07-C08	Salivary glands	0.2	0.2	0.2	0.2	0.2	0.2	0.2	0.2	0.2	0.2	0.2	0.2	0.1	0.2	0.2	0.1	0.2	0.2	0.2	0.1	0.2	0.2	0.2	0.2	0.2	0.2	0.2	0.1	0.2	C07-C08
C09-C10	Oropharynx	0.1	0.1	0.1	0.1	0.2	0.2	0.2	0.2	0.2	0.1	0.2	0.1	0.2	0.2	0.2	0.1	0.2	0.2	0.2	0.1	0.2	0.1	0.1	0.2	0.2	0.2	0.2	0.2	0.2	C09-C10
C11	Nasopharynx	0.1	0.1	0.1	0.1	0.2	0.2	0.1	0.1	0.1	0.1	0.2	0.1	0.1	0.2	0.1	0.2	0.2	0.1	0.2	0.2	0.1	0.2	0.1	0.1	0.1	0.1	0.1	0.1	0.1	C11
C12-C13	Hypopharynx	0.7	0.6	0.5	0.5	0.5	0.6	0.5	0.5	0.4	0.3	0.4	0.3	0.4	0.3	0.4	0.3	0.3	0.2	0.2	0.3	0.3	0.2	0.2	0.2	0.2	0.2	0.2	0.2	0.1	C12-C13
C14	Pharynx unspecified	0.1	0.1	0.1	0.1	0.1	0.1	0.1	0.1	0.1	0.2	0.2	0.2	0.1	0.1	0.1	0.1	0.3	0.2	0.2	0.1	0.2	0.1	0.2	0.1	0.1	0.2	0.1	0.2	0.1	C14
C15	Oesophagus	4.0	3.9	4.1	4.3	4.1	4.3	4.4	4.5	4.3	4.3	4.2	4.4	4.4	4.6	4.5	4.6	4.6	4.5	4.8	4.9	4.8	4.8	5.1	5.0	5.2	5.1	5.1	4.8	5.2	C15
C16	Stomach	14.9	14.9	14.3	13.9	13.6	13.5	12.6	12.7	12.4	11.5	10.9	10.5	10.1	9.9	9.4	8.6	8.6	8.5	8.0	7.6	7.2	7.0	6.6	6.6	6.1	5.5	5.7	5.4	5.0	C16
C17	Small intestine	0.4	0.4	0.3	0.3	0.3	0.3	0.3	0.3	0.3	0.3	0.3	0.3	0.3	0.3	0.3	0.3	0.3	0.3	0.3	0.3	0.3	0.4	0.2	0.2	0.3	0.3	0.3	0.3	0.3	C17
C18-C21	Colorectal	25.5	25.9	25.7	25.7	25.8	25.9	24.5	24.3	23.2	23.2	22.3	21.3	22.0	21.6	21.9	20.6	20.7	20.4	20.1	19.6	19.4	19.3	19.0	18.6	17.8	17.5	16.9	16.2	16.1	C18-C21
C18	Colon	17.6	17.7	17.6	18.0	17.7	17.7	16.6	16.5	15.7	15.6	15.2	14.9	14.7	15.0	15.3	14.3	14.8	14.5	14.3	13.9	13.6	13.9	13.7	13.5	13.0	12.6	12.1	11.6	11.5	C18
C19-C21	Rectum	7.9	8.2	8.1	7.7	8.1	8.2	7.9	7.8	7.6	7.7	7.3	6.5	7.4	6.5	6.7	6.2	5.9	6.0	5.8	5.7	5.8	5.4	5.3	5.1	4.9	4.9	4.8	4.7	4.6	C19-C21
C22	Liver	0.6	0.6	0.6	0.7	0.7	0.5	0.7	0.6	1.0	1.1	1.2	1.3	1.1	1.3	1.4	1.4	1.4	1.5	1.5	1.5	1.6	1.7	1.9	1.9	1.9	1.9	2.1	2.0	2.0	C22
C23-C24	Gallbladder	2.1	2.1	2.0	1.9	1.8	2.1	1.9	1.9	1.7	1.8	1.7	1.5	1.5	1.6	1.6	1.4	1.5	1.3	1.3	1.5	1.2	1.1	1.2	1.0	1.0	0.9	0.9	0.9	0.8	C23-C24
C25	Pancreas	7.2	7.2	7.4	7.7	7.4	7.8	7.7	7.5	7.6	7.6	7.3	7.4	7.3	7.6	7.6	7.6	7.6	7.6	7.7	7.6	7.2	7.3	7.4	7.3	7.4	7.2	7.1	7.0	7.4	C25
C26	Digestive organs etc. unspecified	0.2	0.2	0.1	0.2	0.2	0.2	0.2	0.2	0.6	0.6	0.6	0.8	0.7	0.8	0.8	0.9	1.0	1.0	1.1	1.1	1.2	1.4	1.5	1.7	1.7	1.8	1.8	1.9	2.0	C26
C30-C31	Nasal cavities	0.4	0.3	0.4	0.4	0.4	0.3	0.3	0.3	0.3	0.3	0.3	0.2	0.3	0.3	0.3	0.4	0.2	0.2	0.2	0.2	0.2	0.2	0.2	0.1	0.2	0.1	0.2	0.1	0.1	C30-C31
C32	Larynx	0.4	0.4	0.5	0.5	0.5	0.5	0.5	0.5	0.5	0.6	0.5	0.5	0.5	0.5	0.5	0.4	0.5	0.5	0.4	0.4	0.4	0.5	0.5	0.5	0.5	0.5	0.4	0.4	0.4	C32
C33-C34	Lung	18.3	19.1	19.6	21.0	21.4	21.8	22.6	23.4	24.0	25.0	24.9	26.1	26.2	27.6	27.4	27.9	28.3	29.4	29.3	28.9	29.3	29.1	29.4	29.6	29.4	29.4	28.5	29.1	28.8	C33-C34
C37-C39,C45	Respiratory system	0.3	0.3	0.3	0.3	0.3	0.3	:	:	0.3	0.3	0.2	0.4	0.3	0.3	0.3	0.3	0.3	0.4	0.4	0.3	0.4	0.4	0.3	0.3	0.3	0.3	0.3	0.3	0.3	C37-C39,C45
C40-C41	Bone	0.7	0.7	:	:	0.6	0.7	0.6	0.6	0.3	0.4	0.4	0.4	0.5	0.3	0.4	0.3	0.4	0.3	0.3	0.3	0.3	0.4	0.3	0.2	0.2	0.3	0.3	0.3	0.3	C40-C41
C43	Melanoma of skin	1.4	1.3	1.2	1.3	1.3	1.6	1.5	1.5	1.6	1.7	1.7	1.9	1.6	1.8	1.8	2.0	1.9	2.0	2.0	2.0	1.8	1.8	2.2	2.1	2.0	2.0	2.0	2.1	2.0	C43
C44	Non-melanoma of skin	0.6	0.6	0.6	0.6	0.7	0.6	0.7	0.6	0.5	0.4	0.5	0.4	0.4	0.5	0.5	0.4	0.4	0.4	0.4	0.4	0.4	0.4	0.3	0.4	0.4	0.3	0.3	0.3	0.3	C44
C47,C49	Connective tissue	0.6	0.5	0.5	0.6	0.5	0.4	0.6	0.5	0.7	0.7	0.7	0.7	0.7	0.8	0.7	0.9	0.8	0.8	0.7	0.8	0.8	0.8	0.9	1.0	1.0	0.9	0.9	1.0	0.9	C47,C49
C48	Retroperitoneum and peritoneum	0.6	0.6	0.6	0.5	0.5	0.5	0.5	0.5	0.7	0.3	0.3	0.3	0.3	0.2	0.2	0.3	0.3	0.2	0.2	0.2	0.2	0.2	0.2	0.2	0.3	0.3	0.3	0.3	0.2	C48
C50	Breast	37.9	37.6	38.4	37.7	38.9	39.0	38.9	39.1	39.1	39.3	40.2	35.7	40.0	40.3	40.2	40.5	40.3	40.0	40.3	39.0	38.6	38.0	37.5	36.8	35.7	34.4	33.5	32.6	31.8	C50
C51-C52	Other female genital	1.6	1.5	1.6	1.6	1.5	1.6	1.6	1.3	1.5	1.4	1.3	1.2	1.3	1.3	1.3	1.2	1.3	1.2	1.2	1.2	1.2	1.1	0.9	0.9	1.0	0.9	0.9	0.9	0.9	C51-C52
C53	Cervix	8.3	7.9	8.0	7.1	7.5	7.7	7.5	7.4	7.0	6.9	6.7	6.4	6.5	6.2	6.3	6.6	6.2	6.2	5.7	5.5	5.2	5.0	4.7	4.2	4.1	4.1	3.7	3.5	3.3	C53
C54	Uterus	4.8	4.5	4.8	4.8	4.3	4.4	4.4	4.5	:	4.3	4.2	4.4	3.9	3.9	4.0	3.7	3.7	3.7	3.6	3.5	3.7	3.4	3.3	3.1	3.3	3.2	3.2	3.2	3.0	C54
C55	Uterus unspecified	:	:	:	:	:	:	:	:	1.3	1.2	1.3	1.3	1.2	1.4	1.5	1.4	1.6	1.5	1.4	1.4	1.3	1.5	1.4	1.3	1.5	1.4	1.4	1.3	1.4	C55
C56-C57	Ovary	12.7	12.3	11.6	11.8	12.2	12.4	12.1	12.6	12.0	11.9	11.9	11.3	11.7	12.3	11.7	11.7	11.9	11.5	11.8	11.9	11.4	11.4	11.5	11.4	11.6	12.2	11.5	11.6	11.2	C56-C57
C58	Placenta	0.0	0.0	0.0	0.0	0.0	0.0	0.0	0.0	0.0	0.0	0.0	0.0	0.0	0.0	0.0	0.0	0.0	0.0	0.0	0.0	0.0	0.0	0.0	0.0	0.0	0.0	0.0	0.0	0.0	C58
C64-C66,C68	Kidney	1.9	1.9	2.0	2.0	2.0	2.0	2.0	2.0	2.1	2.1	2.0	2.2	2.3	2.3	2.4	2.3	2.5	2.5	2.5	2.4	2.8	2.7	2.7	2.6	2.7	2.6	2.8	2.8	2.6	C64-C66,C68
C67	Bladder	3.2	3.3	3.3	3.5	3.3	3.4	3.4	3.5	3.4	3.3	3.4	3.2	3.3	3.3	3.3	3.4	3.3	3.2	3.4	3.4	3.2	3.3	3.4	3.4	3.2	3.1	3.1	3.1	3.0	C67
C69	Eye	0.3	0.3	0.3	0.3	0.3	0.2	0.3	0.3	0.2	0.3	0.2	0.2	0.3	0.2	0.1	0.2	0.2	0.2	0.3	0.3	0.2	0.2	0.1	0.1	0.1	0.1	0.1	0.1	0.1	C69
C70,C72	Other nervous system	2.9	2.9	3.0	3.1	2.9	3.0	3.3	3.2	:	:	:	:	:	:	:	:	:	:	:	:	:	:	:	:	:	:	:	:	:	C70,C72
C71	Brain	:	:	:	:	:	:	:	:	3.2	3.2	2.9	3.0	3.3	3.5	3.5	3.3	3.5	3.5	3.6	3.6	3.5	3.6	3.8	3.7	3.7	3.9	3.8	3.9	3.7	C71
C73	Thyroid	0.9	0.9	0.9	0.9	0.9	0.9	0.8	0.8	0.8	0.7	0.7	0.7	0.6	0.6	0.6	0.6	0.6	0.6	0.5	0.6	0.6	0.6	0.5	0.4	0.4	0.5	0.4	0.4	0.5	C73
C74-C75	Other endocrine glands	0.1	0.2	0.2	0.2	0.1	0.2	0.2	0.2	0.3	0.3	0.2	0.2	0.2	0.2	0.3	0.2	0.2	0.3	0.3	0.3	0.3	0.2	0.2	0.3	0.4	0.2	0.3	0.3	0.3	C74-C75
C81	Hodgkin's disease	1.1	1.0	1.1	1.0	0.9	0.9	0.8	0.9	0.8	0.8	0.8	0.6	0.7	0.5	0.6	0.6	0.7	0.6	0.6	0.5	0.5	0.5	0.5	0.4	0.4	0.4	0.3	0.3	0.3	C81
C82-C85,C91.4,C96	Non-Hodgkin's lymphoma	2.5	2.7	2.7	2.5	2.7	2.7	2.8	2.9	3.0	3.1	3.1	3.1	3.4	3.5	3.7	3.7	4.0	4.3	4.2	4.2	4.5	4.6	4.6	4.8	4.8	4.9	4.7	4.6	4.9	C82-C85,C91.4,C96
C88,C90	Multiple myeloma	1.8	1.7	1.7	2.0	1.9	2.0	2.1	2.1	2.3	2.2	2.2	2.3	2.1	2.3	2.4	2.3	2.5	2.5	2.7	2.5	2.5	2.4	2.7	2.5	2.7	2.6	2.4	2.6	2.5	C88,C90
C91-C95 xC91.4	Leukaemia	4.5	4.9	4.3	4.6	4.5	4.6	4.6	4.7	4.6	4.5	4.6	4.6	4.6	4.4	4.5	4.4	4.6	4.4	4.5	4.0	4.1	4.0	4.2	4.1	4.0	4.0	4.2	4.0	4.4	C91-C95 xC91.4
C91 xC91.4	Lymphoid leukaemia	1.5	1.4	1.3	1.3	1.4	1.2	1.4	1.4	1.4	1.3	1.4	1.3	1.4	1.4	1.4	1.4	1.3	1.3	1.4	1.3	1.2	1.2	1.1	1.1	1.2	1.2	1.2	1.0	1.2	C91 xC91.4
C92	Myeloid leukaemia	2.5	2.8	2.5	2.7	2.7	2.9	2.7	2.9	3.0	2.7	2.6	2.7	2.8	2.7	2.8	2.7	3.0	2.8	2.9	2.7	2.7	2.7	2.6	2.6	2.6	2.5	2.7	2.6	2.8	C92
C93	Monocytic leukaemia	0.2	0.2	0.1	0.2	0.1	0.1	0.1	0.1	0.1	0.1	0.1	0.1	0.1	0.1	0.1	0.1	0.1	0.1	0.1	0.0	0.0	0.0	0.0	0.0	0.0	0.0	0.0	0.0	0.0	C93
C94-C95	Other leukaemia	0.2	0.2	0.2	0.2	0.1	0.2	0.2	0.3	0.3	0.3	0.5	0.5	0.4	0.4	0.4	0.2	0.3	0.3	0.4	0.5	0.3	0.2	0.2	0.3	0.3	0.3	0.3	0.3	0.3	C94-C95
C76-C80	Other and unspecified	6.3	6.6	7.0	7.4	7.5	8.2	8.1	8.3	9.4	9.9	11.7	12.8	11.9	12.7	13.2	13.3	13.3	14.1	14.3	14.4	14.7	14.7	16.5	16.3	16.3	16.8	16.5	16.3	16.1	C76-C80

* Directly age-standardised to the European standard population
** Provisional

MORTALITY TO INCIDENCE RATIOS

Appendix D1(a) Mortality to incidence ratios by health region, England and Wales, 1994 - males

ICD10	Site	England and Wales	Northern and Yorkshire	Trent	Anglia and Oxford	North Thames	South Thames	South and West	West Midlands	North West	Wales
C00-C97	All cancers	0.55	0.60	0.56	0.48	0.63	0.65	0.48	0.55	0.52	0.51
C00-C97 xC44	All cancers excluding NMSC	0.64	0.70	0.67	0.64	0.66	0.67	0.56	0.66	0.64	0.61
C00-C14	Lip, mouth and pharynx	0.48	0.52	0.49	0.39	0.53	0.50	0.42	0.54	0.46	0.41
C00	Lip	0.09	0.03	0.11	0.04	0.50	0.40	0.09	0.14	0.09	0.14
C01-C02	Tongue	0.50	0.45	0.56	0.72	0.36	0.48	0.47	0.59	0.53	0.44
C03-C06	Mouth	0.44	0.42	0.40	0.49	0.60	0.43	0.32	0.38	0.50	0.46
C07-C08	Salivary glands	0.43	0.56	0.50	0.38	0.42	0.46	0.39	0.27	0.39	0.64
C09-C10	Oropharynx	0.57	0.70	0.81	0.19	0.74	0.51	0.84	0.73	0.33	0.55
C11	Nasopharynx	0.61	1.55	0.62	0.40	0.47	0.86	0.46	1.00	0.26	0.36
C12-C13	Hypopharynx	0.39	0.44	0.30	0.50	0.43	0.38	0.38	0.52	0.38	0.22
C14	Pharynx unspecified	1.05	1.44	0.71	1.11	1.14	1.00	0.79	1.71	1.13	0.56
C15	Oesophagus	1.00	0.99	0.95	0.93	0.99	1.00	0.92	1.20	0.98	1.09
C16	Stomach	0.75	0.81	0.78	0.83	0.77	0.75	0.70	0.80	0.69	0.68
C17	Small intestine	0.44	0.57	0.48	0.47	0.43	0.70	0.33	0.56	0.26	0.41
C18-C21	Colorectal	0.53	0.55	0.52	0.52	0.58	0.59	0.47	0.53	0.51	0.57
C18	Colon	0.58	0.58	0.58	0.59	0.66	0.64	0.50	0.57	0.56	0.65
C19-C21	Rectum	0.46	0.52	0.44	0.42	0.48	0.52	0.42	0.48	0.44	0.46
C22	Liver	1.03	1.29	1.13	1.13	1.03	0.94	0.85	1.43	0.85	0.85
C23-C24	Gallbladder	0.47	0.45	0.61	0.47	0.36	0.54	0.45	0.39	0.59	0.39
C25	Pancreas	0.97	1.00	1.02	0.96	0.93	0.94	0.93	1.01	1.07	0.87
C26	Digestive organs etc. unspecified	2.72	7.10	2.66	5.15	1.79	1.79	2.11	4.15	3.25	3.00
C30-C31	Nasal cavities	0.34	0.20	0.42	0.28	0.38	0.41	0.16	0.40	0.40	0.50
C32	Larynx	0.37	0.43	0.35	0.24	0.36	0.36	0.33	0.42	0.43	0.30
C33-C34	Lung	0.90	0.93	0.91	0.90	0.92	0.94	0.87	0.92	0.87	0.86
C37-C39,C45	Respiratory system	0.38	0.50	0.48	0.56	0.33	0.24	0.30	0.48	0.33	0.33
C40-C41	Bone	0.61	0.75	0.46	0.81	0.37	0.42	0.58	0.65	0.73	1.00
C43	Melanoma of skin	0.34	0.41	0.42	0.28	0.49	0.51	0.25	0.27	0.22	0.36
C44	Non-melanoma of skin	0.01	0.01	0.01	0.01	0.07	0.07	0.02	0.01	0.01	0.01
C47,C49	Connective tissue	0.53	0.38	0.57	0.46	0.73	0.86	0.38	0.54	0.51	0.68
C48	Retroperitoneum and peritoneum	0.53	0.71	0.42	0.38	0.46	0.69	0.40	0.53	0.61	0.67
C50	Breast	0.46	0.54	0.38	0.71	0.76	0.55	0.36	0.25	0.38	0.31
C60,C63	Penis	0.32	0.28	0.31	0.30	0.27	0.36	0.27	0.53	0.33	0.22
C61	Prostate	0.45	0.51	0.50	0.46	0.43	0.48	0.39	0.44	0.42	0.42
C62	Testis	0.06	0.09	0.09	0.05	0.02	0.07	0.04	0.06	0.03	0.09
C64-C66,C68	Kidney	0.53	0.59	0.61	0.51	0.59	0.53	0.48	0.49	0.49	0.51
C67	Bladder	0.37	0.46	0.38	0.39	0.38	0.39	0.31	0.30	0.40	0.29
C69	Eye	0.18	0.11	0.24	0.10	0.14	0.29	0.07	0.33	0.33	0.08
C70,C72	Other nervous system	0.35	1.17	0.38	0.13	0.21	0.30	0.36	0.40	0.33	0.29
C71	Brain	0.74	0.83	0.67	0.80	0.80	0.76	0.67	0.74	0.75	0.66
C73	Thyroid	0.35	0.28	0.35	0.56	0.45	0.38	0.46	0.28	0.25	0.20
C74-C75	Other endocrine glands	0.64	0.59	0.59	0.17	0.75	1.25	0.37	1.45	0.89	0.75
C81	Hodgkin's disease	0.23	0.21	0.25	0.22	0.28	0.27	0.21	0.19	0.19	0.25
C82-C85,C91.4,C96	Non-Hodgkin's lymphoma	0.50	0.53	0.48	0.52	0.52	0.53	0.44	0.50	0.55	0.50
C88,C90	Multiple myeloma	0.73	0.90	0.74	0.72	0.61	0.69	0.65	0.95	0.72	0.76
C91-C95 xC91.4	Leukaemia	0.61	0.63	0.73	0.62	0.62	0.60	0.52	0.73	0.66	0.43
C91 xC91.4	Lymphoid leukaemia	0.45	0.44	0.57	0.43	0.44	0.43	0.40	0.60	0.54	0.27
C92	Myeloid leukaemia	0.74	0.82	0.84	0.79	0.76	0.74	0.62	0.80	0.73	0.62
C93	Monocytic leukaemia	0.38	:	0.25	0.17	-	0.33	0.80	-	0.40	0.50
C94-C95	Other leukaemias	1.01	1.31	1.88	1.71	0.88	1.07	0.68	1.14	1.17	0.47
C76-C80	Other and unspecified	0.98	0.99	1.04	1.08	0.97	0.96	1.00	1.08	0.95	0.77

Appendix D1(b) Mortality to incidence ratios by health region, England and Wales, 1994 - females

ICD10	Site	England and Wales	Northern and Yorkshire	Trent	Anglia and Oxford	North Thames	South Thames	South and West	West Midlands	North West	Wales	ICD10
C00-C97	All cancers	0.51	0.54	0.53	0.45	0.58	0.61	0.46	0.51	0.48	0.46	C00-C97
C00-C97 xC44	All cancers excluding NMSC	0.59	0.63	0.62	0.57	0.60	0.62	0.53	0.61	0.60	0.53	C00-C97 xC44
C00-C14	Lip, mouth and pharynx	0.48	0.44	0.55	0.41	0.52	0.52	0.48	0.54	0.50	0.37	C00-C14
C00	Lip	0.10	0.07	0.25	-	-	0.30	-	-	0.10	0.25	C00
C01-C02	Tongue	0.55	0.47	0.42	0.56	0.62	0.90	0.45	0.50	0.45	0.56	C01-C02
C03-C06	Mouth	0.49	0.63	0.41	0.60	0.53	0.38	0.61	0.40	0.48	0.32	C03-C06
C07-C08	Salivary glands	0.33	0.15	0.38	0.42	0.38	0.37	0.15	0.50	0.61	0.20	C07-C08
C09-C10	Oropharynx	0.56	0.53	0.83	0.17	0.63	0.56	0.62	0.94	0.52	0.26	C09-C10
C11	Nasopharynx	0.56	0.40	1.67	0.40	0.67	0.33	0.42	0.75	0.67	0.40	C11
C12-C13	Hypopharynx	0.48	0.37	0.82	0.22	0.50	0.24	0.64	0.67	0.43	0.63	C12-C13
C14	Pharynx unspecified	0.90	1.00	0.71	0.67	0.40	1.60	1.33	0.67	0.92	0.60	C14
C15	Oesophagus	0.89	0.93	0.80	0.92	0.75	0.93	0.85	1.08	0.89	0.92	C15
C16	Stomach	0.82	0.78	0.81	0.84	0.82	0.83	0.82	0.82	0.85	0.76	C16
C17	Small intestine	0.38	0.36	0.11	0.50	0.28	0.42	0.34	0.36	0.46	0.62	C17
C18-C21	Colorectal	0.56	0.57	0.61	0.54	0.58	0.58	0.52	0.58	0.56	0.56	C18-C21
C18	Colon	0.60	0.62	0.66	0.57	0.65	0.61	0.57	0.59	0.56	0.60	C18
C19-C21	Rectum	0.49	0.47	0.51	0.45	0.45	0.51	0.42	0.55	0.55	0.47	C19-C21
C22	Liver	1.22	1.38	1.22	1.54	1.41	1.30	0.89	1.74	1.09	0.93	C22
C23-C24	Gallbladder	0.57	0.59	0.58	0.49	0.44	0.65	0.68	0.56	0.71	0.37	C23-C24
C25	Pancreas	0.95	0.97	1.01	0.96	0.92	0.93	0.93	0.94	0.94	1.01	C25
C26	Digestive organs etc. unspecified	2.42	5.50	3.64	2.63	1.39	1.36	2.28	5.80	2.49	2.55	C26
C30-C31	Nasal cavities	0.37	0.29	0.67	0.33	0.20	0.47	0.42	0.33	0.56	0.13	C30-C31
C32	Larynx	0.50	0.49	0.62	0.52	0.69	0.40	0.41	0.59	0.42	0.56	C32
C33-C34	Lung	0.89	0.89	0.88	0.89	0.87	0.95	0.84	0.96	0.88	0.85	C33-C34
C37-C39, C45	Respiratory system	0.36	0.65	0.23	0.59	0.35	0.40	0.21	0.32	0.23	0.24	C37-C39, C45
C40-C41	Bone	0.47	0.92	0.69	0.33	0.43	0.44	0.28	0.08	1.00	0.44	C40-C41
C43	Melanoma of skin	0.26	0.27	0.26	0.23	0.38	0.30	0.21	0.28	0.20	0.30	C43
C44	Non-melanoma of skin	0.01	0.01	0.01	0.00	0.06	0.09	0.01	0.01	0.01	0.01	C44
C47, C49	Connective tissue	0.66	0.49	0.61	0.45	0.86	1.03	0.58	0.72	0.76	0.53	C47, C49
C48	Retroperitoneum and peritoneum	0.49	1.18	0.47	1.60	0.50	0.53	0.11	0.56	0.36	0.17	C48
C50	Breast	0.40	0.44	0.46	0.37	0.39	0.43	0.37	0.41	0.40	0.36	C50
C51-C52	Other female genital	0.39	0.43	0.49	0.57	0.39	0.58	0.31	0.31	0.30	0.25	C51-C52
C53	Cervix	0.43	0.41	0.43	0.50	0.45	0.50	0.41	0.37	0.39	0.50	C53
C54	Uterus	0.17	0.22	0.17	0.14	0.15	0.20	0.17	0.19	0.15	0.16	C54
C55	Uterus unspecified	1.62	0.89	1.19	14.00	9.00	70.00	1.05	1.41	1.46	0.77	C55
C56-C57	Ovary	0.72	0.69	0.73	0.75	0.74	0.77	0.70	0.71	0.68	0.68	C56-C57
C58	Placenta	0.33	-	1.00	-	:	-	0.50	0.50	:	:	C58
C64-C66, C68	Kidney	0.55	0.59	0.51	0.59	0.49	0.65	0.54	0.55	0.51	0.48	C64-C66, C68
C67	Bladder	0.49	0.57	0.49	0.55	0.52	0.53	0.38	0.53	0.50	0.38	C67
C69	Eye	0.27	0.58	0.22	0.32	0.32	0.25	0.21	0.25	0.22	0.17	C69
C70, C72	Other nervous system	0.19	0.33	0.43	0.22	0.20	0.06	0.08	0.27	0.21	0.15	C70, C72
C71	Brain	0.73	0.77	0.65	0.81	0.79	0.80	0.71	0.67	0.72	0.55	C71
C73	Thyroid	0.24	0.27	0.32	0.15	0.25	0.20	0.28	0.27	0.25	0.20	C73
C74-C75	Other endocrine glands	0.68	1.00	0.71	0.58	1.43	0.55	0.35	1.13	0.85	0.33	C74-C75
C81	Hodgkin's disease	0.23	0.25	0.29	0.25	0.21	0.22	0.12	0.33	0.23	0.19	C81
C82-C85, C91.4, C96	Non-Hodgkin's lymphoma	0.54	0.45	0.53	0.50	0.54	0.59	0.51	0.69	0.58	0.46	C82-C85, C91.4, C96
C88, C90	Multiple myeloma	0.73	0.81	0.74	0.89	0.70	0.74	0.57	0.72	0.76	0.74	C88, C90
C91-C95 xC91.4	Leukaemia	0.63	0.62	0.62	0.58	0.66	0.65	0.55	0.80	0.72	0.56	C91-C95 xC91.4
C91 xC91.4	Lymphoid leukaemia	0.42	0.42	0.44	0.29	0.46	0.45	0.34	0.51	0.57	0.41	C91 xC91.4
C92	Myeloid leukaemia	0.79	0.71	0.79	0.88	0.77	0.77	0.75	0.97	0.82	0.65	C92
C93	Monocytic leukaemia	0.80	2.00	-	1.00	1.00	1.00	0.67	0.50	1.00	0.50	C93
C94-C95	Other leukaemias	0.95	1.29	0.69	0.69	1.07	0.79	1.00	1.21	0.95	1.00	C94-C95
C76-C80	Other and unspecified	0.89	0.97	0.96	0.89	0.97	0.87	0.90	0.97	0.91	0.57	C76-C80

QUALITY INDICATORS

Appendix E1 Percentage of zero survival cases by health region, England and Wales, 1994

ICD10	Site	England and Wales Number	England and Wales %	Northern and Yorkshire %	Trent %	Anglia and Oxford %	North Thames %	South Thames %	South and West %	West Midlands %	North West %	Wales %
C00-C97	All cancers	259,628	10.3	5.5	6.4	3.6	23.3	24.6	8.3	5.9	6.4	11.4
C00-C97 xC44	All cancers excluding NMSC	223,079	12.0	6.4	7.6	4.7	24.1	25.2	9.6	7.1	7.9	13.4
C00-C14	Lip, mouth and pharynx	3,626	5.0	1.5	4.1	1.5	13.2	12.3	4.3	0.9	1.2	8.9
C00	Lip	280	1.1	0.0	0.0	0.0	0.0	13.3	0.0	0.0	0.0	3.3
C01-C02	Tongue	774	4.5	1.0	2.7	1.7	12.6	12.2	2.0	0.0	0.0	11.5
C03-C06	Mouth	977	5.9	1.9	7.9	0.0	18.2	12.9	6.3	1.1	0.0	8.9
C07-C08	Salivary glands	420	6.2	1.7	3.7	2.5	14.0	15.3	6.0	4.7	2.0	0.0
C09-C10	Oropharynx	449	5.3	1.7	2.6	0.0	14.3	13.0	2.6	0.0	3.6	10.9
C11	Nasopharynx	179	5.6	0.0	6.3	0.0	11.1	9.1	8.3	0.0	3.1	6.3
C12-C13	Hypopharynx	364	2.5	0.0	0.0	4.3	5.7	3.0	5.0	0.0	0.0	9.7
C14	Pharynx unspecified	183	8.7	8.0	4.8	16.7	10.5	16.7	4.0	5.4	5.6	21.4
C15	Oesophagus	5,969	12.7	5.8	5.6	5.1	30.7	31.0	9.7	5.4	6.4	14.9
C16	Stomach	9,665	15.7	8.7	9.4	6.3	28.9	35.5	16.0	8.9	8.3	21.6
C17	Small intestine	466	9.0	2.0	4.0	6.4	15.8	28.6	5.4	8.0	1.4	20.0
C18-C21	Colorectal	28,770	10.7	6.2	6.4	4.7	22.1	23.6	8.2	7.0	6.1	14.2
C18	Colon	18,218	12.6	7.2	8.0	5.6	26.3	25.8	9.8	8.5	7.7	16.3
C19-C21	Rectum	10,552	7.5	4.6	4.1	3.2	14.6	19.5	5.5	4.3	3.1	10.5
C22	Liver	2,057	28.1	12.0	18.8	18.0	49.2	59.9	19.5	24.8	19.1	27.3
C23-C24	Gallbladder	1,193	14.3	10.3	5.5	12.8	19.6	27.6	13.0	13.3	8.7	21.0
C25	Pancreas	5,992	25.3	11.4	16.9	10.6	48.8	48.5	20.8	17.7	18.2	27.6
C26	Digestive organs etc. unspecified	542	41.1	32.1	33.3	30.0	68.4	66.7	23.8	8.7	23.3	12.1
C30-C31	Nasal cavities	340	5.9	2.1	3.0	3.3	5.9	16.3	9.1	0.0	0.0	11.1
C32	Larynx	2,057	4.7	1.6	2.1	0.6	12.2	11.8	4.0	1.9	2.1	6.4
C33-C34	Lung	35,443	18.2	9.7	10.3	7.7	33.3	37.9	16.2	11.4	13.6	22.3
C37-C39,C45	Respiratory system	1,406	15.0	14.0	14.5	13.0	20.8	25.7	13.5	9.8	7.1	13.8
C40-C41	Bone	364	8.8	2.9	5.4	0.0	24.4	18.4	5.9	6.3	5.3	4.8
C43	Melanoma of skin	4,692	2.7	1.1	1.7	0.2	6.5	7.5	2.7	0.0	0.8	5.8
C44	Non-melanoma of skin	36,549	0.3	0.2	0.2	0.0	2.2	3.5	0.3	0.1	0.2	0.3
C47,C49	Connective tissue	1,061	4.4	4.3	2.7	2.4	11.4	8.0	4.0	1.1	2.1	6.0
C48	Retroperitoneum and peritoneum	283	15.9	10.7	10.3	5.6	26.8	39.4	6.8	16.1	12.5	7.4
C50	Female breast	31,458	6.9	3.1	5.6	2.0	14.0	13.6	6.8	3.2	4.8	7.0
C50	Male breast	188	8.0	0.0	0.0	7.1	35.0	20.0	5.6	0.0	0.0	6.3
C51-C52	Other female genital	1,081	6.8	3.9	8.5	3.6	15.5	16.7	5.0	2.7	3.9	1.3
C53	Cervix	3,143	3.8	1.5	2.2	0.8	11.7	10.6	2.9	1.1	1.9	4.6
C54	Uterus	4,030	5.1	2.2	1.5	1.3	10.3	13.4	3.0	3.6	3.3	5.4
C55	Uterus unspecified	332	21.7	18.8	11.6	25.0	16.7	0.0	23.0	20.5	29.2	26.4
C56-C57	Ovary	5,328	11.6	4.9	7.0	5.8	24.1	21.2	12.1	6.3	7.6	11.7
C58	Placenta	12	8.3	0.0	100.0	0.0	:	0.0	0.0	0.0	0.0	:
C60,C63	Penis	356	4.2	0.0	0.0	0.0	12.2	2.8	7.3	0.0	3.3	11.1
C61	Prostate	19,324	9.5	3.6	6.4	3.5	19.5	22.1	8.3	4.1	4.4	7.8
C62	Testis	1,362	1.2	0.7	1.6	0.0	1.3	2.6	0.5	0.7	1.7	1.3
C64-C66,C68	Kidney	4,741	12.7	6.8	9.2	6.1	24.7	25.0	11.0	7.8	9.7	13.2
C67	Bladder	11,748	6.0	3.4	3.9	2.7	14.0	13.2	3.7	3.3	3.1	6.0
C69	Eye	381	2.9	0.0	8.6	1.9	0.0	8.0	0.0	0.0	4.9	3.3
C70,C72	Other nervous system	185	10.3	0.0	13.3	0.0	8.3	10.7	3.7	6.7	10.0	29.6
C71	Brain	3,474	10.5	5.6	7.5	6.6	25.5	16.2	7.6	8.1	4.4	18.5
C73	Thyroid	1,040	4.7	3.1	0.0	1.0	10.3	7.9	6.4	1.8	2.9	6.3
C74-C75	Other endocrine glands	230	17.8	11.5	19.2	12.5	36.8	21.4	8.7	26.3	22.7	15.0
C81	Hodgkin's disease	1,168	3.5	2.0	2.9	2.4	10.2	5.6	0.6	1.0	2.9	3.8
C82-C85,C91.4,C96	Non-Hodgkin's lymphoma	7,217	10.5	3.5	8.4	3.9	25.5	22.1	5.1	5.7	5.8	9.2
C88,C90	Multiple myeloma	2,863	15.1	4.8	11.7	7.7	37.2	35.0	4.6	10.4	9.0	13.1
C91-C95 xC91.4	Leukaemia	5,523	16.3	7.5	8.0	5.1	37.6	37.3	7.7	13.5	12.0	10.5
C91 xC91.4	Lymphoid leukaemia	2,593	12.8	7.7	5.6	4.1	27.4	27.9	8.8	11.8	11.2	8.0
C92	Myeloid leukaemia	2,598	18.1	6.3	10.6	4.9	43.1	42.9	6.2	12.5	10.8	9.6
C93	Monocytic leukaemia	48	12.5	0.0	0.0	0.0	33.3	40.0	0.0	0.0	12.5	50.0
C94-C95	Other leukaemia	284	33.1	18.5	9.5	21.7	66.7	62.2	11.1	35.7	27.0	29.6
C76-C80	Other and unspecified	14,463	24.0	13.6	15.7	9.6	48.5	46.3	23.3	15.1	15.7	18.3

Appendix E2 Registrations with site unspecified by health region, England and Wales, 1971-1994

	England and Wales Number	England and Wales %	England %	Northern and Yorkshire %	Trent %	Anglia and Oxford %	North Thames %	South Thames %	South and West %	West Midlands %	North West %	Wales %	
1971	1,745	1.2	1.2	1.5	1.3	0.7	1.0	0.8	1.1	1.8	1.3	1.0	1971
1972	1,834	1.2	1.2	1.6	1.8	0.8	1.1	0.7	0.9	1.7	0.9	1.7	1972
1973	2,105	1.3	1.3	1.8	2.0	0.9	1.0	0.7	1.4	1.5	1.2	1.9	1973
1974	2,402	1.5	1.4	1.8	1.9	1.1	1.1	0.8	1.4	1.7	1.5	2.1	1974
1975	2,194	1.3	1.3	1.5	1.7	1.0	1.3	0.5	1.7	1.1	1.5	1.7	1975
1976	2,318	1.4	1.4	1.9	1.9	0.8	1.2	0.6	2.0	1.0	1.4	1.8	1976
1977	2,768	1.6	1.6	1.7	2.0	0.7	1.8	0.8	2.3	1.7	1.6	2.1	1977
1978	3,043	1.8	1.7	1.8	2.0	0.9	2.0	1.0	2.4	1.6	1.8	2.7	1978
1979	3,878	2.2	2.2	2.1	2.3	0.5	2.3	3.2	2.6	2.4	1.8	2.4	1979
1980	4,075	2.3	2.3	2.1	2.2	0.7	2.7	3.7	2.8	2.4	1.4	2.1	1980
1981	3,800	2.1	2.0	2.7	2.2	0.9	2.8	1.0	2.5	2.6	1.6	2.6	1981
1982	4,288	2.3	2.3	2.6	2.4	1.0	2.6	2.1	2.6	3.0	1.8	3.3	1982
1983	4,802	2.6	2.6	2.7	2.5	1.2	3.2	3.4	2.9	3.1	1.5	2.4	1983
1984	4,744	2.5	2.5	2.9	2.2	1.1	3.1	2.8	3.2	2.7	1.7	2.3	1984
1985	5,338	2.7	2.6	3.0	2.4	1.5	3.4	2.6	3.1	3.0	1.9	3.0	1985
1986	5,111	2.6	2.6	3.1	2.7	1.8	3.0	2.1	2.8	3.3	2.1	2.4	1986
1987	6,993	3.4	3.5	3.0	2.9	2.3	3.9	3.9	3.2	6.7	2.1	3.2	1987
1988	6,482	3.1	3.1	3.0	2.7	2.4	3.2	2.9	2.8	6.2	2.2	2.7	1988
1989	6,466	3.1	3.1	3.2	2.5	2.8	2.8	2.6	2.9	6.2	2.4	2.8	1989
1990	6,436	3.1	3.1	3.6	2.6	3.1	2.8	2.4	2.5	6.0	2.4	2.8	1990
1991	6,981	3.3	3.3	3.7	2.5	3.1	2.9	2.6	2.7	6.7	2.4	3.0	1991
1992	6,820	3.1	3.1	3.3	2.6	3.7	3.0	2.5	2.2	6.0	2.3	2.5	1992
1993	7,040	3.2	3.2	3.8	2.7	3.2	3.2	3.0	2.4	5.2	2.5	3.5	1993
1994	6,859	3.1	3.1	3.7	2.5	2.8	3.5	3.7	2.3	3.1	2.8	3.1	1994

POPULATION ESTIMATES

Appendix F   Mid-year population estimates, England and Wales, 1971, 1981, 1991 and 1998, and European and World standard populations

	1971 Males	1971 Females	1981 Males	1981 Females	1991 Males	1991 Females
Under 1	401,500	380,100	324,400	309,600	359,700	342,400
1-4	1,626,100	1,543,700	1,217,600	1,153,900	1,400,900	1,327,500
5-9	2,079,000	1,976,700	1,642,000	1,554,300	1,657,000	1,565,600
10-14	1,877,600	1,771,800	1,996,600	1,892,100	1,573,600	1,484,700
15-19	1,711,200	1,632,600	2,114,400	2,015,400	1,685,000	1,586,500
20-24	1,903,900	1,869,100	1,896,400	1,847,300	2,025,200	1,940,600
25-29	1,653,500	1,613,900	1,700,100	1,671,400	2,162,700	2,083,300
30-34	1,475,100	1,421,700	1,869,000	1,845,800	1,902,600	1,859,300
35-39	1,424,300	1,380,000	1,605,900	1,579,400	1,685,200	1,672,800
40-44	1,467,000	1,464,600	1,418,000	1,392,700	1,853,800	1,844,500
45-49	1,549,900	1,576,500	1,370,000	1,352,700	1,562,900	1,559,200
50-54	1,427,500	1,497,700	1,393,800	1,415,200	1,356,300	1,355,300
55-59	1,436,500	1,545,400	1,414,500	1,486,800	1,279,700	1,293,400
60-64	1,337,000	1,516,300	1,226,800	1,379,900	1,233,800	1,319,400
65-69	1,075,800	1,351,800	1,124,400	1,365,500	1,148,900	1,326,500
70-74	702,100	1,098,400	895,400	1,233,800	878,500	1,152,600
75-79	416,100	791,700	564,700	959,300	652,800	1,014,800
80-84	220,500	497,100	260,100	604,500	382,400	760,500
85+	112,400	325,300	126,000	414,600	194,200	615,500
All ages	23,897,000	25,255,000	24,160,100	25,474,200	24,995,200	26,104,400

1971   ■ Males ■ Females

1981

1991

	1998			EUROPEAN		WORLD
	Males	Females				
Under 1	324,200	308,800	Under 1	1,600	Under 1	2,400
1-4	1,341,600	1,273,600	1-4	6,400	1-4	9,600
5-9	1,771,200	1,685,600	5-9	7,000	5-9	10,000
10-14	1,712,800	1,623,700	10-14	7,000	10-14	9,000
15-19	1,653,600	1,565,100	15-19	7,000	15-19	9,000
20-24	1,579,400	1,504,700	20-24	7,000	20-24	8,000
25-29	1,993,900	1,889,500	25-29	7,000	25-29	8,000
30-34	2,196,200	2,097,900	30-34	7,000	30-34	6,000
35-39	2,053,200	1,982,100	35-39	7,000	35-39	6,000
40-44	1,749,700	1,730,000	40-44	7,000	40-44	6,000
45-49	1,703,100	1,700,700	45-49	7,000	45-49	6,000
50-54	1,746,300	1,753,800	50-54	7,000	50-54	5,000
55-59	1,345,200	1,364,200	55-59	6,000	55-59	4,000
60-64	1,224,000	1,265,900	60-64	5,000	60-64	4,000
65-69	1,100,000	1,214,700	65-69	4,000	65-69	3,000
70-74	939,600	1,146,100	70-74	3,000	70-74	2,000
75-79	728,300	1,052,900	75-79	2,000	75-79	1,000
80-84	384,900	704,700	80-84	1,000	80-84	500
85+	269,400	747,300	85+	1,000	85+	500
All ages	25,816,700	26,611,200	All ages	100,000	All ages	100,000

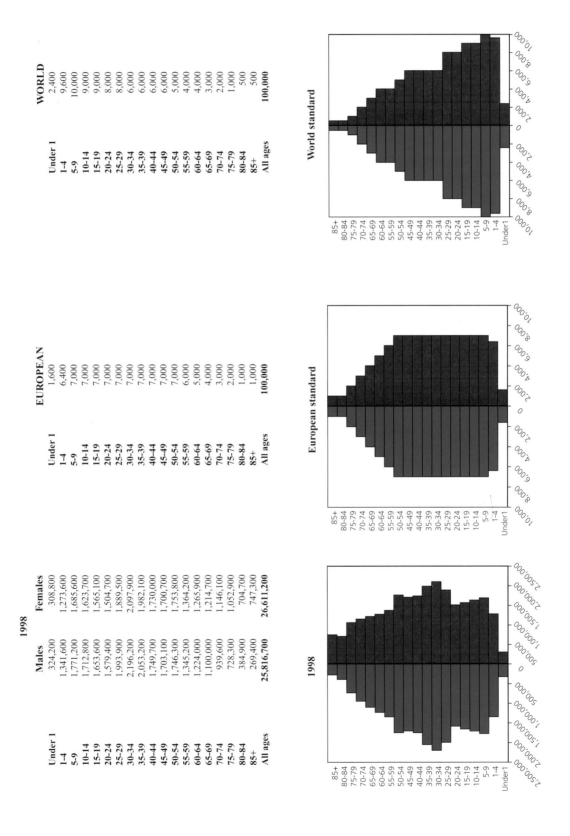

1998

European standard

World standard

Appendix G

THE CANCER REGISTRATION SYSTEM IN ENGLAND AND WALES

This appendix contains a brief history of the cancer registration system in England and Wales including an outline of the role of the Office for National Statistics (ONS), and details of the United Kingdom Association of Cancer Registries.

Background and early history

Cancer registration is the process of maintaining a systematic collection of data on the occurrence and characteristics of malignant neoplasms and certain non-malignant tumours. The procedure is widely established throughout the world and generally follows guidelines established by bodies such as the International Union Against Cancer (UICC), the International Agency for Research on Cancer (IARC), the International Association of Cancer Registries (IACR) and the World Health Organisation (WHO)[1,2].

The great and increasing suffering due to cancer was of concern to the Ministry of Health in the early 1920s and with the introduction of radium treatment, a system was initiated in parts of England and Wales to follow the outcome of treated patients. Both the Radium Commission of 1929 and the Cancer Act of 1939 incorporated the principle that statistical information about cancer patients was essential for planning and operating cancer care services. In 1945, the Radium Commission was designated as the Statistical Bureau to which the data should be sent for final analysis. This work was taken over by the General Register Office in 1947; and the Cancer Act was repealed in 1948 when the National Health Service Act came into force. From that time the General Register Office and its successors OPCS and, more recently, ONS, have collected and processed data forwarded under voluntary arrangements. Since January 1993, it has been mandatory for the NHS, including Trusts, to provide the core items listed in the cancer registration minimum data set to the regional cancer registries; and for the registries to send these data to ONS.

The 1960s

In February 1963 a conference was held at the Ministry of Health for the purposes of paving the way for 100% registration of cancer patients and for seeking means of improving the cancer registration scheme. A Working Party agreed on the regional and national objectives of the cancer registration scheme. At the **regional** level, the objectives were to improve the service to the cancer patient through good record keeping and efficient follow-up; and to provide information for local research into the value of treatment and for epidemiological studies, for the planning and assessment of the cancer service, and for the production of national statistics. At the **national** level, the objectives were to produce national statistical analyses likely to assist in the management of the disease and the understanding of it; to co-operate with other Government Departments and outside bodies in any survey aimed at furthering knowledge of the disease; and to participate, by supplying statistical data as required, in the work of international cancer organisations established to carry out research into the cause and course of cancer.

The Working Party spent a considerable amount of time determining what information should be obtained for analysis at the national level, but it was agreed that the information requested should be kept to a minimum – with the intention of obtaining a more complete record and a greater degree of accuracy. The Working Party's report (unpublished) also discussed and agreed recommendations on desirable national and regional tabulations; the elimination of duplicate activity (in data processing); duplicate registrations; dissemination of information; and the unique difficulties of the (then) Metropolitan Regional Hospital Board areas which are now covered by the Thames Registry and the South and West Cancer Intelligence Unit based in Winchester (formerly the Wessex Registry).

Advisory Committee Report 1970

Following discussions in 1969 between the Department of Health and Social Security (DHSS) and the Registrar General, an Advisory Committee on Cancer Registration was set up. It was requested simply "to consider and advise on matters of policy and method relating to the national cancer registration scheme", and its members included several eminent epidemiologists in addition to representatives from the DHSS, the registries and (the then) OPCS.

The Committee reviewed the existing scheme, in which each case of cancer was registered first of all on a registration form and the data subsequently transferred onto an abstract card. These cards were to be updated and re-submitted to OPCS after five, ten and fifteen years. Each registry received, through the machinery of the general system of vital registration and statistics, details of any death in its area where cancer was mentioned on the death certificate (this is known as the 'green card' system after the colour of the paper onto which the death certificate information was copied). Much difficulty had been caused at OPCS by the late submission of abstract cards, and – even worse – of follow-up cards. The quality of data varied considerably among the regions and even the best fell 'rather short' of 100 per cent accuracy in all particulars. The Committee felt that some of the data collected (for example on treatment) were of doubtful value and placed an unnecessary workload on the registries[3].

There was, however, unanimous agreement that some form of national cancer registration scheme was necessary in order not only to establish national incidence rates and monitor them for purposes of logistic planning and general epidemiological research, but also to permit prospective studies of cancer in selected groups of the population. In addition, information at the international level for comparison with experience in other countries made a valuable contribution to the understanding of the disease.

Revised scheme

A revised scheme was proposed[3] covering the definition of cases to be registered; the documentation (a revised and shortened abstract card); a nominal index for use by research workers; national tabulations (to be produced by OPCS); and death notifications

Figure G1 The cancer registration system of England and Wales

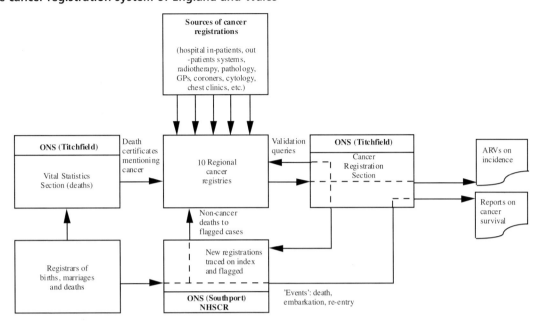

(green cards). Probably the most important change suggested was that the system of five, ten and fifteen year follow-up abstract cards should be stopped. Instead, cancer registrations would be 'flagged' in the records maintained by the National Health Service Central Register (NHSCR) – another part of OPCS (in Southport) – in the same way that deaths were. As non-cancer deaths to persons flagged as cancer-registered could be notified routinely to the registries, this, together with the green cards, would relieve them of the expensive and laborious task of tracing patients clerically (for example by using hospital records or writing to GPs). This revised scheme was introduced in 1971, backdated to cover all registrations whose incidence date fell on or after 1st January 1971. The essential features of the system (illustrated in Figure G1) have now remained unchanged for almost 30 years.

Advisory Committee Report 1980
The revised scheme was reviewed some ten years later when the Advisory Committee was reconvened. Its report[4] presented a large amount of national statistics on cancer incidence, survival, prevalence and mortality. It also highlighted the growing demands for information for clinical research; planning, organising and evaluating services for the prevention and treatment of cancer; epidemiological research; and education of the public.

Many of the Committee's comments on areas where problems were being experienced are still relevant today. The Committee re-emphasised the great value of recording the NHS number, and stressed that personal identification data were essential – for the elimination of duplicate notifications; to enable follow-up and calculation of survival rates; and to enable registration data to be linked (with suitable safeguards) to other data about the same person. They found a substantial degree of variation among the regions in the excess of registrations over deaths; although difficult to interpret, this suggested an equivalent variation in the degree of ascertainment. The report discussed the three main methods of collection: peripatetic staff, hospital staff and the Hospital Activity

Analysis (HAA) system. HAA data were often considered to be insufficiently reliable, but the Committee noted that the three regions which used HAA as their primary source were not those which had low numbers of registrations compared with deaths. The use of information from pathology departments, to increase not only the accuracy but also the completeness of ascertainment, was encouraged. As well as being complete, the data needed to be up to date and here the Committee found grave shortcomings since the inception of the revised scheme.

While the average cost of registering one patient with cancer was only a very small fraction of the total cost of the management of the patient's illness, it was noted that (in England) the regional registries were funded by the regional health authorities, with no direct financial input from the DHSS or OPCS. It was possible that registration might not be given the necessary resources at regional level where priorities were decided autonomously.

The Committee concluded that cancer registration covering the whole of England and Wales should continue and be improved in several areas for the following reasons: preventative action was usually based on information from epidemiological studies (using the national register linked to the NHSCR); changes in incidence needed to be monitored because of public, political and medical concern, and improvements in treatment were making mortality data increasingly unreliable as an index of trends; changes in survival needed to be monitored; and reliable and up-to-date data on incidence were essential for the planning and operating of services for cancer detection and treatment.

Medical Advisory Committee review 1990
In 1989 a Working Group of the Registrar General's Medical Advisory Committee (MAC) was set up to review the operation of the cancer registration system, particularly the regional and national data collection methods, the quality and timeliness of the statistics produced, the uses made of the regional and national

registers, and the growing tendency to treat cancers in out-patient departments or privately. It was also asked to consider the implications of changes in demand for information and developments in information technology, and the priorities and level of resources required to maintain adequate registers. The potential implications of the recommendations of the White Paper *Working for Patients*[5] were also considered.

The Working Group noted that in addition to the traditional uses of cancer registration (monitoring of time trends and geographical variation in incidence) the system had become vital in several other areas[6]. These included the management of the substantial resources required for the preventative, curative and laboratory services for cancer; the planning and evaluation of services, particularly the screening programmes for breast and cervical cancer; the planning and evaluation of clinical management and treatment based on accurate and unbiased survival data and clinical trials; research into causes of cancer, involving case-control studies and the flagging of cohorts at the NHSCR; and information for health education and health promotion for both professionals and the public. Future uses of cancer registration (especially if linked with other databases) were identified, including evaluating programmes of care, quality assurance, and relating costs to clinical outcome.

The seventeen recommendations made by the Working Group for improvements to the system fell into several categories, relating to the organisation of the system; the collection, processing, quality, timeliness and completeness of the data; and the safeguarding of the necessary data release in view of the impending NHS changes and the use of the private sector.

One of the six recommendations in the 'organisational' area was that a Steering Committee should be established to oversee national cancer registration, with representation from the registries, OPCS, regional and district health authorities, the UK Co-ordinating Committee of Cancer Research, the Health and Safety Executive and the private health sector. This Steering Committee, which was chaired by Dr J Metters, the Deputy Chief Medical Officer at the Department of Health, held its first meeting in June 1991 and met subsequently at approximately six monthly intervals. This committee was re-formed as the Advisory Committee on Cancer Registration in 1997; it is chaired by Dr S Atkinson of the NHS Executive.

Three recommendations involved both the registries and OPCS: an expanded national core data set; co-operation with the private health sector; and the establishment of guidelines for the handling and release of data. These were discussed at several consultative meetings with the registries. Work on three other recommendations, relating to the provision of timely estimates of incidence at the national and regional level, quality control checks and the provision of up-to-date anonymous and summary data, was carried forward at ONS which in 1995 completed the redevelopment of its longstanding computer system to a new database environment (see below).

The role of ONS in cancer registration

The Office for National Statistics was formed by the merger of OPCS and the Central Statistical Office (CSO) in 1996. The Director of ONS, Mr Len Cook, is also the Registrar General for England and Wales. The National Cancer Intelligence Centre (NCIC) at ONS includes part of the Demography and Health Division in London which co-ordinates all the work on cancer registration and carries out a wide range of secondary analysis and research; part of the Population and Vital Statistics Division in Titchfield which conducts the primary data processing of registry data; and a section at the NHSCR in Southport which the flags the cancer registrations on the central register. Much of the secondary analysis and research, which is carried out by a statistician and three researchers, supported by a medical epidemiologist and the Deputy Chief Medical Statistician who is also Professor of Epidemiology and Vital Statistics at the London School of Hygiene and Tropical Medicine (LSHTM), is done in collaboration with academic and external researchers, for example at the LSHTM, the Cancer Screening Evaluation Unit at the Institute for Cancer Research, and the Small Area Health Statistics Unit at Imperial College.

Most registries collect a large amount of information about the patient, the tumour and the treatment. The registries carefully collate all the data for any one patient to avoid duplication of records. This is not a quick process, as information is often not made available to the registry until the main course of treatment is finished. A sub-set of the data, as defined in the cancer registration minimum data set[7] is sent to the national registry at the ONS office in Titchfield, near Southampton. The data items are:

Core	Optional
Record type (new registration, amendment, deletion)	Country of birth
	Ethnic origin*
Identity number (unique)	Patient's occupation
Patient's name	Patient's employment status
Patient's previous surname	Patient's industry
Patient's address	Head of household's occupation
Post code	Head of household's employment status
Sex	Head of household's industry
NHS number	Registration from screening*
Marital status	
Date of birth	
Date of death (if dead)	
Incidence date	
Site of primary growth	
Type of growth	
Behaviour of growth	
Multiple tumour indicator	
Basis of diagnosis*	
Death certificate only indicator*	
Side (laterality)*	
Treatment(s) (indicators)*	
Stage†	
Grade†	

* From incidence year 1993

† From incidence year 1993; phased introduction – initially only for breast and cervix.

The data are loaded onto the new person-based database (see below) and validated. The extensive checks include the compatibility of the cancer site and the associated histology; these checks are closely based on those promulgated by IARC[2]. Once all the expected records for any one incidence year have been received and validated at ONS, detailed tables are published on the numbers and rates of all types of cancer by age and sex, and by region of residence[8].

All the work on processing in Titchfield and flagging at the NHSCR in Southport has, since 1993, been paid for by the Department of Health (DH). A service level agreement (SLA) has been negotiated between DH and ONS. Work on the key targets and outputs established in the relevant ONS divisional business plans and the SLA is monitored continuously. ONS makes formal progress reports to the six-monthly meetings of the Joint (DH and ONS) Cancer Regsistration Management Board, and to the Advisory Committee on Cancer Registration.

Redevelopment of the ONS cancer registration computer system

Beginning in 1990, over 20 of the major computer processing systems at the (then) OPCS – including births, deaths, cancer registrations, the Longitudinal Study (1% linked sample from the censuses), marriages and divorces – were redeveloped onto a modern database environment. The two main objectives of the redevelopment of the cancer registration computer system were to have an effective and efficient processing system; and a person-based database (rather than annual files of tumours). To meet the timetable for introducing the new system, it was necessary to convert the 21 annual tumour files (1971 to 1991 inclusive) to a person-based database before the new system began operation. From among the 4.5 million records, those which were either duplicates or were true multiple primary records for the same person were linked together by a probability matching process[9] based on those successfully operated by the Oxford Record Linkage Study, Statistics Canada, and the Information and Statistics Division (ISD) of the Scottish Health Service[10,11,12]. Information on linked registrations was sent to the cancer registries for the deletion or amendment of records as appropriate. The essential structure of the cancer registration system in England and Wales, shown in Figure G1 above, has remained unchanged; but the identification, and the sending to the regional cancer registries, of the death certificates mentioning cancer and the non-cancer deaths to flagged cases, is now done by the new system in Titchfield. In addition, all validation errors are now returned to the appropriate registry for resolution.

In parallel with the work on the redevelopment of the system at ONS, a very large amount of data enhancement work was completed. This included 13,000 new registrations, amendments and cancellations; amendments to about 40,000 records from the probability matching exercise; 15,000 updates of date of death; 25,000 date of birth and date of death discrepancies; 7,000 no trace indicators added to the database; and smaller numbers of trace and event rejects, multiple primary cancer queries from registries, mis-traced Welsh records, "dead" now known to be alive, sex discrepancies, partial or invalid postcodes, and embarks. In addition, 36,000 queries from NHSCR about possible multiple primary cancers were dealt with.

The backlog of over 600,000 records which had built up in the registries during the time that the person-based database was being constructed was successfully processed by the NCIC in Titchfield. Priority for the processing of amendments resulting from validation errors was given to data for incidence years 1990 and 1991. At the same time, the NCIC worked steadily through the remaining problems – some left over from the old computer system, and some new ones. These included amendments to the way the system handled the notifications to the registries of death certificates containing a mention of cancer; corrections to records with duplicate identity numbers; re-numbering of some records for one regional registry; and improvements to postcodes. In addition, the revalidation – to the higher standards embedded in the new system – of all the data which had previously been processed on the old computer system has been carried out, queries sent to the regional registries, and records amended. The new NHS numbers for flagged cases, together with any dates of death, have been sent from the NHSCR to Titchfield, and passed to the cancer registries. This information has enabled both ONS and the registries to amend records for the "immortals" – cases registered alive but whose death was not previously linked to the cancer registration.

The backlog of records which had been processed in Titchfield was sent to the NHSCR in Southport once the testing of the module of their new computer system which deals with the flagging of cancer cases had been completed. It was known that about 65,000 of these were for people who had died before 1991 when the computerised index was assembled, and so they would not be on the database at NHSCR. These records were therefore stripped off the Titchfield database and sent separately to Southport on paper. Of the remaining records, which were sent on electronic media, it was expected that about 300,000 would match automatically on the system. It was planned to do the batch runs in order, ie the earliest registrations first, to facilitate the determination of true multiple cancers and duplicates. The flagging of the stockpiled registrations for incidence years 1971 to 1990 was completed in January 1997; and the resulting trace and event (death, embark, re-entry) data were sent to Titchfield and added to the database. All flagging for records up to incidence year 1996 which have been received at ONS and have passed the validation checks (about 85% of the expected totals for 1995 and 1996) has been completed [June 2000] and work has begun on cases diagnosed in 1997. At the same time, ONS is attempting to keep earlier incidence years up to date by processing and flagging any "late" registrations received from the cancer registries.

Proposed extension to the cancer registration minimum data set

A conflict exists between the number of data items collected and data quality. This has been recognised by the three reviews of the national system described above[3,4,6]. The minimum data set is currently being reviewed. There is great pressure to include the stage of disease for all cancers, and details of treatment. This will require the information on stage to be made explicit by clinicians. Although the private sector is not covered by the minimum data set, members of the Independent Healthcare Association have generally been very co-operative; however, the growth of private pathology laboratories is a concern.

Processing problems

There have been three main problems with the cancer registration process. First, the timeliness of national data based on the full set of individual records depends on the speed of the slowest registry in completing its submissions to ONS. In the past, there has always been (at least) one registry which, for a variety of excellent reasons at the time, has lagged considerably behind the others. The most timely complete results were those for 1982 and 1983 which were published in 1985 and 1986 and were therefore only two years out of date. With the co-operation of the registries, however, it was possible to produce provisional results for 1990 to 1993 well in advance of the corresponding reference volumes. The latest available full results are those for 1994[8].

Second, the database is "live" or "dynamic" in the sense that records may be modified or deleted if new information is obtained. The information from "trace back" of a death certificate may result in a case being registered many years after the true incidence date. This, together with the general timeliness problem, meant that any attempts in the past to bring forward the publication of national results has artefactually reduced the numbers of cases reported in the OPCS and ONS reference volumes. For several incidence years in the mid-1980s, there are now around 10% more cases on the national register that when the reference volumes were published (see Figure H1 in Appendix H). Recently, however, several registries have redeveloped their computer systems (as has ONS) and their timeliness has improved dramatically. The availability of complete information for incidence years up to 1996 from half of the registries enabled ONS to produce in 1999 reliable provisional results for 1994 to 1996 for the 20 or so major cancer sites[13]. This volume contains similar provisional results up to 1997[14].

Third, cancer registration is not statutory, and ONS has no organisational, managerial or financial control over the regional registries. In 1994, the registries passed from regional control to lead purchasers. Local needs for up to date information have in some areas resulted in considerable improvements in timeliness. On the other hand, although safeguards, and quality and timeliness standards, for national data were included in the national core contract[15,16], the requirements of lead purchasers who hold the registries' budgets sometimes take priority over the supply of data to ONS. In short, there was a power vacuum which, together with chronic underfunding of registries over a long period, means that it has been difficult to obtain timely, accurate and comparable data at the national level.

Advisory Committee on Cancer Registration review 1999/2000

In recent years, and particularly since the publication in 1995 of the Calman-Hine report on cancer services[17], the role of cancer registries has been extended. Cancer registries have contributed to studies on the variations in the outcomes for cancer patients across the UK and in the investigations into the underlying causes of these variations. Cancer registries were also increasingly being asked to provide data to support the planning and monitoring of cancer service delivery, including the national breast and cervical screening programmes. For these purposes, more extensive data sets are needed and the timeliness of information is of great importance. For the purposes of clinical governance, data on the patterns of care and outcomes for specified sub-groups of patients, for example defined by extent of disease or "stage", are needed.

This expansion of the traditional role of cancer registries led to renewed interest in them, but drew attention to the variable quality of the service that individual registries provided. Concerns were expressed about their capacity to provide up to date, complete and accurate data.

Despite the changes implemented following the three national reviews described above, these concerns had persisted, and in April 1999 the Advisory Committee on Cancer Registration, on behalf of the Department of Health, commissioned Professor Charles Gillis, Director of the West of Scotland Cancer Surveillance Unit, to undertake a further review of cancer registration in England.

The review found that due to the history of the cancer registries, which had grown up more or less autonomously since before the second world war, there were considerable variations among them in terms of organisational structures; type of host institution (hospital, health authority, academic); title; data collection process (predominantly manual or electronic); range of tumours registered; data items collected; IT systems; research activity; and significant variations in completeness, accuracy and timeliness of data submission to ONS. The budgets per head of population served and the cost per case registered appeared to vary considerably, although those for the majority of registries clustered closely around the average.

The timeliness of data acquisition by some of the registries had been poor, with the knock on effect that they were, in turn, slow in submitting data to ONS for national collation. For example, it was only in August 1997 that provisional figures were published for cancers diagnosed in 1992 – so at first sight national cancer registration data looked five years out of date – and confirmed registrations for 1991 were only published in December 1997. But as noted above, the timeliness of several registries improved dramatically during the late 1990s following redevelopment of their computer systems, and the provisional results up to incidence year 1996 were only two years out of date (and two years behind the available mortality data).

The issue of timeliness was addressed through the allocation by the Department of Health of £500,000 from the Public Health Development Fund, with the aim of ensuring a measurable improvement in the timeliness and quality of national cancer incidence and survival data. The target was that through this investment, all cancer registries would submit complete data up to and including 1997, to the quality standard in the national core contract, to ONS by the end of September 2000.

The review noted that data quality varied between registries. The editors of *Cancer Incidence in Five Continents Volume VII*[18] assess the quality of data submitted by individual cancer registries. It was a matter of concern that not all cancer registries in England provided data acceptable to the editors of this standard work.

Most cancer registries collect far more data than required for the national minimum data set. The review found tensions regarding the priority given to local and national need for data. In some cases, national priorities were unduly neglected. Some cancer registries had not complied with the requirement to submit data to ONS within the timescales specified in the national core contract. Data on variables relating to stage of disease and treatment were variably collected. Registries generally only collected information on treatment given within six months of diagnosis, as specified in the core contract, and so surgical, radiotherapy and chemotherapy treatments given later in the course of a patient's illness would have been excluded.

The review concluded that the credibility of the data for comparisons of the risks of cancer over time, and of outcomes within some cancer registry areas was well established. But the reliability of inter-regional comparisons was doubtful and the requirement for data of a uniform high standard in all parts of England, for the purposes of public health and clinical governance was certainly not being met.

The review made a number of key recommendations for how cancer registries should be strengthened, so that they would be able to contribute fully to the cancer modernisation agenda by providing robust data to support the planning and monitoring of cancer service delivery and identify the scope for NHS intervention in relation to deprivation and cancer.

The report by Professor Gillis defined an enhanced role for cancer registries in directly informing and improving national and local cancer strategies through the expert use of timely, complete and accurate data. Specifically, it recommended that:

- Cancer registration should be provided by a new framework driven by and accountable to a new National Cancer Registration Policy Executive, chaired by the National Cancer Director. The role of the National Cancer Registration Policy Executive would be to ensure that all the cancer registries operate to the same high standards and ensure that they contribute to the national cancer agenda.

- The National Cancer Registration Policy Executive should have the support of an advisory group to aid in development of issues where detailed professional input is required.

- The directors of cancer registries should be accountable (through Regional Directors of Public Health) for the conduct of the work programmes set by the National Cancer Registration Policy Executive.

- Information on clinical treatment and stage of disease at diagnosis is critical to the development of cancer strategies. The performance of cancer registries in obtaining and disseminating such data, and of trusts in making it available, should be subject to regional appraisal.

- Data collection should continue on the basis of the existing geographical boundaries of cancer registries. However, their outputs must relate to the new regional structure of the NHS in England.

- Provision of up-to-date, complete and accurate basic data is critical to the users of cancer data. The National Cancer Registration Policy Executive should set and monitor performance standards in these areas, particularly to address the issue of improving timeliness.

The implementation of the main recommendations of the Gillis review by the Department of Health will ensure that the cancer registries are strengthened so that they not only fulfil their longstanding role in collating information on the incidence of and survival from cancer, but also provide complete, accurate and timely information on cancer services.

The United Kingdom Association of Cancer Registries
In the early 1990s, the regional cancer registration system in the UK was subject to rapid change. With the development of information technology, the pace of change in registration practice quickened, and increasing demands for accurate and timely information were made on the cancer registration system. Changes in the organisation of the health service and in the methods of health care delivery contributed to an increased interest from various authorities and scientists. There were new uses which could and should be made of registration data, such as medical audit and quality assurance of health care, as well as the routine uses which have been made of these data in the past, such as estimation of incidence and evaluation of survival and mortality.

There was widespread awareness both of the need to improve the quality and completeness of cancer registration data, and of the opportunities to do so through the use of information technology. Together with the increased interest from external bodies in using the data, this led to the creation of several groups bringing together cancer registry staff and personnel from OPCS (as it then was) to discuss and resolve matters of common interest.

The longest standing of these was the *Cancer Registries' Consultative Group* which concerns itself essentially with issues of data collection, including coding and data quality. It now has representation from all cancer registries in the UK and Eire, and its members are for the most part registry managers and others closely involved in the day-to-day business of data collection. The *Cancer Surveillance Group* was set up in 1989 to meet a perceived need for a forum bringing together those with an interest in the use of cancer data. It has a loose, open and informal membership and structure. Its members include epidemiologists and statisticians, as well as other registry staff. The *Cancer Registries' Information*

Technology Group brings together technical experts from the various registries. Education and training was another area of activity thought to be of such importance that it could justify the establishment of another group. There was, however, no forum which brought together registry directors on a regular basis. There was a danger, therefore, with so many different perspectives and forums in which different points of view could be expressed, that the cancer registries might fail to speak with a united voice when, for example, making representations or giving advice to government. With no coherent framework of organisation, there would be a strong possibility of duplication of effort and inadequate communication between the various groups.

It was therefore proposed that a United Kingdom Association of Cancer Registries (UKACR) be established. Following preliminary meetings at which almost all of the UK registries were represented, the Association was brought into being on 2nd April 1992 in Cardiff.

The Association has a federal structure. All affiliated population-based cancer registries in the United Kingdom, ONS, the Information and Statistics Division of the NHS in Scotland and the Northern Ireland Cancer Registry are full members with their representative, usually the director, having a vote on the Executive Committee. Associate (non-voting) members currently (September 2000) comprise the Childhood Cancer Research Group in Oxford, the CRC Paediatric and Familial Cancer Research Group in Manchester, the Republic of Ireland national cancer registration scheme and the Marie Curie Cancer Care charity. Since the formation of the UKACR, a *Quality Assurance Group* has been set up to standardise the methodology for, and report on, various registry performance indicators included in the national core contract[15,16] such as timeliness and the percentage of registrations made solely from a death certificate. A *Training Group* and a *Coding and Classification Group* have been established to oversee and co-ordinate the implementation of developments in those particular aspects of cancer registries' work. And a *Clinical Effectiveness Group* is taking forward issues relating to the registries' expanding role in clinical audit and performance monitoring on cancer. The Chairs of the various sub-groups, are invited, as appropriate, to attend Executive Committee meetings as observers.

The current officers are: Chair – Dr M Roche, Director of the Oxford Cancer Intelligence Unit; Vice-Chair – Professor D Forman, Director of Information and Research at the Northern and Yorkshire Cancer Registry and Information Service; and Treasurer – Mr P Needham, Deputy Director of the Trent Cancer Registry. It was agreed that ONS was the most appropriate body to provide secretariat facilities; Dr M J Quinn (Director of the NCIC) was nominated by ONS to be the Association's Executive Secretary.

The UKACR provides:
- a focus for national initiatives in cancer registration;
- a coherent voice for representation of cancer registries in the United Kingdom;
- a channel for liaison between registries and for agreeing policy on matters connected with cancer registration;
- a framework to facilitate the operation of special interest groups and regional registries; and
- a means of stimulating the development of cancer registration, of information procedures and practices, and of research based on cancer registry data[19,20].

The UKACR represents the views of its members to government and other bodies operating at national level on issues concerned with data quality, the definition of information requirements, and the development of health information systems where these have implications for cancer registration, in particular where matters of overall policy are concerned. The Association is represented on the National Advisory Committee on Cancer Registration. The establishment of such close links is very important given the intimate ties many regional registries have with NHS information systems, and the potential importance of cancer registration to NHS functions such as medical audit and contracting.

The UKACR has, through consensus, examined and improved coding and classification issues; agreed the complex interface document for transmission of data to and from ONS; developed performance indicators; produced a training manual and cancer-specific training packs for registry staff; developed guidelines for the release of data, including for the rapidly expanding field of genetic counselling; developed guidelines for standardisation of reported results; and established a forum for sharing the latest epidemiological research. Consensus may be slower to achieve than coercion, but may in practice be stronger and more valuable as there is often a better chance that an agreed procedure will actually be followed. Even near consensus requires those disagreeing to continually justify their minority position.

Appendix H

DATA AND METHODS

1 DATA

Cancer registrations

Care is required in the interpretation of cancer registration statistics, particularly when addressing either trends over time or differences between regions or countries.

Registration of cases of cancer is a dynamic process in the sense that the data files both in the regional cancer registries and at the Office for National Statistics (ONS) are always open. Cancer records may be amended – for example, the site code may be modified should later, more accurate, information become available. The date of death is added for cases registered when the person was alive. Records may be cancelled, although this is relatively unusual. Also, complete new "late" registrations may be made after either the regional cancer registry, or ONS, or both, have published what were thought at the time to be virtually complete results for a particular year.

Consequently, the figures for registrations published by a regional cancer registry may be different from those published by ONS, which will generally have been produced at a different (usually later) time. In addition, both sets of published figures will differ again from the numbers of registrations currently on the files. Further differences between regional registry and ONS figures may arise if records which have been rejected by the validation process at ONS have not been corrected by the registry concerned before the corresponding tables for publication are produced.

ONS has been advised both by expert epidemiologists and by members of the Advisory Committee on Cancer Registration, that non-melanoma skin cancer (ICD10 C44) is greatly under-registered. Registration varies widely depending on a registry's degree of access to out-patient records and general practitioners. This under-registration of non-melanoma skin cancer is not just a problem for the registries of England and Wales. *Cancer Incidence in Five Continents Volume VII*[1] reports that cancer registries in the United States, Australia, and parts of Europe, also collect very limited information on these skin cancers. Figures given in this book for "all cancers" therefore **exclude** non-melanoma skin cancer.

Cancer incidence data for England and Wales for 1995-1997 are provisional[2]. Figures are based on the complete submissions of records from five of the ten cancer registries. For the five registries for which complete data were not available, it was assumed that the age and sex specific incidence rates changed, compared with the previous year, by the same amount as the average change in those regions for which data were available.

Quality of cancer registration data

For the purposes of the national cancer registration scheme the term "cancer" includes all malignant neoplasms and the reticuloses, that is conditions listed under site code numbers C00

to C97 of the Tenth Revision of the International Statistical Classification of Diseases and Related Health Problems[3]. In addition, all carcinoma *in situ* and neoplasms of uncertain behaviour are registered. Benign neoplasms and neoplasms of unspecified nature of bladder and brain, including the pineal and pituitary glands, are also registered, together with hydatidiform mole.

It should be noted that some cancer registries are not always able to collect complete information about benign, uncertain and unspecified neoplasms and therefore these registration rates are almost certainly underestimates of the true incidence. In particular this should be noted when interpreting regional differences.

A brief history of cancer registration in England and Wales is given in Appendix G. The essential features of the current system have remained unchanged for nearly 30 years. The main flows of information to and from the regional registries and ONS, including the National Health Service Central Register (NHSCR), are illustrated in Figure G1 in Appendix G. Some aspects of the system which are relevant to the interpretation of the data have been discussed in considerable detail by Swerdlow[4]. These and others including geographic coverage; methods of data collection; ascertainment (or completeness of registration); completeness of recording of data items; validity; accuracy; late registrations, deletions and amendments; duplicate and multiple registrations; registrations from information on death certificates; clinical and pathological definitions and diagnoses; changes in coding systems; completeness of flagging at NHSCR; changes in definition of resident population; and error are discussed below.

Over the years, changes have occurred to the number of registries and to their geographic coverage. In 1950 there were 74 centres registering cancer in England and Wales, but the system was progressively simplified and by 1958 ten regions were covered by regional cancer registries; full coverage of England and Wales (but not 100% ascertainment of cases – see below) was achieved in 1962. Some registries covered more than one RHA: the current Thames Registry was formed in 1985 with the merger of the North West, North East and South Thames registries (the last of these covered both the South West and South East Thames RHAs). Wessex was separated from the South Thames registry in 1973; this coincided with a change in the method of data collection and a substantial increase in numbers of registrations for some parts of the Wessex region. Following reorganisations at the regional level in the NHS in 1996, the former South Western and Wessex RHAs were covered by the South and West Cancer Intelligence Unit based in Bristol and Winchester. The former Yorkshire RHA and part of the former Northern RHA are now covered by the Northern and Yorkshire Cancer Registry and Information Service based in Leeds (the remainder of the former Northern RHA, South Cumbria, is now covered by the North Western registry). Further reorganisations at the regional level in the NHS occurred in 1999, but no corresponding major changes have been made to the areas covered by the regional cancer registries. Some registries received reports from several centres in their region – at various times five regional centres existed in Trent, two in South Western, and three in East Anglian.

The independent regional registries differ considerably in their methods of data collection; some employ peripatetic clerks, others use hospital record staff to extract data for the registry, and several rely heavily on other organisations' computer systems including those in hospitals and pathology laboratories. The registries probably also differ in the level of ascertainment of their data (that is the degree to which reportable incident cases of cancer in the population are actually recorded in the registry) but the best are known to have very high levels. Direct measures are only available from occasional special studies[5,6]. That by Hawkins and Swerdlow [5] estimated that the under-ascertainment of registration of childhood cancers by the regional registries was just under 5%; under-ascertainment may be greater for adults, for whom registration and record linkage (in the registries and at NHSCR) may be more difficult, than for children. General indications of ascertainment levels can be obtained from comparisons of the numbers of registrations and deaths in a period. The figures for deaths are those coded to a particular type of cancer as the underlying cause of death in residents of the same geographical area. Such mortality to incidence ratios by sex and site for 1994 are presented in Appendix D. These ratios have several limitations, but there are variations between regions (and over time) which would be difficult to explain unless there were similar variations in ascertainment.

It may be difficult to interpret any apparent trends in cancer registrations because the registries are continually striving to increase their levels of ascertainment of cases. Any particularly large increases from year to year in the numbers of registrations for an individual registry are most likely to have arisen because of this. For example, the recorded incidence for residents in some parts of the Thames Regional Health Authorities was unusually high in 1992, and unusually low in 1993, as a result of a one-off exercise by the Thames Cancer Registry in 1993 to find further information for people with cancer mentioned on their death certificate[7].

Completeness is the extent to which all appropriate data items have been recorded in the registry database. Some data items are essential; if high proportions of such items are missing, this is an indicator of poor quality. For example, for cases that have been registered solely from the information on a death certificate (DCO) the incidence date is unknown and has to be taken as the date of death and the case may well be recorded against the wrong calendar year. A high DCO rate also implies under-ascertainment[8] because patients are being missed by the registry while they are alive and not all cancer patients die of their disease (in which case, cancer is not mentioned on the death certificate). Other quality indicators are the proportion of cases where the primary site is unknown, and the proportions where important information such as the age of the patient or their postcode, is missing. Tables giving the proportions of registrations by region that have zero survival (which include both DCO cases and patients who were known to have died on the day of diagnosis – true zero survival) are given in

Appendix E1. Tables giving the proportions of registrations by region with site unspecified are given in Appendix E2.

The agreed procedures to be followed by the regional cancer registries and ONS when submitting and processing data are set out in the "Registry/ONS Interface Document"[9]. When a registry's submission is loaded onto the database at ONS, a large number of validity checks are carried out. There are over 40 checks on individual data items. These include that dates are valid, or that an "indicator" is either 0 or 1 (or "&" if not known). There are around 20 cross checks between data items. These include the consistency of dates, for example that the incidence date is not after the date of death, and that the cancer site and histology are compatible. These latter cross checks are based closely on those promulgated by the International Agency for Research on Cancer (IARC)[8] and used by them when verifying data for inclusion in "*Cancer Incidence in Five Continents*"[1]. Combinations of site and histology are checked against three lists:

(i)   histology codes which will be accepted in combination with any site code;

(ii)   histology codes which will only be accepted if the site code is in the appropriate group (of which there are over 50); and

(iii)   histology codes which will not be accepted in combination with any of the sites in a group (of which there are two).

If a record passes all the checks and cross checks, it is given a quality status of 1. If a record fails any one of a small number of vital checks and cross checks, for example if the date of birth is invalid, thus making it impossible either to include the data in an output table in the ONS annual reference volume[7] or to flag the person concerned at the NHSCR (see Appendix G), it is given a quality status of 3. If a record passes all the vital checks and cross checks but fails one or more other checks, it is given a quality status of 2, and along with records that have a quality status of 1, can be used in outputs and sent to the NHSCR for flagging. Information about all records which fail any of the validation checks is sent to the registries for them to investigate and submit corrections.

The national core contract for cancer registries[10,11] requires that when a registry's data for a particular year are complete, no more than 0.5% of records should have a quality status of 3. When OPCS redeveloped its cancer registration computer processing system in the early 1990s, all the previously submitted records were re-validated using the more stringent checks[9] incorporated in the new system. The quality status of all the records on the database at the National Cancer Intelligence Centre from 1971 up to 1994 (the latest complete year) is shown in Table H1.

As with completeness, the accuracy of the data (that is the proportion of cases recorded with a given characteristic that truly have the attribute) is only occasionally known directly from special studies. Various indirect measures, however, suggest that there is considerable variation between regions. A report of a project to

Table H1 Number of newly diagnosed cases of cancer* by quality status, as at May 2000, England and Wales, 1971-1994

| Year | Total submitted | Quality | | | Status 3 as |
		Status 1	Status 2	Status 3	% of total
1971	150,455	147,965	2,176	314	0.2
1972	155,144	152,686	1,963	495	0.3
1973	160,081	157,377	2,178	526	0.3
1974	165,845	164,423	803	619	0.4
1975	167,156	165,780	598	778	0.5
1976	168,163	165,805	1,464	894	0.5
1977	171,950	170,824	172	954	0.6
1978	171,837	170,730	176	931	0.5
1979	173,981	173,013	302	666	0.4
1980	178,947	177,880	337	730	0.4
1981	184,827	182,325	1,596	906	0.5
1982	186,415	183,653	1,809	953	0.5
1983	188,347	184,997	2,308	1,042	0.6
1984	190,049	185,719	3,165	1,165	0.6
1985	201,867	199,202	1,619	1,046	0.5
1986	197,231	194,131	1,938	1,162	0.6
1987	204,135	200,522	2,538	1,075	0.5
1988	210,951	206,742	2,879	1,330	0.6
1989	210,686	205,836	3,355	1,495	0.7
1990	210,882	189,709	20,925	248	0.1
1991	213,934	207,913	5,715	306	0.1
1992	222,846	217,115	5,141	590	0.3
1993	220,917	215,641	4,527	749	0.3
1994	225,119	224,322	385	412	0.2

* All malignant neoplasms excluding non-melanoma skin cancer

audit the quality and comparability of cancer registration data in the UK, carried out under the aegis of the United Kingdom Association of Cancer Registries (see Appendix G) was published in 1995[12]. Variations among the registries were found in data quality for diagnostic factors, incidence date, stage of disease, treatment information, and use of death information. A study at the Merseyside and Cheshire Registry[13] also found that data quality within a registry varied by the age of the patient, the cancer site, and area of residence. However, a substantial audit of Scottish cancer registry data[14], in which information was re-abstracted from the available records, found that severe discrepancies had occurred in under 3% of cases. The review[12] concluded that although comparisons between the various published studies was difficult, cancer registry records were largely complete, accurate and reliable. The review found that the quality of cancer registry data

Figure H1 Number of registrations (thousands) published in ARVs and currently (May 2000) on the NCIC database

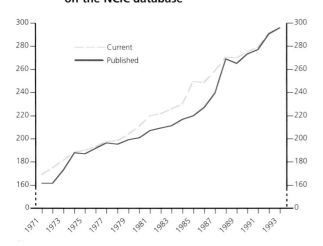

depended heavily on the competence and experience of staff in the registry; on maintaining good relationships with clinicians, staff in health authorities, and scientists and on the registry's active involvement in research.

The point in time at which ONS, in consultation with the regional registries, decides to produce the tables for the reference volume is necessarily a compromise between two principal considerations – the need to minimise the delay between the relevant data year and the publication of the detailed results, and the requirement to obtain a very high level of completeness of the data and hence minimise the number of <u>late registrations</u>. The gap between the data year and production of tables has varied considerably; as a result there are currently varying proportions of additional cancer registrations held on the computer files at ONS compared with the numbers published in the corresponding reference volume, as shown in Figure H1. Over the twenty three year period the differences have averaged around 3% although the differences for 1985, 1986 and 1987 are larger as a result of the problems with the transmission of data between the Thames Registry and ONS[15]. The overall figures contain within them some substantial variations among the regions. For example, a problem at OPCS (as it was then) with the processing of one data tape for 1985 from the North Western registry resulted in a shortfall in the published figures of around two thousand registrations. Although this made a difference of less than 1 per cent to the total for England and Wales, it represented a shortfall of around 10 per cent for the North Western region. <u>Late deletions and amendments</u> to data are in general a much smaller problem than late new registrations.

A CD-ROM containing anonymised records of new cases of cancer – including all the "late" registrations – for incidence years 1971 to 1992 has been produced by ONS[16]; the data are geographically coded to regional health authority level. Also included are anonymised records of deaths from cancer for 1971-1997; and the relevant mid-year population estimates to enable the calculation of incidence and mortality rates.

While late registrations result in the figures published in the reference volume being too low, <u>duplicate registrations</u> can artificially inflate them. Such duplication may arise if a patient is resident in one region but treated in another; this is particularly so for those resident in North Wales and treated in Liverpool, and for those resident around London who are treated in central London. Duplications are prevented firstly by the regional registries which hold alphabetic indexes of names and carry out computer searches; and secondly by the flagging at NHSCR, where if on flagging, a previous registration is found for the individual, the registrations are examined to see if they are duplicates or <u>true multiple primary</u> cancers. The rules for decisions on duplicates/multiples have changed over time, particularly for 1978 registrations which led to a 13 per cent decrease in registrations for Welsh residents. Currently, with the agreement of the regional registries, all such cases are referred back to them by ONS, and decisions taken according to an agreed set of rules[9].

Since the early 1960s, copies of information from all <u>death</u>

Table H2 Definition of cancer codes, ICD6-ICD10, 1950-1999

Site	ICD6	ICD7	ICD8	ICD9	ICD10
All cancers	140-205	140-205	140-207	140-208	C00-C97
All cancers excluding NMSC	140-205 x191	140-205 x191	140-207 x173	140-208 x173	C00-C97 xC44
Lip	140	140	140	140	C00
Tongue	141	141	141	141	C01-C02
Salivary glands	142	142	142	142	C07-C08
Gum	143	143	143	143	C03
Floor of mouth	143	143	144	144	C04
Unspecified parts of mouth	144	144	145	145	C05-C06
Oropharynx	145	145	146	146	C09-C10
Nasopharynx	146	146	147	147	C11
Hypopharynx	147	147	148	148	C12-C13
Pharynx unspecified	148	148	149	149	C14
Oesophagus	150	150	150	150	C15
Stomach	151	151	151	151	C16
Small intestine	152	152	152	152	C17
Colon	153	153	153	153	C18
Rectum	154	154	154	154	C19-C21
Liver	155-156 part	155,156 x155.1	155	155	C22
Gallbladder	155 part	155.1	156	156	C23-C24
Pancreas	157	157	157	157	C25
Retroperitoneum and peritoneum	158	158	158	158	C48
Digestive organs etc. unspecified	159	159	159	159	C26
Nasal cavities	160	160	160	160	C30-C31
Larynx	161	161	161	161	C32
Trachea, bronchus and lung	162-163 part	162,163 x162.2	162	162	C33-C34
Pleura	162-163 part	162.2	163.0	163	C38.4, C45
Thymus, heart and mediastinum	164	164	163.1	164	C37-C38 xC38.4
Respiratory system	:	165	163.9	165	C39
Bone	196	196	170	170	C40-C41
Connective tissue	197	197	171	171	C47, C49
Melanoma of skin	190	190	172	172	C43
Non-melanoma of skin	191	191	173	173	C44
Female breast	170	170	174	174	C50
Male breast	170	170	174	175	C50
Uterus unspecified	174	174	182.9	179	C55
Cervix	171	171	180	180	C53
Placenta	173	173	181	181	C58
Uterus	172	172	182.0	182	C54
Ovary	175	175	183	183	C56-C57
Other female genital	176	176	184	184	C51-C52
Prostate	177	177	185	185	C61
Testis	178	178	186	186	C62
Penis	179	179	187	187	C60, C63
Bladder	181	181	188	188	C67
Kidney	180	180	189	189	C64-C66, C68
Eye	192	192	190	190	C69
Brain	193 part	193.0, 193.9	191	191	C71
Other nervous system	194 part	193.1-193.8	192	192	C70, C72
Thyroid	194	194	193	193	C73
Other endocrine glands	195	195	194	194	C74-C75
Other and ill-defined site	199 part	199 part	195	195	C76
Secondary & unspecified of lymph nodes	199 part	199 part	196	196	C77
Secondary - respiratory & digestive system	199 part	199 part	197	197	C78
Secondary - other specified sites	199 part	199 part	198	198	C79
Unspecified site	199 part	199 part	199	199	C80
Non-Hodgkin's lymphoma	200, 202, 205	200, 202, 205	200, 202	200, 202	C82-C85, C91.4, C96
Hodgkin's disease	201	201	201	201	C81
Multiple myeloma	203	203	203	203	C88,C90
Leukaemia	204	204	204-207	204-208	C91-C95 xC91.4

certificates mentioning cancer have been sent by ONS to the registry covering the RHA in which the death occurred. Any cancers registered solely from the information on the death certificates were not included in the published information prior to 1974, at which point an abrupt increase occurred. Registries use the death certificate information in different ways. For example, some check the data by reference to clinical notes or other local data sources, but others simply enter the death as a registration (with the year of death as the incidence year).

Inaccuracies and incompleteness may arise from diagnostic practice, and changes in it, although such errors and changes come from outside the cancer registration system and are not under its control. Misclassification of cancers is more likely to occur when there is no opportunity to obtain histological confirmation of disease, or if the tumour has a pre-malignant stage which can be confused with invasive carcinoma. Misclassification may also result from mistakes in the collection, abstraction or coding of information both before and after it reaches the registry. Also, clinical and pathological (and registry) definitions of cancer may change over time and between places, particularly for borderline malignant conditions.

Changes in coding systems may cause discontinuities in published data. For the national data held by ONS, for incidence years 1971 to 1978 site is coded to ICD8 and histology by the Manual of Tumor Nomenclature and Coding (MOTNAC) 1968 edition[17]; for incidence years 1979 to 1994, site is coded to ICD9 and histology to ICD-O[18]; and from incidence year 1995 onwards site is coded to ICD10 and histology to ICD-O2[3]. Details of the effect of the changes between the ICD revisions on mortality statistics have been published[19]; these give an indication of their likely effect on cancer registrations. Table H2 gives the site codes for the ICD6 to ICD10 classifications, according to the ICD9 definition of sites. In addition, there have been some minor changes in ONS coding and classification rules[5]. Over time the submission of data from the registries to ONS on abstract cards was superseded by computer media (punched cards, magnetic tape and diskettes). Abstract cards were coded at ONS whereas magnetic tapes and diskettes were coded by the registry before being sent to ONS. Thus a change to magnetic tape (the last region to do so was Oxford in 1985) may have been accompanied by changes in interpretation of coding.

In addition, the completeness of flagging of registrations by NHSCR is important for cohort studies. The proportion of cancer registrations received by ONS which were successfully linked to an NHSCR record was on average about 96 per cent from 1971 up to 1989. With the computerisation at NHSCR and improvements in data quality by the regional cancer registries, this has risen to over 99% for data for 1993 and subsequent years. The importance for any particular study of the records not traced will depend upon any biases by region, site or other main factors of interest[20].

Rates of cancer incidence are dependent not only on the accuracy of the cancer registration data but also on that of the population denominator data. Recent censuses are believed to have been very accurate overall: under-enumeration in 1981 was estimated to be 0.5 per cent (240,000 people) and in 1991 to be 1.1 per cent (572,000 people), but this varied by age and by geographic area. Annual mid-year estimates of population, based on census data together with information on births, deaths and migration (see above) also appear to be very accurate on a national basis, although errors of several per cent have been found for some counties, districts and London boroughs. There may also be differences between the definitions of 'place of residence' used for cancer registrations and for population estimates. For the former, the address used is 'the usual place of residence as given by the patient', whereas the census definition is not so straightforward, particularly when a person lives at more than one address throughout the year [21]. This may lead to biases in analyses of data for small areas which include large numbers of students, armed forces or people living in institutions.

Finally, in published data on the scale of the national cancer registration system it is almost inevitable that straightforward errors will occur, for example in the transcription and printing of tables. Corrections to known errors have been published.

Mortality
Most deaths are certified by a medical practitioner. The death certificate is then usually taken to a registrar of births and deaths by a person known as an informant – usually a near relative of the deceased. In certain cases, deaths are referred to, and sometimes then investigated by, a coroner who sends information to the registrar of deaths which is used instead of that from the medical practitioner. In some cases, additional information from the coroner's certificate is forwarded to ONS by the registrar. Thus the information used in ONS mortality statistics may have come from one of four sources: the doctor, the informant, a coroner, or derived from one or other of the above (for example, the age of the dead person is derived from date of birth and date of death).

In the early 1990s, OPCS redeveloped its deaths registrations computer processing system. The main changes affecting the data included the progressive computerisation of local offices of registrars of births and deaths, and the automation of cause of death coding.

A full set of notes and definitions for mortality data has been published by ONS[22]. This includes: base populations; occurrences and registrations; areal coverage; death rates and standardisation; certification of cause of death; coding the underlying cause of death; analysis of conditions mentioned on the death certificate; amended cause of death; accelerated registrations; legislation on registration of deaths and the processing, reporting and analysis of mortality data; and historical changes in mortality data including the introduction of the Ninth Revision of the International Classification of Diseases[18] in 1979, industrial action taken by registration officers in 1981-82, and the amendment by OPCS in 1984 of WHO Rule 3 (one of the rules used to select the underlying cause of death).

Further information is provided in reference 22 about the redevelopment of the deaths computer processing system, the use of "medical enquiries" to the certifier of death for further information in order to assign a more definite code, and the use of WHO Rule 3. There is also advice on using cause of death from 1993 onwards.

The main change in introducing automated cause of death coding was in the interpretation of WHO Rule 3. The death certificate is set out in two parts; part I gives the condition or sequence of conditions leading to death, while part II gives details of any associated conditions. Rule 3 states that "if the condition selected by the General rules or Rules 1 and 2 can be considered a direct sequel of another reported condition, whether in part I or part II, select this primary condition"[18]. The interpretation of Rule 3 was broadened by OPCS in 1984 so that certain conditions which were often terminal, such as bronchopneumonia or pulmonary embolism, could be considered a direct sequel of any more specific condition reported. The more specific condition would then be regarded as the underlying cause. This change in interpretation meant that the numbers of deaths from certain conditions such as pneumonia fell suddenly in 1984, while deaths from conditions often mentioned in part II of the certificate rose.[23] The change in 1993 was thus a move back to the internationally accepted interpretation of Rule 3 operating in England and Wales before 1984.

Information on the effects of moving back to this earlier interpretation of Rule 3 have been published[24,25]. The expected effects were based on the assumption that any allowance for them was the same in 1993 as it was in 1984 (which is unlikely to be exactly true). But the effects of the change appear to be generally in the opposite direction to those of 1984 and of a similar magnitude[22].

All mortality data given in this book have been adjusted to allow for the impact of automated cause coding and related changes.[25] The mortality data also exclude deaths in people who were not resident in England and Wales. Mortality data for 1999 are provisional: they are registrations (not occurrences) and the rates are based on mid-year populations for 1998 (as those for 1999 were not available).

Quality of mortality data

As explained above, mortality statistics in England and Wales are derived from the registration of deaths certified by a doctor or coroner. The data pass through a number of processes before becoming usable for analysis. These processes are complex, and involve a wide range of people, organisations and computer systems. The scope for error is correspondingly wide. ONS aims to produce mortality statistics with the highest achievable quality given the available resources.

The quality checks and validations carried out at the various stages in the creation of mortality statistics are described in detail in reference 22. These include: writing the medical certificate of death; registration of the death; entry of data in the computer system used by registrars of births and deaths; other checks made by the registration service; receipt of death registration data at ONS; validation processes; routine checks by ONS; the automated cause coding system; checks before and after extraction of data for analysis; checks on routine outputs; and analysis of ill-defined causes of death.

Advantages and disadvantages of incidence and mortality data

In 1981, Doll and Peto[26] compared the quality and utility of incidence and mortality data in the USA. The incidence data came from two "one off" national cancer surveys in 1947/48 and 1969-71, and from continuous collection up to 1977 by the Surveillance, Epidemiology and End Results (SEER) cancer registries (which operated in various cities and states and in total covered about 10% of the US population). They showed that mortality data were largely reliable and stable over time. But examples for a few major sites such as breast (in females) and prostate indicated that there were discrepancies with incidence that were too large to be explained without there being serious upward biases in the trends in cancer registration data, and that mortality data were generally more trustworthy.

These conclusions do not apply to cancer registration data in the UK. As noted above, a recent review of the quality of UK cancer

Table H3 Advantages and disadvantages of incidence and mortality data

Incidence	Mortality
Advantages	Disadvantages
• high quality coding	• diagnostic accuracy less certain than for incidence
• both cancer site and histology	• site only, no histology
• very low proportion site unspecified	• around 10% site unspecified
• incidence date known (except for small proportion registered solely from a death certificate)	• deaths in any one year result from cases diagnosed over a long previous period
Disadvantages	Advantages
• may not be complete	• virtually 100% complete
• may not be sufficiently timely	• timely (within months of the end of a data year)
• national coverage not achieved until 1962; evidence of under-ascertainment in the early 1970s	• very long time series (if not affected by ICD or other coding changes[19])

Table H4 Definition of childhood malignancies

Site	ICD9 code	Morphology code
Leukaemia	204-208	9800-9804, 9810, 9820-9825, 9830, 9840-9842, 9850, 9860-9866, 9870, 9880, 9890-9894, 9900-9940
Brain and spinal cord	191, 192, 194.3, 194.4	8000-8004, 9990
	140-208	8270-8281, 8300, 9350-9362, 9380-9384, 9390-9394, 9505, 9530-9539
	140-199	9400-9441, 9442-9460, 9470-9481
Lymphomas	201	9650-9662
	140-200, 202	9590-9642, 9690-9701, 9710-9722, 9730-9750
Soft tissue sarcoma	140-199	8810, 8811, 8813-8920, 8990, 8991, 9040-9044, 9120-9170, 9251, 9540, 9560, 9581
	140-169, 171-199	8800-8840
	140-169, 171-195	9240, 9260
Neuroblastoma	140-208	9490-9500
Wilms' tumour	140-199	8960

registry data[12] concluded that results were largely complete, accurate and reliable. The data on cancer registration "quality indicators" – mortality to incidence ratios, zero survival cases, and unspecified site – given in Appendices D and E demonstrate that although there is some variability within England and Wales, the overall ascertainment and reliability is good. And the trends in incidence and mortality illustrated for the major cancer sites in Chapter 2 clearly confirm that, although there may have been some under registration, particularly for lung and stomach cancer, in the early 1970s, from the late 1970s onwards the trends in incidence are consistent with those for mortality and the recorded improvements in survival[27,28].

Mortality data are generally more timely than incidence (the current gap is two years – the latest mortality data[29] are for 1999, while provisional incidence data are available up to 1997). This is largely because there is a statutory requirement to register a death within five days, and for the large majority of deaths there is only one source document. As explained in Appendix G, cancer registration is not statutory and collating information from the necessary wide variety of sources is time consuming, and ONS cannot produce final results for England and Wales until data have been received from all registries. But trends in mortality give only a delayed indication of trends in new cases, because for cancers with moderate or good survival, those dying in any one year may have been diagnosed and treated many years earlier. Even in the 1970s, five-year survival from many of the major cancers, for example breast (in females), cervix, larynx, melanoma of skin, testis and uterus, was in the range 50-70% and since then there have been notable improvements in survival for almost all except the highly fatal cancers (lung, oesophagus, pancreas)[27,28]. This has made incidence data increasingly more important for early monitoring of trends, and for assessment of major public health interventions such as breast and cervical screening[30-33].

Mortality data never were free from bias or criticism[34]. Death is not always correctly certified, nor is the underlying cause always correctly coded, even for cancer. Many studies have shown wide variability in certification and coding, particularly between countries[35-46]. Although the mortality data are virtually 100% complete, while cancer registration data may not be, around 10% of deaths in England and Wales are coded to "site unspecified"[22] whereas the corresponding proportion for incidence data is only 3% (see Appendix E2). These and other advantages and disadvantages of incidence and mortality data are summarised in Table H3 (see also Chapter 1).

Cancer mortality trends are therefore an imperfect and fuzzy indicator of trends in the efficacy of treatment – they reflect earlier trends in both incidence and survival and cannot be interpreted sensibly without them. Incidence and survival trends from the national cancer registry, based on data from the regional cancer registries, provide additional insight into the complex problems of cancer control. None of these indicators is perfect, and none is adequate on its own[47].

International Classifications of Diseases

Over the period 1950-1999 for which cancer trends are presented in this volume, five revisions of the International Classification of Diseases (ICD) were in force at various times:

ICD6	1950 – 1957	
ICD7	1958 – 1967	
ICD8	1968 – 1978	
ICD9	1979 – 1994	for incidence
	1979 onwards	for mortality
ICD10	1995 onwards	for incidence

With the exception of mortality coding remaining in ICD9 after

Table H5 NHS Regions and health authorities (1998)

Northern and Yorkshire
Bradford
Calderdale and Kirklees
County Durham
East Riding
Gateshead and South Tyneside
Leeds
Newcastle and North Tyneside
North Cumbria
North Yorkshire
Northumberland
Sunderland
Tees
Wakefield

Trent
Barnsley
Doncaster
Leicestershire
Lincolnshire
North Derbyshire
North Nottinghamshire
Nottingham
Rotherham
Sheffield
South Humber
Southern Derbyshire

Anglia and Oxford
Bedfordshire
Berkshire
Buckinghamshire
Cambridge and Huntingdon
East Norfolk
North West Anglia
Northamptonshire
Oxfordshire
Suffolk

North Thames
Barking and Havering
Barnet
Brent and Harrow
Camden and Islington
Ealing, Hammersmith and Hounslow
East and North Hertfordshire
East London and The City
Enfield and Haringey
Hillingdon
Kensington and Chelsea and Westminster
North Essex
Redbridge and Waltham Forest
South Essex
West Hertfordshire

South Thames
Bexley and Greenwich
Bromley
Croydon
East Kent

East Surrey
East Sussex, Brighton and Hove
Kingston and Richmond
Lambeth, Southwark and Lewisham
Merton, Sutton and Wandsworth
West Kent
West Surrey
West Sussex

South and West
Avon
Cornwall and Isles of Scilly
Dorset
Gloucestershire
Isle of Wight
North and East Devon
North and Mid Hampshire
Portsmouth and South East Hampshire
Somerset
South and West Devon
Southampton and South West Hampshire
Wiltshire

West Midlands
Birmingham
Coventry
Dudley
Herefordshire
North Staffordshire
Sandwell
Shropshire
Solihull
South Staffordshire
Walsall
Warwickshire
Wolverhampton
Worcestershire

North West
Bury and Rochdale
North Cheshire
South Cheshire
East Lancashire
North West Lancashire
South Lancashire
Liverpool
Manchester
Morecambe Bay
St Helens and Knowsley
Salford and Trafford
Sefton
Stockport
West Pennine
Wigan and Bolton
Wirral

Wales
Bro Taf
Dyfed Powys
Gwent
Morgannwg
North Wales

the introduction of ICD10 in the NHS in 1995, all incidence and mortality data were coded by the cancer registries and ONS (and its predecessors) to the revision of ICD current at the time.

All the site specific information in this volume is labelled with ICD10 codes. Table H3 contains lists of the relevant codes for each of ICD6 to ICD10 for all the major cancers.

Cancers in childhood (ages 0-14 years) have been grouped according to *Cancer Survival Trends in England and Wales* and the *International Incidence of Childhood Cancer Volume II* as shown in Table H4.[27,48] They attempted to follow the concepts underlying the Birch and Marsden classification together with subsequent advances in scientific knowledge and methodological development.

Populations

The population figures used to calculate incidence rates given in this volume are the ONS mid-year estimates of the population resident in England and Wales. The mid-1990 estimates and those for later years are not directly comparable with those produced for years before 1981: residents who were outside Great Britain on census night are now included whereas overseas visitors to Great Britain are now excluded.

Appendix F contains the population estimates for 1971, 1981, 1991 and 1998, by sex and age. Users requiring further information on these estimates should contact the Population Estimates Unit at: Office for National Statistics, Segensworth Road, Titchfield, Fareham, Hants, PO15 5RR.

Occasional Paper No. 37 describes methods used by ONS to produce annual mid-year estimates of the population of local and health authority areas in England and Wales. It includes historical background and methods used in the 1980s. Details are given of the components of change (births, deaths and migration), and of methods used to estimate some special groups in the population, such as students and armed forces. Methods for re-basing the estimates for the 1990s, incorporating the results of the 1991 Census, are also included. The paper is available from the ONS Direct at: Office for National Statistics, D140, Government Buildings, Cardiff Road, Newport, South Wales, NP9 1XG.

NHS Regions

Regional incidence data in this book are presented by the patient's health region of usual residence. The 1998 NHS regions as defined in terms of the constituent health authorities were as shown in Table H5.

Some regional cancer registry publications present regional statistics based on the number of patients treated in the cancer registry area. Statistics in some regional cancer registry reports may therefore differ from the analyses by region of residence given in this volume.

2 METHODS

Incidence

The number of newly diagnosed patients registered in a given period (usually a calendar year). The incidence rate for a particular sex and age group is the number of new cases divided by the corresponding sex and age-specific mid-year population; the incidence rate is usually expressed per 100,000 population.

Mortality

The number of deaths in a given period divided by the sex and age specific mid-year population; the death rate or mortality is usually expressed per 100,000 population.

Age-standardised rates

Differences in the age structure of populations between geographical areas or over time need to be controlled to give unbiased comparisons of incidence and mortality. Except for the trends by birth cohort, this has been achieved through direct age standardisation in which the age and sex specific rates are multiplied by the corresponding number of people in the European (or World) standard population (see Appendix F), and then summed to give an overall rate per 100,000 population (ASR)[49,50].

Thus the directly age standardised incidence rate using the European standard population is given by

$$I(ESR) = \{\sum_k i_k P_k\} / \sum_k P_k$$

where i_k = observed incidence rate in age group k
   k = 1, ... , 19 and the 19 age groups are 0, 1-4, 5-9, ... , 80-84, and 85 and over
   P_k = European standard population in age group k

The cohort incidence and mortality ratios presented in Figures 6 and 8 in chapters 3-22 were derived using indirect age standardisation[50,51]. The ratios compare incidence or mortality in each cohort with that in a "reference" cohort (1929 for incidence, 1916 for mortality). The ratio is of observed to "expected" cases or deaths, where the expected number is that which would have occurred if the age specific incidence or mortality rates of the reference year had applied.

Thus the indirectly age-standardised cohort incidence ratio is given by

$$SCIR = \{\text{observed cases in age groups m to n} / \text{expected cases}\} \times 100$$

where   expected cases $= \sum_{j=m}^{n} i_j P_j$

   i_j = incidence rate in age group j in the (five year) reference cohort
   P_j = population in age group j in the (five year) birth cohort corresponding to the observed cases

and j is an index whose range depends upon which age groups have data for both the reference cohort and the birth cohort in question. For example, for a cohort of people born in the five years 1950-54, the oldest people for whom there exist mortality data are those age 45-49; thus (as the reference cohort is chosen such that it has data for all age groups) j would range from 1 to 11.

In line with established practice[52] the standardised cohort incidence and mortality ratios were not calculated if the observed number of cases or deaths in any particular cohort was fewer than 20.

Appropriate confidence intervals for the directly age-standardised rates and the indirectly standardised cohort ratios were calculated according to the methods of Breslow and Day[50].

Cumulative lifetime risk

The risk of a person developing cancer during their lifetime. It is obtained by applying the sex and age-specific incidence rates to the person years at risk derived from the numbers of survivors free from cancer in a hypothetical cohort based on an England and Wales life table. It gives the percentage of the cohort that would develop cancer should the current age and sex-specific rates be experienced throughout the lifetime of the cohort[53]. It can also be expressed as the odds of developing the disease during a person's lifetime.

Socio-economic deprivation

The Carstairs index[54] was used to measure deprivation. The value of the index for each enumeration district (ED) is calculated using four variables from the 1991 census: % unemployed (economically active) males; % people in overcrowded households (density over 1 person per room); % residents in households with no car; and % residents in households with an economically active head of household in social classes IV or V. The values of each variable for each ED are "normalised" (by subtracting the average value for Great Britain and dividing by the standard deviation of the distribution) and then summed. EDs with low values are termed "affluent" and those with high values "deprived". EDs were categorised into fifths or twentieths based on the distribution of values for Great Britain. Values of the index were assigned to each individual case of or death from cancer using a directory which links postcode to ED – thus everyone living in a particular ED is assigned the same level of socio-economic deprivation.

Prevalence

The prevalence of cancer in a population – the number (or proportion) of cases alive on a given date who were diagnosed with a particular cancer within a specified period of time before that date – is a crucial indicator for health care planning and resource allocation[55]. The records for all cases diagnosed in residents of England and Wales from 1971 to 1992 were checked to determine the vital status of the patient. For those patients not lost to follow up, the percentage who were still alive on 1st January 1993 was calculated. This percentage was then applied to the total number of cases to give the estimated number of prevalent cases. [A detailed description of the follow up process through the National Health Service Central Register is given in reference 20. For incidence years up to the late 1980s, 3-4% of cases were lost to follow up; by 1993 this had fallen to only $1/_2$%.]

Survival

A brief outline of the methods used for cancer survival analysis is given in Chapter 1, and a description of the processing system at ONS which links cancer cases with national death registrations at the NHSCR in Appendix G.

The methods used to analyses survival trends for cases diagnosed in the period 1971-90 and followed up to the end of 1995 were described in the *Cancer Survival Trends* volume[27]. The methods used for cases diagnosed in 1991-93 and followed up to the end of 1998 were based closely on those used earlier[28]. All adult residents of England and Wales aged 15-99 years who were diagnosed with a first primary invasive malignant neoplasm (except non-melanoma skin cancer) during the period 1971-93 were eligible for analysis. Benign and *in situ* tumours were excluded. Incidence data submitted to ONS by all the regional cancer registries in England and Wales were linked with death registrations and emigrations at the NHSCR. The data sets were frozen for analysis on 6th July 1997 for cases diagnosed during 1971-90, and on 31st January 2000 for cases diagnosed during 1991-93, when death linkage was considered acceptably complete up to and including 31st December 1995 and 31st December 1998, respectively.

Patients were excluded from analysis for various reasons including: aged under 15 or 100 or more years at diagnosis (less than 0.1% for most cancers); duplicate registration (none); sex not known or incompatible with the cancer (less than 0.1%); and invalid date(s) or sequence of dates (less than 0.1%). Cancers diagnosed in the period 1991-93 were linked to the entire data set for 1971-90 to detect any previous tumour in the same patient. Patients were retained in the analysis only if the tumour diagnosed during 1991-93 was the first invasive malignant tumour (excluding non-melanoma skin cancer) in that patient. For synchronous tumours in the same organ or bilateral tumours, only one record was retained for analysis.

Tumours registered from a death certificate only (DCO) have unknown survival because the incidence date is unknown (it is taken as the date of death). For most of the period 1991-93, these cases could not be reliably distinguished in the national data from patients who were known to have died on the day of diagnosis (true zero survival). Patients with zero or unknown survival were also excluded form analysis. For 1991-93 data, these comprised 4-15% (overall 9%) of eligible patients – up to 3% points less (depending on the cancer) than for patients diagnosed during 1986-90.

When the record of a patient diagnosed with cancer during 1991-93 could not be traced at the NHSCR, the person's cancer record could not be linked with death or emigration records; such persons were excluded from analysis because their vital status was therefore unknown. Fewer than 1% of patients were excluded for this reason, however, compared with 5.3% for the period 1986-90. This represents a considerable improvement in the efficiency of linkage between cancer registrations and deaths. Overall, the percentage of patients diagnosed during 1991-93 excluded from analysis for all the above reasons fell by almost half compared with the corresponding values for those diagnosed during 1986-90.

Survival time was calculated as the number of days between the dates of diagnosis and death (or emigration or the relevant end of follow-up, whichever was the earlier) divided by 365.25. Crude survival is simply the cumulative probability of survival at a particular period after diagnosis. Relative survival for each age group and sex is the ratio of the crude survival and the expected survival in the corresponding group of the general population. Expected survival was estimated from the England and Wales life tables. These were obtained from the Government Actuary's Department for each sex and single year of age up to 106 years, both for the peri-censal period 1990-92 (used for deaths up to 1994) and the period 1995-97 (used for deaths in 1995-98). Mortality is, however, known to differ between groups of patients defined by their region of residence or their degree of socio-economic deprivation. Using the same life table for all deprivation groups can lead to bias in estimating survival in each group. If affluent groups have better relative survival, the true gradient in survival between affluent and deprived groups will be exaggerated. This is because the life table represents background mortality averaged across all socio-economic groups and it ignores the known gradient in mortality between them, which favours the more affluent[27]. Tables of mortality rates for England and Wales by single year of age at death up to age 99, for every combination of NHS region, deprivation category and sex, for each of two time periods, were therefore constructed. Full details are given in the *Cancer Survival Trends* volume[27].

Computations were done with a STATA algorithm using the method of Estève et al[56]. Relative survival was estimated for each single year of age, but results are presented for age ranges 15-39, 40-49, 50-59, 60-69, 70-79 and 80-99 years; 95% confidence intervals were also calculated. Overall (all ages) survival rates for patients diagnosed during 1991-93 were not age standardised, but any changes in the age distribution of cancer patients compared with those diagnosed during 1986-90 would have only a trivial effect on the survival estimates.

Appendix I

LIST OF PUBLICATIONS USING CANCER REGISTRATION DATA

This appendix contains lists of publications from 1990 onwards which used cancer registration data and/or involved staff of the cancer registries in the UK, the Childhood Cancer Research Group, Oxford, the Small Area Health Statistics Unit (SAHSU) at Imperial College, London, or the Office for National Statistics.

The cancer registries in England appear in alphabetical order, followed by those in Wales, Scotland and Northern Ireland. For each registry, the publications are given in three groups: the latest incidence and/or survival report; any special reports (for example, the "Lung Cancer Bulletin: a framework for action" produced by the Merseyside and Cheshire Registry); and papers in peer reviewed journals. The lists of references were supplied by the registries and ONS cannot vouch for their accuracy; if further information is required the registry concerned should be contacted (addresses, telephone numbers and e-mail addresses of all the registries are given in Appendix G).

East Anglian Cancer Registry

Cancer Incidence in Eastern Region 1997, 1999.

Cancer incidence in East Anglia, 1996, 1998.

Cancer incidence and survival in East Anglia, 1995, 1997.

Years of life lost due to cancer in East Anglia 1990-94, 1996.

East Anglian study on survival by hospital 1989-93, 1997

The quality of East Anglian Cancer Registry data 1997, 1998.

Davies TW, Prener A, Engholm G.Body size and cancer of the testis Acta Oncologica 1990; 29 Fasc 3: 287-290.

Mori M, Davies TW, et al. Maternal factors of Testicular Cancer. A case-control study in Japan. Jpn. J. Oncol 1990; 20: 72-77

Cheng KK, Day NE , Davies TW. Oesophageal cancer mortality in Europe: paradoxical time trend in relation to smoking and drinking. British J Cancer (1992), 65 613-617

Cheng KK, Davies TW, Sham JS, Hedley, AJ. A starting point in the evaluation of care; an example using cancer registry information J Hong Kong Med Assoc 1994; 46(1):53-55

Day NE, McCann J, Camilleri-Ferrante C, Britton P, Hurst G, Cush S, Duffy S. Monitoring interval cancers in breast screening programmes: the East Anglian experience. Journal of Medical Screening 1995;2:180-185.

Day NE, Davies TW. Cancer registration: integrate or disintegrate? [editorial] BMJ 1996;313:896.

Davies TW, Palmer CR, Ruja E, Lipscombe JM. Adolescent milk, dairy product and fruit consumption and testicular cancer. British Journal of Cancer1996;74:657-60.

Stockton D, Davies TW, Day NE, McCann J. Retrospective study of reasons for improved survival in patients with breast cancer in East Anglia: earlier diagnosis or better treatment? BMJ 1997;314:472-475 [published erratum appears in BMJ 1997;314:721]

Gibson L, Spiegelhalter DJ, Camilleri-Ferrante C, Day NE. Trends in invasive cervical cancer incidence in East Anglia from 1971 to 1993. Journal of Medical Screening 1997;4:44-48.

Badrinath P, Day NE, Stockton D. Seasonality in the diagnosis of acute lymphocytic leukaemia British Journal of Cancer 1997;75:1711-1713.

Badrinath P, Day NE, Stockton D. Population-based survival trends for leukaemia in East Anglia, United Kingdom Journal of Public Health Medicine 1997;19:403-407.

McCann J, Wait S, Séradour B, Day NE. A comparison of the performance and impact of breast cancer screening programmes in East Anglia, UK, and Bouches Du Rhône, France. European Journal of Cancer 1997;33:429-435.

Stockton D, Cooper P, Lonsdale R N. Changing incidence of invasive adenocarcinoma of the uterine cervix in East Anglia. Journal of Medical Screening 1997;4:40-43.

McCann J, Stockton D, Day NE. Breast cancer in East Anglia: the impact of the breast screening programme on stage at diagnosis. Journal of Medical Screening 1998;5:42-48.

Thomas R, Brown C, Dalton L, Welton S, Stockton D. Anxiety and depression – effect on patients preferences for information following a diagnosis of cancer. Annuls of Oncology (ESMO) 1998;vol 9 suppl 4:141.

Thomas R, Dreary A, Kaminiski E, Stockton D, De Zeeuw N. Patients' preferences for video cassette recorded information. Effect of age, sex and ethnic group. *European Journal of Cancer Care* 1998;8.

Sala E, Warren R, McCann J et al. Mammographic parenchymal patterns and mode of detection: implications for the breast screening programme Journal of Medical Screening 1998;5:207-212.

Sala E; Warren R; McCann J; Duffy S; Luben R; Day-N High-risk mammographic parenchymal patterns and anthropometric measures: a case-control study.Br-J-Cancer 1999; 81(7): 1257-61.

Hunt DPJ. Predicting Meningioma Recurrence: Clinical Oncology 1999;11:398-404

Post P, Stockton D, Davies TW, Coebergh JW. Prostate cancer below 60 years of age: increasing incidence and worsening prognosis? *British Journal of Cancer* 1999;79:13-17.

Badrinath P, Day N E, Stockton D. Geographical clustering of acute adult leukaemia in the East Anglian region of the United Kingdom: a registry-based analysis. *British Journal of Cancer* 1999;53:317-318.

Pharoah PD, Easton DF, Stockton DL, Gayther S, Ponder BA. Survival in familial, BRCA1-associated, and BRCA2-associated epithelial ovarian cancer. United Kingdom Co-ordinating Committee for Cancer Research (UKCCCR) Familial Ovarian Cancer Study Group Cancer Research 1999;59:868-71.

Stockton D, Davies TW. Multiple cancer site comparison of adjusted survival by hospital of treatment: an east Anglian Study *Br J Cancer* 2000; 82(1): 208-12.

Sala E, Solomon L, Warren R, McCann J, Duffy S, Luben R, Day NE. Size, node status and grade of breast tumours: association with mammographic parenchymal patterns. Eur. Radiol. 2000:10:157-61.

Sala E, Warren R, McCann J, Duffy S, Luben, R & Day NE. Smoking and high-risk mammographic parenchymal patterns: a case-control study. Breast Cancer Research 2000;

Britton PD, McCann J, O' Driscoll D, Hunnam G & Warren R. Interval cancer peer review in East Anglia; implications for monitoring doctors as well as the breast screening programme. Clinical Radiology.

Godward-S.Screening for congenital dislocation of the hip [letter] Lancet.2000; 355: 231-2; discussion 232-3.

Merseyside and Cheshire Cancer Registry

Littler J. Chester K. Ashurst D. Williams E.M.I. Cancer Incidence in Merseyside and Cheshire and its constituent health districts 1990-1995. Merseyside and Cheshire Cancer Registry. December 1998

Williams EMI. Youngson JH. Ashby A. Donnelly R J. Lung Cancer Bulletin: a framework for action. Merseyside and Cheshire Cancer Registry: Liverpool, 1993

Williams EMI. Maudsley G. Developments in the Mersey Regional Cancer Registry . In: The changing health of Mersey 1948-94: The Report of the Regional Director of Public Health. Mersey Regional Health Authority Annual Public Health Report: 1994

Williams EMI. Maudsley GM. Friedmann, PS. Grey P. Skin Cancer Bulletin: establishing the baseline. Merseyside and Cheshire Cancer Registry: Liverpool, 1994

Williams EMI. Somerville M. Youngson JH. Smith C. White FE. Leinster S.J. Zakhour HD. Breast Cancer Bulletin: stimulating the debate. Merseyside and Cheshire Cancer Registry: Liverpool, 1994

Williams EMI. Cancer section in: The Health of the North West of England. The report of the Regional Director of Public Health 1995. North West Regional Health Authority, 1995

Williams EMI. Newton M. Bean D. Cancers in children and young people. (14 pages) In: Health prospects for young citizens in the North West. A special report for the Regional Director of Public Health, 1998 ed. Barrett S. NHS Executive North West, 1998.

Rivers H. Williams EMI. Cancer in the Older Person Section (11 pages) In: Health Opportunities for Older People in the North West. A special report of the Regional Director of Public Health 1998 ed. Flanagan C, Barnes G.NHS Executive North West, 1998

Maudsley G.M. Williams EMI. 'Inaccuracy' in death certification – where are we now? Journal of Public Health Medicine 1996; 18:59-66

Maudsley G. Williams EMI. Variability of skin cancer registration practice in the United Kingdom. Journal of Epidemiology and Community Health 1997; 51:337-338

Seddon DJ. Williams EMI. Data quality in population-based cancer registration: an assessment of the Merseyside and Cheshire Cancer Registry. *British Journal of Cancer* 1997; 76(5):667-674

Nandapalan V. Roland NJ. Helliwell TR. Williams EMI. Hamilton JW. Jones AS. Mucosal melanoma of the head and neck. Clinical Otolaryngology. 1998, 23, 107-116

McCulloch PG. Williams EMI. McCulloch C. Mullins D. Can we improve the uptake of gastroscopy in the population at risk of gastric cancer? The effect of home letter information. Journal of the Royal College of Surgeons of Edinburgh 43, 1998 385-389

Dolan K. Walker SJ. Sutton R. Morris AI. Campbell F. Williams EMI. New classification of oesophageal and gastric carcinoma derived from changing patterns in epidemiology. *British Journal of Cancer* 80, 1999, 834-42.

Maudsley, G. Williams, EMI. What lessons can be learned for cancer registration quality assurance from data users? Skin cancer as an example. International Journal of Epidemiology 1999; 28: 809-815. 1999.

North Western Cancer Registry

Woodman CBJ et al. Trends in Cancer Incidence and Survival. Centre for Cancer Epidemiology, Manchester 1998. (This includes cases incident in 1996)

Woodman CBJ, Parry J, Scott N. Management of colorectal cancer in Greater Manchester and Lancashire. Centre for Cancer Epidemiology, Manchester, 1998.

Woodman CBJ, Dey P. Reorganisation of gynaecological cancer services. Centre for Cancer Epidemiology, Manchester 1997.

Dey P, Woodman CBJ. Monitoring of quality indicators for breast cancer units. Centre for Cancer Epidemiology, Manchester 1997.

Woodman CBJ and Donnelly B. Geographical variation in the un-met need for radiotherapy. In: review of cancer services. Eds: Selby, Cumisky S. North West Regional Health Authority, 1994.

Wilson S, Hare L, Donnelly B and Woodman CBJ. Skin cancer in the North West. Centre for Cancer Epidemiology, Manchester, 1994.

Wilson S, Hare L, Woodman CBJ et al. Progress towards regional cancer health of the nation targets. Public Health Cancer Resource Centre, Centre for Cancer Epidemiology, Machester, 1994.

Wilson S, Smith A, Woodman CBJ et al. Cervical cancer in the North Western Region, Centre for Cancer Epidemiology, Manchester, 1994

Wilson S, Woodman CBJ, Cooke G et al, Breast cancer in the North West Region, Centre for Cancer Epidemiology, Manchester, 1993.

T Moran, S Collins, A Gibbs, CBJ Woodman. Survival of patients with colon cancer in Europe: a cautionary tale. Colorectal Disease 2000; 2:190-92

Dimitriv B. Cyclic patterns of incidence variations for stomach cancer in the North-Western region of England. Croat Med J 2000;41:197-202

Parry JM, Collins S, Mathers J, Scott NA, Woodman CBJ Influence of volume of work on the outcome of treatment for patients with colorectal cancer. Brit. J. Surg. 1999,86;475-481

Clarke F, Dey P and Collins S. A population based survey of the management of women with cancer of the cervix. Brit. J. Cancer. 1999,80(12);1958-1961

McGaughran J, Harris DI, Donnai D, Teare D, McLeod R, Kingston H, Super M, Harris R, Evans DGR. A clinical study of type 1 neurofibromatosis in North West England. J Med Genet 1999;36:197-203

Ellis D, Greenman J, Hodgson S, McCall S, Lalloo F, Cameron J, Izatt L, Scott G, Jacobs C, Watts S, Chorley W, Perrett C, MacDermott K, Mohammed S, Evans G, Mathew C. Low prevalence of germline BRCA1 mutation in early onset sporadic breast cancer. J Med Genet (in press)

Prior P, Woodman CBJ, Collins S. International differences in survival from colon cancer: more effective care versus less complete cancer registration. Brit J Surg 1998,85;101-104

Boer R, DeKoning H, Threlfall A, Warmerdam P, Street A, Friedman E, Woodman CBJ. Cost effectiveness of shortening the screening interval or extending the age range of the NHS breast screening programme: computer simulation study. BMJ 1998;317:376-79

Collins S, Woodman CBJ, Threlfall A, Prior P. The National Health Service Breast Screening Programme: no evidence to suggest a worse outcome for interval cancers. BMJ 1998;316:832-33

Byrne J, Attwood S, Parry J, Woodman CBJ. The epidemiology of adenocarcinoma of the oesophagus and gastric cardia in the north west of England. In: The Esophagogastric Junction. 3-7 Eds: Giuli R, Galmiche JP, Jamieson GG, Scarpignato C. London: John Libbey 1998

Lalloo F, Boggis C, Evans DGR, Shenton A, Threlfall A, Howell A. Screening by mammography, women with a family history of breast cancer. European Journal of Cancer 1998;34: 6:937-940

Woodman CBJ, Baghdady A, Collins S, Clyma JA. What changes in the organisation of cancer services will improve the outcome for women with ovarian cancer? Br J

Obstet and Gyn 1997; 104:135-9

Threlfall A, Woodman CBJ, Prior P. Breast Screening Programme: should the interval between tests depend on age? Lancet 1997: 349:472

Dey P, Woodman CBJ, Coyne J, Gibbs A. Completeness of reporting on prognostic factors for breast cancer – A regional survey. J Clin Path. 1997;50(10):829-31

Moran A, Collins S, Evans DGR, Davies R. Risk of subsequent primary cancers in patients with carcinoma of the Ampulla of Vater. *British Journal of Cancer* 1997; 76(9): 1232-1233

Maddock IR, Moran A, Maher E et al. A genetic register for von Hippel-Lindau disease. Journal of Medical Genetics 1996; 33:120-27

Prior P, Woodman CBJ, Wilson S, Threlfall A. The reliability of underlying incidence rates for assessing the effect and efficiency of screening for breast cancer. J Med Screening 1996; 3:119-122

Woodman CBJ, Coyne J, Balldam A, Singleton J. Management of premenopausal women with primary breast cancer. Br J Surgery 1996; 83:1256-1257

Asbury D, Boggis C, Sheals D, Threlfall A, Woodman CBJ. NHS Breast Screening programme: Is the high incidence of interval cancers inevitable? BMJ. 1996;313:1369

Woodman CBJ, Threlfall A, Boggis C, Prior P. Is the 3 year breast screening interval too long? – Occurrence of interval cancers in the National Health Service Breast Screening Programme's north west region. BMJ 1995: 310:224-6

Woodman CBJ, Prior P, Joseph R, Watson A. Prospects for the secondary prevention of colorectal cancer: screening by flexible sigmoidoscopy. J Med Screening 1995;2:71-78

Lancaster G, Moran A, Woodman CBJ. Towards achieving the health of the nation target for cervical cancer: Accuracy of cancer registration. J Pub Hlth Med 1994;16 No 1:50-52

Kehoe S, Powell J, Wilson S and Woodman CBJ. The influence of the operating surgeon's specialisation on patient survival in ovarian carcinoma. *Br J Cancer.* 1994; 70:1014-1017

Gillen CD, Walmsley RS, Prior P. Ulcerative colitis and Crohn's disease: a comparison of the colorectal cancer risk in extensive colitis. GUT 1994; 35:1590-92

Gillen CD, Andrews HA, Prior P, Allan RN. Crohns disease and colorectal cancer. GUT. 1994, 35:651-56

Parry J, Cummins C, Redman V, Wilson S, Woodman CBJ. Incidence and survival of malignant partoid tumours in the West Midlands Region 1977-1986. Clin Oncology. 1993, 5:150-153

Parry J, Wilson S, Cummins C, Redman V, Woodman CBJ. A review of partoid pleomorphic adenomas in the West Midlands Region 1977-1986. Clin. Oncology. 1993. 5:147-149

Wilson S, Prior P, Woodman CBJ. Use of cancer surveillance data for comparative analysis. J Pub Health Med. 1992. 14, No 2:151-156

Finn C, Luesley D, Nuxton E, Blackledge C, Dunn J, Wilson S. Is stage 1 epithelial ovarian cancer over-treated both surgically and systemically? Results of a five year cancer registry review. Brit J Obs & Gynae. 1992,99:54-58

Northern and Yorkshire Cancer Registry and Information Service

Cancer in Yorkshire (1996) Cancer Registry Report, Cancer Statistics 1989-93, (Forman D & Rider PL, eds) Yorkshire Cancer Organisation, Leeds.

NYCRIS Annual Report 1998/99. Includes Cancer Statistics for Yorkshire 1994-96, Northern 1994.

Lung Cancer in Yorkshire (1994) Cancer in Yorkshire, Special Report Series No. 1, (Joslin C & Rider PL, eds) Yorkshire Cancer Organisation, Leeds.

Head and Neck Cancers in Yorkshire (1994) Cancer in Yorkshire, Special Report Series No. 2, (Joslin C, Rider PL & Crellin A. eds) Yorkshire Cancer Organisation, Leeds.

Breast Cancer in Yorkshire (1995) Cancer in Yorkshire, Special Report Series No. 3, (Forman D & Rider PL, eds) Yorkshire Cancer Organisation, Leeds.

Cancer Treatment Policies, Delay in Referral and Effects on Survival: 1. Central Nervous System (1998) Northern and Yorkshire Cancer Registry and Information Service, Leeds.

Cancer Treatment Policies, Delay in Referral and Effects on Survival: 2.Lung cancer (1999) Northern and Yorkshire Cancer Registry and Information Service, Leeds.

Cancer Treatment Policies, Delay in Referral and Effects on Survival: 3. Malignant melanoma (1999) Northern and Yorkshire Cancer Registry and Information Service, Leeds.

Cancer Treatment Policies, Delay in Referral and Effects on Survival: 4. Pancreas cancer (2000) Northern and Yorkshire Cancer Registry and Information Service, Leeds.

Lung Cancer Referral Patterns – The Yorkshire Experience. (2000) Northern and Yorkshire Cancer Registry and Information Service, Leeds.

The Provision of Bereavement Support Services – A Pilot Study. (2000) Northern and Yorkshire Cancer Registry and Information Service, Leeds.

Sagar PM, Gauperaa T, Sue-Ling H, McMahon MJ, Johnston D. (1994) An audit of the treatment of cancer of the oesophagus. Gut 35:941-945.

Crawford SM, Atherton F. (1994) Lung Cancer: histological aspects of diagnosis in England and the south east Netherlands. Journal of Epidemiology and Community Health 48: 420-421.

Muers MF. (1994) Pre-operative screening for metastases in lung cancer. Thorax 48: 1-2.

Sainsbury R, Haward R, Rider L, Johnston C, Round C. (1995) Influence of clinician workload and patterns of treatment on survival from breast cancer. Lancet. 345: 1265-1270.

Sainsbury R, Rider L, Smith A, McAdam A. (1995) Does it matter where you live? Treatment variation for breast cancer in Yorkshire. *British Journal of Cancer* 71: 1275-1278.

Varghese C, Barrett JH, Johnston C, Shires M, Rider L, Forman D. (1996) High risk of lymphomas in children of Asian origin: ethnicity or confounding by socioeconomic status? *British Journal of Cancer* 74: 1503-5,

Selby P, Gillis C, Haward RA (1996). Benefits from specialised cancer care. Lancet 348:13-18.

Connolly CK, Crawford SM, Rider PL, Smith ADM, Johnston CF, Muers MF. (1997) Carcinoma of the bronchus in the Yorkshire region of England 1976-1990: Trends since 1984. European Respiratory Journal 10: 397-403.

Muers MF, Haward RA. (1997) Management of lung cancer. Thorax 51: 557-60.

Barrett JH, Parslow RC, McKinney PA, Law GR, Forman D (1998). Nitrate in drinking water and the incidence of gastric, oesophageal and brain cancer in Yorkshire, England.Cancer Causes and Control 9: 153-159.

Faivre J, Forman D, Esteve J, Gatta G and the Eurocare Study Group (1998). Survival of patients with oesophageal and gastric cancers in Europe. European Journal of Cancer 34: 2167-2175.

Faivre J, Forman D, Esteve J, Obradovic M, Sant M and the Eurocare Study Group (1998). Survival of patients with primary liver cancer, pancreative cancer and biliary tract cancer in Europe. European Journal of Cancer 34: 2184-2190.

Janssen-Heijnen, Gatta G, Forman D, Capocaccia R, Coebergh JWW and the Eurocare Study Group (1998). Variation in survival of patients with lung cancer in Europe, the Eurocare Study. European Journal of Cancer 34: 2191-2196.

Winter H, Cheng KK, Cummins C, Maric R, Silcocks P, Varghese C (1999). Cancer incidence in the south Asian population of England (1990-92). *British Journal of Cancer* 79: 645-654.

Sainsbury JRC, Johnston C , Haward RA (1999). Effect on survival of delays in referral of patients with breast symptoms: a retrospective analysis. Lancet 353: 1132-35.

Haward RA, Sainsbury JRC, Johnston C (1999). Achieving consistency in breast cancer care: An analytical tool to measure change. Cancer Strategy 1: 65-72.

Turner NJ, Haward RA, Mulley GP, Selby PJ (1999). Cancer in old age: Is it inadequately investigated and treated? British Medical Journal 319:309-12.

Murphy M, Johnston C, Whelan PP, Rider PL,Lloyd SN (1999). Changing Trends in Prostate Cancer. British Journal of Urology 83:786-91.

Cummins C, Winter H, Cheng K-K, Maric R, Silcocks P, Varghese C (1999). An assessment of the Nam Pehchan computer program for the identification of names of south Asian ethnic origin. Journal of Public Health Medicine 21:401-406.

Velikova G, Wright EP, Smith AB, Cull A, Gould A, Forman D, Perren T, Stead M, Brown J, Selby PJ (1999). Automated collection of quality of life data: a comparison of paper and computer – touchscreen questionnaires. Journal of Clinical Oncology 17:998-1007.

Oxford Cancer Intelligence Unit

Cancer Incidence in the Four Counties

Incidence in 1998 and survival for years 1991 to 1993

Targeting melanoma – A Priority for the Health of the Nation

Trends in Incidence and Mortality 1981-1993

Cervical Cancer in Berkshire, Buckinghamshire, Northamptonshire and Oxfordshire

Trends in Cancer Survival in Berkshire, Buckinghamshire, Northamptonshire and Oxfordshire (1979-1988)

Incidence of melanoma in four English counties, 1989-92. – British Medical Journal 1995, Vol. 310, Pages 502-503

The incidence and prognosis of nail apparatus melanoma. A retrospective study of 105 patients in four English regions. – British Journal of Dermatology 1998; 139: 276-279.

Improving Survival of Melanoma Patients in Europe Since 1978 – European Journal of Cancer Vol. 34, No. 14, Pages 2197-2203, 1999

South and West Cancer Intelligence Unit

Herbert A, Breen C, Bryant TN, Hitchcock A, Macdonald H, Milward-Sadler H and Smith JAE. Invasive cervical cancer in Southampton and South West Hampshire : effect of introducing a comprehensive screening programme. J Med Screening 1996 : 3, (1) : 23-28.

Etherington DJ, Pheby DFH, Bray FI . An Ecological Study of Cancer Incidence and Radon Levels in South West England European Journal of Cancer (1996) 32A. 7. 1189-1197

Thorne P, Etherington D, Burchall MA. Head and Neck Cancer in the South West of England: Influence of Socio-economic Status on Incidence and Second Primary Tumours European Journal of Surgical Oncology 1997 23: 503-508

Preece AW, Iwi GR, Etherington DJ. Radon Skin Cancer and Interaction with Power Line Sources

Annual Review of Research on Biological Effects of Electric and Magnetic Fields from the Delivery, Generation and Use of Electricity, San Antonio, Texas, 1996, pp. 43-44.

Moss SM, Smith JAE and Nicholas DS. The quality of histopathological data in a computerised cancer registry system: implications for future audit of care. Public Health 1997 : 111 :101-106

Herbert A, Bryant T, Campbell MJ and Smith J. Investigation of the effect of occult invasive cancer on progress towards successful cervical screening. J Med Screening 1998 : 5 (2): 92-98

Smith JAE, Whatley PM and Redburn JC. Improving survival from melanoma in Europe since 1978. European J Cancer 1999;. 34 (14): 2197-2203.

Aitken RJ, Thompson MR, Smith JAE, Radcliffe AG, Stamatakis JD and Steele RJC. Training in large bowel cancer surgery: observations from three prospective regional United Kingdom audits. BMJ 1999 : 318(7185) 702-3.

Herbert A and Smith JAE. Cervical intraepithelial neoplasia grade III (CIN III) and invasive cervical carcinoma: the yawning gap revisited and the treatment of risk. Cytopathology 1999, 10, 161-170.

Cancer in the South & West. South & West CIU Annual Report; 1997 registration Data. Sept 2000

No.1 Colorectal Cancer in the South and West (1996)

No.2 Malignant Melanoma in the South and West (1996)

No.3 Prostrate Cancer in the South and West (1996)

No.4 Breast Cancer in the South and West (1996)

No.5 An Overview of Common Epidemiological Terms (1996)

No.6 Lung Cancer in the South and West (March 1997)

No.7 Bladder Cancer in the South and West (March 1997)

No.8 Ovarian Cancer in the South and West (March 1997)

No.9 Cervical cancer in the South and West (March 1997)

No.10 Childhood Cancer in the South and West (September 1997)

No.11 Testicular Cancer in the South and West (September 1997)

No.12. Head and Neck Cancer in the South and West (September 1997)

No.13. Radiotherapy Treatment (September 1997)

Hospital In-Patient Care and Cancer Patients. (Eastern sector data) March 1996

Pan European Comparison of Cancer Incidence and Mortality April 1998

Cancer in the South &West 1988-96. South and West Cancer Intelligence Unit Sept 1999

Annual Report (1996 registration data)

Cervical Cancer in the South and West May 1999

Cancer Incidence Reports for the Channel Islands May 1999

Cancer Incidence in the SE Region (1997 registration data) May 1999

Thames Cancer Registry

Bell CMJ, Coleman MP. Extremely low frequency (ELF) electromagnetic fields and leukaemia in children. *Br J Cancer* 1990;62:331-2

Coleman MP, Cardis E, Boyle P, Saracci R, Ahlbom A, Feychting M et al. Extremely low-frequency electric and magnetic fields and risk of human cancer. Bioelectromagnetics 1990;11:91-9

Johnson NW,Warnakulasuriya KAAS. Oral cancer: is it more common than cervical? Br Dental J 1990;170:170-1

Kaldor JM, Day NE, Clarke EA, van Leeuwen FE, Henry-Amar M, Fiorentino MV et al. Leukemia following Hodgkin's disease. N Engl J Med 1990;322:7-13

Kaldor JM, Day NE, Pettersson F, Clarke EA, Pedersen D, Mehnert W et al. Leukemia following chemotherapy for ovarian cancer. N Engl J Med 1990;322:1-6

Murphy M, Goldblatt P, Thornton-Jones H, Silcocks PBS. Survival among women with cancer of the uterine cervix: influence of marital status and social class. J Epidemiol Community Health 1990;44:1-6

Murrell DS, Helm CW, Bourne HM. Carcinoma of the cervix in women up to 35 years of age Clin Oncol 1990;2:260-3

Scholefield, JH, Thornton-Jones H, Cuzick J, Northover JMA. Anal cancer and marital status. *Br J Cancer* 1990;62:286-8

Stiller CA, Bunch KJ. Trends in survival for childhood cancer in Britain diagnosed 1971-85. *Br J Cancer* 1990;62:806-15

Tait DM, Thornton-Jones H, Bloom HJG, Lemerle J, Morris-Jones P. Adjuvant chemotherapy for medulloblastoma: the first multi- centre control trial of the International Society of Paediatric Oncology (SIOP I). Eur J Cancer 1990;26:464-9

Thornton-Jones H. Cancer registry data. *Br J Cancer* 1990;61:942

Boffetta P, Cardis E, Vainio H, Coleman MP, Kogevinas M, Nordberg G et al. Cancer risks related to electricity production. Eur J Cancer 1991;27:1504-19

Boffetta P, Cardis E, Vainio H, Coleman MP, Kogevinas M, Nordberg G et al. Cancer risks related to different energy sources. Énergies Santé 1991;2:385-401

Gulliford MC, Petruckevitch A, Burney PGJ. Hospital case notes and medical audit: evaluation of non-response. Br Med J 1991;302:1128-9

Gulliford MC, Petruckevitch A, Burney PGJ. Survival with bladder cancer, evaluation of delay in treatment, type of surgeon, and modality of treatment. Br Med J 1991;303:437-40

Vainio H, Coleman MP, Wilbourn J. Carcinogenicity evaluations and ongoing studies: the IARC databases. Environ Hlth Persp 1991;96:5-9

Bell CMJ. Temporal trends in the incidence of childhood leukaemia and in radiation exposure of parents. J Radiol Prot 1992;12:3-8

Brada M, Ford D, Ashley S, Bliss JM, Crowley S, Mason M et al. Risk of second brain tumour following conservative surgery and radiotherapy of pituitary adenoma. Br Med J 1992;304:1343-6

Coleman MP, Muir CS, Ménégoz F. Confidentiality in the cancer registry. *Br J Cancer* 1992; 66:1138-49

Coleman MP, Reiter RJ. Breast cancer, blindness and melatonin. Eur J Cancer 1992;28:501-3

Freni SC, Gaylor DW. International trends in the incidence of bone cancer are not related to drinking water fluoridation. Cancer 1992;70:611-8

Staunton MD, Bourne HM. Thyroid cancer 1932-1972: outcome in 492 patients. *Eur J Surg Oncology* 1992;18:469-74

Walker A, Petruckevitch A, Bourne HM, Burney PGJ. Contributions of incidence and case fatality to mortality from bladder cancer in the South Thames Regions. J Epidemiol Community Health 1992;46:387-9

Bell CMJ, Bourne HM. A plea for better coding rules for bladder cancer (authors' reply) *Br J Cancer* 1993;68:1255

Bell CMJ. Trends in the incidence of childhood leukaemia between 1961 and 1985 and trends in radiation exposure in parents. Health Reports 1993;5:111-5

Bourne HM. Patient confidentiality. J Roy Soc Med 1993;86:618

EUROGAST Study Group. An international association between Helicobacter pylori infection and gastric cancer. Lancet 1993;341:1359-62

EUROGAST Study Group. Epidemiology of, and risk factors for, Helicobacter pylori infection among 3194 asymptomatic subjects in 17 populations. Gut 1993;34:1672-6

Forman D, Møller H, Coleman MP. International association between Helicobacter pylori and gastric cancer. Lancet 1993;342:120-1

Gulliford MC, Barton JR, Bourne HM. Selection for oesophagectomy and postoperative outcome in a defined population. 1993;2:17-20

Gulliford MC, Bell CMJ, Bourne HM, Petruckevitch A. The reliability of cancer registry records. *Br J Cancer* 1993;67:819-21

Gulliford MC, Bell CMJ, Bourne HM, Petruckevitch A. Reply to letter by Drs Pollock and Vickers. *Br J Cancer* 1993;68:1046

Haylock BJ, Murrell DS, Bourne HM, Acworth P. Stage I endometrial carcinoma: the role of neoadjuvant progesterone therapy. Clin Oncol 1993;5:102-6

Parkin DM, Cardis E, Masuyer E, Friedl HP, Hansluwka H, Bobev D et al. Childhood leukaemia following the Chernobyl accident: the European Childhood Leukaemia-Lymphoma Incidence Study (ECLIS). Eur J Cancer 1993;29A:87-95

Pollock AM, Vickers N. Reliability of cancer registry records. *Br J Cancer* 1993;68:1045

Staunton MD, Bourne HM. Thyroid cancer in the 1980s – a decade of change. Ann Acad Med 1993;22:613-6

Vickers N, Pollock AM. Incompleteness and retrieval of case notes in a case note audit of colorectal cancer. 1993;2:170-4

Bell CMJ, Chouillet AM. Management of breast cancer. (Letter). Br Med J 1994;308:1508

Basnett I, Pollock AM, Gill M. Collecting data on cancer. Br Med J 1994;308:791

Chouillet AM, Bell CMJ, Hiscox JG. Management of breast cancer in south-east England. Br Med J 1994;308:168-71

EUROGAST Study Group. The epidemiology of low serum pepsinogen A levels and an international association with gastric cancer rates. Gastroenterology 1994;107:1335-44

EUROGAST Study Group. O6-methylguanine in blood leucocyte DNA: an association with the geographic prevalence of gastric cancer and with low levels of serum pepsinogen A, a marker of severe chronic atrophic gastritis. Carcinogenesis 1994;15:1815-20

Horwich A, Bell CMJ. Mortality and cancer incidence following radiotherapy for seminoma of the testis. Radiother Oncol 1994;30:193-8

Lutz J-M, Coleman MP. Trends in primary cerebral lymphoma. *Br J Cancer* 1994;70:716-8

Pollock AM, Vickers N. The impact on colorectal cancer survival of cases registered by 'death certificate only': implications for national survival rates. *Br J Cancer* 1994;70:1229-31

Pollock AM. The future of cancer registries. Br Med J 1994;309:821-2

Warnakulasuriya KAAS, Acworth P, Bell CMJ, Johnson NW. Incompleteness of oral cancer registration in South East England, 1971-1987. *Br J Cancer* 1994;70:736-8

Bell CMJ, Lawrence G, Pheby DFH, Smith J, Coleman MP. The role of cancer registries. Clin Oncol 1995;7:143-4

Ford D, Easton DF, Peto J. Estimates of the gene frequency of BRCA1 and its contribution to breast and ovarian cancer incidence. Am J Human Genetics 1995;57:1457-62

Kaldor JM, Day NE, Kittelmann B, Pettersson F, Langmark F, Pedersen D et al. Bladder tumours following chemotherapy and radiotherapy for ovarian cancer: a case-control study. Int J Cancer 1995;63:1-6

Pollock AM, Benster R, Vickers N. Why did treatment rates for colorectal cancer in South East England fall between 1982 and 1988? The effect of case ascertainment and registration bias. J Pub Health 1995;17:419-28

Pollock AM,.Vickers N. Why are a quarter of all cancer deaths in south-east England registered by death certificate only? Factors related to death certificate only registrations in the Thames Cancer Registry between 1987 and 1989. *Br J Cancer* 1995;71:637-41

Pollock AM,Vickers N. The reliability of Thames Cancer Registry data on 673 cases of colorectal cancer: the effect of the registration process. *Qual Health Care* 1995;4:184-9

Schrijvers CTM, Mackenbach JP, Lutz J-M, Quinn MJ, Coleman MP. Deprivation and survival from breast cancer. *Br J Cancer* 1995;72:738-43

Schrijvers CTM, Mackenbach JP, Lutz J-M, Quinn MJ, Coleman MP. Deprivation, stage at diagnosis and cancer survival. Int J Cancer 1995;63:324-9

Garvican L, Littlejohns P. An evaluation of the prevalent round of the breast screening programme in South East Thames, 1988-1993: achievement of quality standards and population impact. J Med Screening 1996;3:1-6

Richards MA, Wolfe CDA, Tilling K, Barton J, Bourne HM, Gregory WM. Variations in the management and survival of women under 50 years with breast cancer in the South East Thames region. Br Med J 1996;73:751-7

Wolfe CDA, Tilling K, Bourne HM, Raju KS. Variations in the screening history and appropriateness of management of cervical cancer in south east England. Eur J Cancer 1996;32:1198-204

Cosford P, Garrett C, Turner K. Travel times and radiotherapy uptake in two English counties.

Ma M, Bell J, Campbell S, Basnett I, Pollock A, Taylor, Clinical Advisory Panel. Breast cancer management: is volume related to quality? *Br J Cancer* 1997;75:1652-9

Pollock AM, Vickers N. Management of colorectal cancer in three South Thames District Health Authorities. Public Health 1997;111:165-70

Pollock AM,Vickers N. Breast, lung and colorectal cancer survival in South East England, 1987-1992: the effect of deprivation. J Pub Health Med 1997;78:288-94

Garvican L, Littlejohns P. Comparison of prognostic and socio-economic factors in screen-detected and symptomatic cases of breast cancer. Public Health 1998;112:15-20

Little MP, de Vathaire F, Shamsaldin A, Oberlin O, Campbell S, Grimaud E et al. Risks of brain tumour following treatment for cancer in childhood: modification by genetic factors, radiotherapy and chemotherapy. Int J Cancer 1998;78:269-75

Rustin GJS, Newlands ES, Lutz J-M, Holden L, Bagshawe KD, Hiscox JG et al. Combination but not single-agent methotrexate chemotherapy for gestational trophoblastic tumors increases the incidence of second tumors. J Clin Oncol 1998;14:2769-73

Sant M, Capocaccia R, Verdecchia A, Estève J, Gatta G, Micheli A et al. Survival of women with breast cancer in Europe: variation with age, year of diagnosis and country. Int J Cancer 1998;77:679-83

Best N, Wakefield J. Accounting for inaccuracies in population counts and case registration in cancer mapping studies. Journal of the Royal Statistical Society 1999;162:363-82

De Vathaire F, Hawkins M, Campbell S, Oberlin O, Raquin MA, Schliegner JY, Shamsaldin A, Diallo I, Bell J, Grimaud E, Hardimen C, Lagrange JL, Daly-Schveitzer N, Panis X, Zucker JM, Sancho-Garnier H, Eschwège, Chavaudra J and Lemerie J. Second malignant neoplasms after a first cancer in childhood: temporal pattern of risk according to type of treatment. *British Journal of Cancer* 1999;79:1884-93

Edmondson PC, Curley RK, Marsden RA, Robinson D, Allaway SL, Willson CD. Screening for malignant melanoma using instant photography. Journal of Medical Screening 1999;6:42-6

Macdonald N, Sibley K, Rosenthal A, Menon U, Jayarajah A, Oram D, Jacobs I. A comparison of national cancer registry and direct follow up in the ascertainment of ovarian cancer. *British Journal of Cancer* 1999;80:1826-7

Warnakulasuriya KAAS, Johnson NW, Linklater KM, Bell J. Cancer of the mouth, pharynx and nasopharynx in Asian and Chinese immigrant in Thames regions. Oral Oncology 1999;35:471-5

Winter H, Cheng KK, Cummins C, Maric R, Silcocks P, Varghese C. Cancer incidence in the south Asian population of England (1990-92). *British Journal of Cancer* 1999;79:645-54

Bell CMJ, Ma M, Campbell S, Basnett I, Pollock A, Taylor I. Methodological issues in the use of regional guidelines and audit to improve clinical effectiveness in breast cancer. European Journal of Surgical Oncology 2000;26:130-6

Bullard J, Coleman MP, Robinson D, Lutz JM, Bell CMJ, Peto J. Completeness of cancer registration: a new method for routine use. *British Journal of Cancer* 2000;82:1111-6

Kocher HM, Patel S, Linklater K, Ellul JP. Increase in the incidence of oesophagogastric carcinoma in the South Thames region: an epidemiological study. British Journal of Surgery 2000;87:362-73

Pollock AM, Vickers N. Reducing DCO registrations through electronic matching of cancer registry data and routine hospital data. *British Journal of Cancer* 2000;82:712-7

Retsas S, Mohith A, Horward N, Bell J, Alexander H. Melanoma and additional primary cancers. Melanoma Research 2000;10:145-52

Roy P, Vaughan Hudson G, Vaughan Hudson B, Esteve J, Swerdlow AJ. Long-term survival in Hodgkin's disease patients. A comparison of relative survival in patients in trials and those recorded in population-based cancer registries. European Journal of Cancer 2000;36:384-9

Trent Cancer Registry

Trent Cancer Registry Third Report, 1999

Bright NM, Hollingsworth L, Silcocks PB, Ainsworth IM, Chattle B, Hethershaw S, Powell R, Tingle J (edited by Botha JL, Needham EP). Recording tumour stage for cancer registration purposes: feasibility of tumour staging from medical records by non-clinical staff. Occasional Report 99/1 July 1999.

Silcocks PB. Survival analysis sample size calculations: a brief outline and a guide to using a spreadsheet programme. Occasional Report 99/2 July 1999.

Bright NM, Botha JL. Relative survival: a spreadsheet approach. Occasional Report 99/3 November 1999.

Read C, Silcocks, PB (edited by Botha, JL). A cancer cluster in a Sheffield school? Occasional Report 99/4 November 1999.

Glencross J, Robinson M, Silcocks PB, Botha JL. Audit of soft tissue sarcoma management in Trent. Occasional Report 99/5 November 1999.

Choyce A, McAvoy BR. Cervical cancer screening and registration-are they working? J Epidemiol Community Health 1990 Mar;44(1):52-4

Smith SJ, Muir KR, Wolstenholme JL, Thornhill KG, Zamorski A, Tolley K, Logan RF, Chilvers CE . Continued inadequacies in data sources for the evaluation of cancer services. Br J Cancer 1997;75(1):131-3

Sokol RJ, Booker DJ, Stamps R. Erythrocyte autoantibodies, autoimmune haemolysis and carcinoma. J Clin Pathol 1994;47:340-343.

Miller JM, Slater DN. Do histopathology reports of primary cutaneous melanoma contain enough essential information? J Clin Pathol 1996 Mar;49(3):202-4

Key TJA, Silcocks PB, Davey GK, Appleby PN, Bishop DT. A case-control study of diet and prostate cancer. Brit J Cancer 1997;76:678-87.

Andalib AR, Lawry J, Ali SA, Murray AK, Sisley K, Silcocks P, Herlyn M, Rees RC. Cytokine modulation of antigen expression in human melanoma cell lines derived from primary and metastatic tumour tissues. Melanoma Research 1997;7:32-42

Darby,S, Whitley E, Silcocks P, Thakrar B, Green M, Lomas P, Miles J, Reeves G, Fearn T, Doll R. Risk of lung cancer associated with residential radon exposure in South-west England: a case-control study. Br J Cancer 1998;78:394-408.

Winter H, Cheng KK, Cummins C, Maric R, Silcocks P, Varghese C. Cancer incidence

in the South Asian population of England (1990-1992). Br J Cancer 1999;79:645-654.

Gillespie AM, Tidy J, Bright N et al. Primary gynaecological management of gestational trophoblastic tumours and the subsequent development of persistent trophoblastic disease. Br J Obs Gynae 1998; 105(suppl 17): 95.

Dobson L.S, Hancock H, Bright N, Robinson M.H, Hancock BW. Localised non-Hodgkin's lymphoma: The Sheffield Lymphoma Group experience (1970-1995). Int J Onc 1998;13:1313-1318.

Nevin J, Laing D, Kaye P, McCulloch T, Barnard R, Silcocks P, Blackett T, Paterson M, Sharp F, Cruse P. The significance of C-erb-B2 staining in cervical cancer, Gynecol Oncol 1999;73:354-358.

Silcocks P, Needham P, Hemsley F. Audit of prostate cancer: lessons learnt for current clinical practice, surrogates for quality of care and standardisation and quality assurance. Public Health 1999;113:161-164.

Silcocks P, Needham P, Hemsley F. Audit of prostate cancer: validity and feasibility of registry-based staging. Public Health 1999;113:157-160.

Cummins C, Winter H, Cheng KK, Maric R, Silcocks P, Varghese C. An assessment of the Nam Pehchan computer program for the identification of names of South Asian ethnic origin. J Pub Health Med 1999; 21: 401-406.

West Midlands Cancer Intelligence Unit

Burkinshaw L, Battle G, Somervaille L. (1999) West Midlands Cancer Intelligence Unit 1999 Report.

Burkinshaw L, Somervaille L. (2000) Workshop on Colorectal Cancer Care – Moving forward on Quality Indicators and Targets.

Cummins C, Kirk A, Saunders PJ, Stevens A, Wilson RC, Smith R, Somervaille L, Tonks A, Wyldes M. (1998) Key Health Data for the West Midlands. Dept of Public Health and Epidemiology, University of Birmingham, 3.

Bell CMJ, Lawrence GM, Pheby DFH, Smith J, Coleman MP. (1995) The Role of Cancer Registries. Clinical Oncology, 7, 143-144.

Cooper D. (2000) Decision Support for Health Research Policy Implementation in the West Midlands Implementation in the West Midlands – A GIS. BURISA, 200, 2-7.

Cooper D. (1999) Association for Geographical Information Health Special Interest Group. AGI Conference Proceedings 1999, 8.12.

Cooper D. (1999) Looking Below the Surface. Mapping Awareness, 13(7), 22–25.

Cooper D. (1998) Examples of GIS in Cancer Surveillance. AGI Conference Proceedings 1998,8.10.

Faux AM, Lawrence GM, Wheaton ME, Wallis MG, Jeffery CL, Griffiths RK. (1998) Slippage in the NHS breast screening round is being achieved. Journal of Medical Screening, 5(2), 88-91.

Faux AM, Richardson DC, Lawrence GM, Wheaton ME, Wallis MG. (1997) Interval breast cancers in the NHS Breast Screening Programme: does the current definition exclude too many? Journal of Medical Screening, 4, 169-173.

Gilman E, Sorahan T, Lancashire RJ, Lawrence GM, Cheng KK. (1998) Seasonality in the presentation of acute lymphoid leukaemia. British Journal of Cancer, 77, 676-678.

Harrison R, Leung P, Somervaille L. (1999) Analysis of incidence of childhood cancer in the West Midlands of the United Kingdom. Occupational and Environmental Medicine, 56, 774-780.

Parsons NP, Somervaille L. (2000) Estimation and projection of population lung cancer trends (United Kingdom). Cancer Causes and Control, 11.

Philips A, Lawrence G. (1999) Resection rates in Lung Cancer. Thorax, 54, 374.

Smith R. (1998) Geographic Information within Public Health and Environmental Health. Parliamentary Information Technology Committee Journal, GIS Supplement.

Smith R, Jarvis C. (1998) Just the Medicine, GIS in the NHS. Mapping Awareness, 12(7), 30-33.

Smith R. (1999) Co-ordinating healthcare – taking a geographic perspective. British Journal of Healthcare Computing & Information Management, 16(10), 40.

Winter H, Cheng KK, Cummins C, Maric R, Silcocks P, Varghese C. (1999) Cancer Incidence in the South Asian Population of England (1990-2). British Journal of Cancer, 79, 645-654.

Welsh Cancer Intelligence and Surveillance Unit

Cancer Registration in Wales 1974 – 1990, WCISU 1998

Cancer Registration and Survival in Wales 1985 – 1994, WCISU 1999

CROPS – Cancer Registration through On-line Pathology Systems

Evaluation of the Impact of Pathology Guidelines and a Minimum Dataset on the Quality of Cancer Registration in Wales. Steward JA et al , Welsh Office R&D 1998.

Results Of A Preliminary Study To Test The Irish Sea Proximity Hypothesis of Busby et al Steward JA, Jones DA, Beer H, John G, National Assembly of Wales, July 1999.

International Conference on Ecological Studies and Disease Clustering, ar Royal Statistical Society December 1999. An ecological investigation of environmental concerns raised over the incidence of childhood cancer on the Irish Sea coast of Wales. Steward J and John G (submitted to JRSS(A))

UKACR Conference, Cambridge November 1999 Bayesian Mapping – Leukaemia Incidence in Wales Dunstan F, Cornelius V (UWCM) Steward J, Beer H (WCISU)

Scottish Cancer Intelligence Unit

Black RJ, MacFarlane GJ, Maisonneuve P, Boyle P. Cancer Incidence and Mortality in Scotland 1960-89. Edinburgh: ISD Publications, 1995.

Black RJ, Storm HH, Simonato L, Démaret E. Automated data collection in cancer registration. IARC Technical Report No. 32. Lyon: International Agency for Research on Cancer, 1998.

Information & Statistics Division. Cancer. In: Scottish Health Statistics 1999. (pp 73-93) Edinburgh: ISD Scotland Publications, 2000.

Scottish Cancer Intelligence Unit. Trends in Cancer Survival in Scotland 1971 – 1995. Edinburgh: Information & Statistics Division, 2000.

Harris V, Sandridge A, Black RJ, Brewster DH, Gould A. Cancer Registration Statistics Scotland, 1986 – 1995. Edinburgh: ISD Publications, 1998.

Brewster D, Crichton J, Muir CS. Accuracy of 1990 Cancer Registration Data in Scotland. Information & Statistics Division of the Scottish Health Service: Edinburgh, 1994.

Storm HH, Clemmensen I, Black RJ. Survey of Cancer Registries in the European Union. IARC Technical Report No. 28. Lyon: International Agency for Research on Cancer, 1998.

Alexander FE, Wray N, Boyle P, Bring J, Coebergh JW, Draper G, Levi F, McKinney PA, Michaelis J, Peris-Bonet R, Petridou E, Pukkala E, Storm H, Terracini B, Vatten L. Clustering of childhood leukaemia: a European study in progress. J Epidemiology Biostats 1996; 1: 13-24.

Arrundale J, Bain M, Botting B, Brewster D, Cartwright R, Chalmers J, et al. Handbook and Guide to the Investigation of Clusters of Diseases. Leeds: Leukaemia Research Fund, 1997.

Bain M, Harvey JC, Muir CS. Cancer in Older Persons in Scotland. Health Bulletin 1996; 54: 375-389.

Bain M, Harvey JC, Muir CS. Epidemiology Research in Ageing: Perspectives and Limitations. In: Balducci L, Lyman GH, Ershler WB (eds). Comprehensive Geriatric Oncology. Amsterdam: Harwood Academic Publishers, 1998.

Bain MRS, Chalmers JWT, Brewster DH. Routinely collected data in national and regional databases – an underused resource. J Public Health Med 1997; 19: 413-418.

Black RJ, Sharp L, Finlayson AR, Harkness EF. Cancer incidence in a population potentially exposed to Radium-226 at Dalgety Bay, Scotland. Br J Cancer 1994; 69: 140-143.

Black RJ, Sharp L, Harkness EF, McKinney PA. Leukaemia and non-Hodgkin's lymphoma: incidence in children and young adults resident in the Dounreay area of Caithness, Scotland in 1968-91. J Epidemiol Community Health 1994; 48: 232-236.

Black RJ. Clarke JA, Warner JM. Malignant melanoma in Scotland: trends in incidence and survival, 1968-87. A review prepared by the CSA Information & Statistics Division for the Health Monitoring Group. Health Bulletin 1991; 49: 97-105.

Black RJ, Sharp L, Urquhart JD. An analysis of the geographical distribution of childhood leukaemia and non-Hodgkin lymphomas in Great Britain using areas of approximately equal population size. In: Geographical epidemiology of childhood leukaemia and non-Hodgkin lymphomas in Great Britain, 1966-83. OPCS Studies on Medical and Population subjects No. 53. HMSO, 1991.

Black RJ, Urquhart JD, Kendrick SQ, Bunch KJ, Warner J, Adams Jones D. Incidence of leukaemia and other cancers in birth and schools cohorts in the Dounreay area. BMJ 1992; 304: 1401-1405.

Black RJ, Sharp L, Urquhart JD. Analysing the spatial distribution of disease using a method of constructing geographical areas of approximately equal population size. In: Alexander FE, Boyle P (eds). Methods for Investigating Localised Clustering of Disease. IARC Scientific Publications No. 135. Lyon: International Agency for Research on Cancer, 1996.

Black RJ, Bray F, Ferlay J, Parkin DM. Cancer incidence and mortality in the European Union: cancer registry data and estimates of national incidence for 1990. Eur J Cancer 1997; 33: 1075-1107.

Black RJ, Sankaranarayanan R, Parkin DM. Interpretation of cancer survival data. In: Sankaranarayanan R, Black RJ, Parkin DM (eds). Cancer survival in developing countries. IARC Scientific Publications No. 145. Lyon: International Agency for Research on Cancer, 1999.

Black RJ. Cancer mortality in Italian migrants to Scotland. In: Geddes M, Parkin DM, Khlat M, Balzi D, Buiatti E (eds). Cancer in Italian migrant populations. IARC Scientific Publications No. 123. Lyon: International Agency for Research on Cancer, 1993.

Brewster D, Black RJ, Muir CS. Bladder cancer in Scotland: Recent trends in incidence, survival and geographical distribution. Health Bulletin 1994; 52: 248-259.

Brewster D, Black RJ. Breast, lung and colorectal cancer incidence and survival in South Thames Region, 1987-1992: the effect of social deprivation (letter). J Publ Hlth Med 1998; 20: 236-237.

Brewster D, Crichton J, Muir CS. How accurate are Scottish cancer registration data? Br J Cancer 1994; 70: 954-960.

Brewster D. Improving the quality of cancer registration data. J R Soc Med 1995; 88: 268-271.

Brewster D, Muir CS, Crichton J. Registration of lung cancer in Scotland: an assessment of data accuracy based on review of medical records. Cancer Causes Control 1995; 6: 303-310.

Brewster D, Muir CS, Crichton J. Registration of colorectal cancer in Scotland: an assessment of data accuracy based on review of medical records. Public Health 1995; 109: 285-292.

Brewster D, Muir C, Crichton J. Registration of non-melanoma skin cancer in Scotland: how accurate are site and morphology codes? Clin Exp Dermatol 1995; 20: 401-405.

Brewster D, Everington D, Harkness E, Gould A, Warner J, Dewar JA, Arrundale J. Incidence of and mortality from breast cancer since the introduction of screening (letter). BMJ 1996; 312: 639-640.

Brewster D, Crichton J, Harvey JC, Dawson G, Nairn ER. Benefits and limitations of pathology databases to cancer registries. J Clin Pathol 1996; 49: 947-949.

Brewster D, Crichton J, Harvey JC, Dawson G. Completeness of case ascertainment in a Scottish Regional Cancer Registry for the year 1992. Public Health 1997; 111: 339-343.

Brewster DH, Bain MRS, Chalmers JWT, Gould A, Dewar JA, George WD. Explicit consent is not needed for studies using medical records (letter). BMJ 1998; 317: 948.

Brewster DH, Harris V, Black RJ, Goldberg DJ. Epidemiology of Kaposi's sarcoma in Scotland, 1976-1996. Br J Cancer 1999; 79: 1938-1942.

Brewster DH, Fraser LA, Harris V, Black RJ. Rising incidence of prostate cancer in Scotland: increased risk or increased detection? BJU Int 2000; 85: 463-472.

Brewster DH, Fraser LA, McKinney PA, Black RJ. Socioeconomic status and risk of adenocarcinoma of the oesophagus and cancer of the gastric cardia in Scotland. Br J Cancer 2000; 83: 387-390.

Carnon A, Hole D, Gillis CR, Brewster D. Incidence of and mortality from breast cancer since the introduction of screening (letter). BMJ 1996; 312: 640.

Cheng KK, Sharp L, McKinney PA, Logan RFA, Chilvers CED, Cook Mozaffari P, Ahmed A, Day NE. A case-control study of oesophageal adenocarcinoma in women: a preventable disease. Br J Cancer 2000; 83: 127-132.

Clarke K, Howard GCW, Elia MH, Hutcheon AW, Kaye SB, Windsor PM, Yosef HMA. Referral patterns within Scotland to specialist oncology centres for patients with testicular germ cell tumours. Br J Cancer 1995; 72: 1300-1302.

Clinical Outcome Indicators Working Group. Survival from selected cancers by Health Board. In: Clinical Outcome Indicators. pp 23-35. Edinburgh: The Scottish Office, 1996 (July).

Colonna M, Grosclaude P, Faivre J, Revzani A, Arveux P, Chaplain G, Tretarre B, Launoy G, Mace-Lesech J, Raverdy N, Schaffer P, Buemi A, Ménégoz F, Black RJ. Cancer registry data based estimation of regional cancer incidence: application to breast and colorectal cancer in French administrative regions. J Epidemiol Community Health 1999; 53: 558-564.

Crosher R, McIlroy R. Multiple primary malignancies in patients with oral cancer in Scotland. Br J Oral Maxillofac Surg 1998; 36: 58-62.

Deans H, Everington D, Cordiner C, Kirkpatrick AE, Lindsay E. The Scottish experience of double reading in the Scottish National Breast Screening Programme. The Breast 1998; 7: 75-79.

Dewar JA, Duncan W, Eremin O, Kaye SB, Muir CS, Gould A. The Scottish Cancer Therapy Network: the first year. Health Bulletin 1994; 52: 492-495.

Dewar J, Twelves CJ, Thomson CS. Breast cancer in Scotland: changes in treatment and workload. Clin Oncol 1999; 11: 52-54.

Everington D, Gilbert FJ, Tyack C, Warner J. The Scottish Breast Screening Programme's experience of monitoring interval cancers. J Med Screening 1999; 6: 21-27.

Ferlay J, Black RJ, Valdivieso MT, Pisani P, Parkin DM. CI5VII: Electronic Database of Cancer in Five Continents Vol. VII. IARC CancerBase No. 2. Lyon: International Agency for Research on Cancer, 1998.

Gould A, Muir CS, Sharp L. Prognosis and Survival in Prostate Cancer. In: Garraway M (editor). Epidemiology of Prostate Disease. pp 214-24. Berlin Heidelberg: Springer-Verlag, 1995.

Higginson J, Muir CS, Munoz N. Human Cancer: Epidemiology and environmental causes. Cambridge Monographs on Cancer Research. Cambridge University Press, 1992.

Howard GCW, Clarke K, Elia MH, Hutcheon AW, Kaye SB, Windsor PM, Yosef HM, Sharp L. A Scottish national mortality study assessing cause of death, quality of and variation in management of patients with testicular non-seminomatous germ-cell tumours. Br J Cancer 1995; 72: 1307-1311.

Howard GCW, Clarke K, Elia MH, Hutcheon AW, Kaye SB, Windsor PM, Yosef HMA. A Scottish national audit of current patterns of management for patients with testicular non-seminomatous germ-cell tumours. *Br J Cancer* 1995; 72: 1303-1306.

MacFarlane GJ, Sharp L, Porter S, Franceschi S. Trends in survival from cancers of the oral cavity and pharynx in Scotland: a clue as to why the disease is becoming more common? *Br J Cancer* 1996; 73: 805-808.

McKinney PA, Ironside JW, Harkness EF, Arango JC, Doyle D, Black RJ. Registration Quality and Descriptive Epidemiology of Childhood Brain Tumours in Scotland 1975-1990. *Br J Cancer* 1994; 70: 973-979.

McKinney PA, Smith K, Findlay E. The Scottish Case Control Study of Childhood Leukaemia and Cancer: Methodology and Environmental Measures of Exposure. Health Bulletin 1995; 53: 222-229.

McKinney PA, Sharp L, MacFarlane GJ, Muir CS. Oesophagus and gastric cancer in Scotland, 1960-90. *Br J Cancer* 1995; 71: 411-415.

McKinney PA, Juszczak E, Findlay E, Smith K. Case-control study of childhood leukaemia and cancer in Scotland: findings for neonatal intramuscular vitamin K. BMJ 1998; 316: 173-177.

McKinney PA, Juszczak E, Findlay E, Smith K, Thomson CS. Pre- and perinatal risk factors for childhood leukaemia and other malignancies: a Scottish case control study. *Br J Cancer* 1999; 80: 1844-1851.

Marson LP, Stevenson J, Gould A, Aitken RJ. A prospective colorectal cancer audit appears to be improving outcome. Br J Surg 1997; 84 (suppl): 24.

Ménégoz F, Black RJ, Arveux P, Magne V, Ferlay J, Buéme A, Carli PM, Chapelain G, Faivre J, Gignoux M, Grosclaude P, Mace-Lesech J, Reverdy N, Schaffer P. Cancer incidence and mortality in France 1975-95. Eur J Cancer Prev 1997; 6: 442-466.

Miñarro R, Black RJ, Martínez C, Navarro C, Garau I, Izarzugaza I, Ardanaz E, Vergara A, Galcerán J, Alonso R, Mateos A, Rodríguez M. Cancer Incidence and Mortality in Spain. IARC Technical Report No. 36. Lyon: International Agency for Research on Cancer, 2000.

Muir CS. Epidemiology, basic science, and the prevention of cancer: implications for the future. Cancer Res 1990; 50: 6441-6448.

Muir CS. Geographical patterns of cancer: Role of Environment. In: Macieira-Coelho A, Nordenskjold B (eds). Cancer and Aging. Florida: CRC Press, 1990.

Muir CS. The Burden of Cancer in Europe. Eur J Cancer 1990; 26: 1111-1113.

Muir C, Nectoux J, Staszewski J. The epidemiology of prostatic cancer. Geographical distribution and time-trends. Acta Oncol 1991; 30: 133-140.

Muir CS, Sasco AJ. Prospects for cancer control in the 1990s. Ann Rev Public Health 1990; 11: 143-163.

Muir CS. Cancer Registry in Cancer Control: an overview with a Scottish dimension. Health Bulletin 1993; 51: 208-229.

Muir CS, McKinney PA. A strategy for cancer control in Scotland. Glasgow: Scottish Forum for Public Health Medicine, 1994.

Muir CS, Fraumeni JR, Doll R. The interpretation of time trends. In: Cancer Surveys volume 19/20: Trends in Cancer Incidence and Mortality. New York: Imperial Cancer Research Fund, 1994.

Muir CS, Storm HH, Polednak A. Brain and other nervous system tumours. In: Cancer Surveys volume 19/20: Trends in Cancer Incidence and Mortality. New York: Imperial Cancer Research Fund, 1994.

Muir CS, Weiland LH. Cancer of the head and neck. Cancer 1995; 75 (suppl): 147-153.

Muir CS. Cancer of unknown primary site. Cancer 1995; 75 (suppl): 353-356.

Muir CS. International patterns of cancer. In: Greenwald P, Kramer BS, Weed DL (eds). Science and Practice of Cancer Prevention and Control. New York: Marcel Dekker, 1995.

Muir CS, Harvey JC. Cancer of the Stomach (Review). GI Cancer 1996; 1: 213-25.

Muir CS, Harvey JC. Cancer of the stomach: overview. In: Sugimura T, Sasako M (eds). Gastric Cancer. Oxford: Oxford University Press, 1997.

Parkin DM, Cardis E, Masuyer E et al. Childhood leukaemia following the Chernobyl accident: the European Childhood Leukaemia-Lymphoma Incidence Study (ECLIS). Eur J Cancer 1992; 29A: 87-95.

Parkin DM, Muir CS, Whelan SL, Raymond L, Young J. Cancer Incidence in Five Continents, volume VII. IARC Scientific Publications No.120. Lyon: International Agency for Research on Cancer, 1992.

Parkin DM, Clayton D, Black RJ, et al. Childhood leukaemia in Europe after Chernobyl: 5 year follow-up. *Br J Cancer* 1996; 73: 1006-1012.

Percy C, Young JL, Muir CS, Ries L, Hankey BF, Sobin LH, Berg JW. Histology of cancer, Incidence and Prognosis: SEER Population-based data 1973-1987. Cancer 1995; 75 (suppl): 139-421.

Richards MA, Stockton D, Babb P, Coleman MP. How many deaths have been avoided through improvements in cancer survival? BMJ 2000; 320: 895-898.

Robertson AG, Robertson C, Perone C, Clarke K, Dewar J, Elia MH, Hurman D,

MacDougall RH, Yosef HMA. Effect of gap length and position on results of treatment of cancer of the larynx in Scotland by radiotherapy: a linear quadratic analysis. Radiotherapy Oncol 1998; 48: 165-173.

Sankaranarayanan R, Black RJ and Parkin DM. Cancer Survival in Developing Countries. IARC Scientific Publication No. 145. Lyon: International Agency for Research on Cancer, 1998.

Scott N, Gould A, Brewster D. Laryngeal cancer in Scotland, 1960-1994: trends in incidence, geographical distribution and survival. Health Bulletin 1998; 56: 749-756.

Scottish Breast Cancer Focus Group, Scottish Cancer Trials Breast Group, Scottish Cancer Therapy Network. Scottish Breast Cancer Audit 1987 & 1993. Edinburgh: SCTN, 1996.

Sinclair DJ, Gilbert FJ, Best JK, Clarke K. An audit into the use of intravascular contrast media in Scotland. Health Bulletin 1996; 54: 371-374.

Sharp L, Black RJ, Muir CS, Warner J, Clarke JA. Trends in cancer of the testis in Scotland, 1961-90. A paper prepared by the Information & Statistics Division for the Health Monitoring Group. Health Bulletin 1993; 51: 255-268.

Sharp L, Finlayson AR, Black RJ. Cancer survival and deprivation in Scotland. J Epidemiol Community Health 1995; 49 (suppl 2): S79.

Sharp L, Finlayson AR, Black RJ. Cancer incidence and deprivation in Scotland. J Epidemiol Community Health 1995; 49 (suppl 2): S79-S80.

Sharp L, Black RJ, Muir CS, Gemmell I, Finlayson AR, Harkness EF. Will the Scottish Cancer Target for the year 2000 be met? The use of cancer registration and death records to predict future cancer incidence and mortality in Scotland. *Br J Cancer* 1996; 73: 1115-1121.

Sharp L, Black RJ, Harkness EF, McKinney PA. Incidence of childhood leukaemia and non-Hodgkin's lymphoma in the vicinity of nuclear sites in Scotland, 1968-93. Occ Environ Med 1996; 53: 823-831.

Sharp L, McKinney PA, Black RJ. Incidence of childhood brain and other non-haematopoietic neoplasms near nuclear sites in Scotland, 1975-1994. Occup Envir Med 1999; 56: 308-314.

Sharp L, Brewster D. The epidemiology of lung cancer in Scotland: a review of trends in incidence, survival and mortality and prospects for prevention. Health Bulletin 1999; 57: 318-331.

Sharp L, Gould A, Harris V, Harkness EF, McKinney PA, Brewster DH, Black RJ. United Kingdom, Scottish Cancer Registry, 1981-1990. In: Parkin DM, Kramárová E, Draper GJ, Masuyer E, Michaelis J, Neglia J, Qureshi S, Stiller CA (eds). International Incidence of Childhood Cancer, Volume II. pp. 369-371. IARC Scientific Publications No. 144. Lyon: International Agency for Research on Cancer, 1998.

Smans M, Muir CS, Boyle P. Atlas of cancer mortality in the European Economic Community. IARC Scientific Publications No. 107. Lyon: International Agency for Research on Cancer, 1992.

Stroner PL, Brewster DH, Dewar JA, Eremin O, Gould A, Howard GCW, Kaye SB. In pursuit of excellence for patients with cancer: the Scottish Cancer Therapy Network model. *Br J Cancer* 1999; 79: 1641-1645.

Swerdlow AJ, dos Santos Silva I, Reid A, Qiao Z, Brewster DH, Arrundale J. Trends in cancer incidence and mortality in Scotland: description and possible explanations. *Br J Cancer* 1998; 77 (suppl 3): 1-16.

Thomas E, Brewster DH, Black RJ, Macfarlane GJ. The Risk of Malignancy Amongst Patients with Rheumatic Conditions. Int J Cancer 2000; (in press).

Twelves CJ, Thomson CS, Young J, Gould A. Entry into clinical trials in breast cancer: the importance of specialist teams. Eur J Cancer 1998; 34: 1004-1007.

Twelves CJ, Thomson CS, Gould A, Dewar J. Variation in the survival of women with breast cancer in Scotland. *Br J Cancer* 1998; 78: 566-571.

Twelves CJ, Thomson CS, Dewar JA. Deprivation and emergency admissions for cancers. Social factors affect patterns of referral for breast cancer [letter]. BMJ 1999; 318: 326.

Urquhart J. Studies of disease clustering: problems of interpretation. In: Elliott P, Cuzick J, English D, Stern R (eds). Geographical and Environmental Epidemiology: Methods for Small-Area Studies. Oxford: Oxford University Press, 1992.

Urquhart JD, Black RJ, Muirhead MJ, Sharp L, Maxwell M, Eden OB, Adams Jones D. Case control study of leukaemia and non-Hodgkin's lymphoma in children in Caithness near the Dounreay nuclear installation. BMJ 1991; 302: 687-692.

Van der Esch EP, Muir CS, Nectoux J, et al. Temporal change in diagnostic criteria as a cause of the increase of malignant melanoma over time is unlikely. Int J Cancer 1991; 27: 483-490.

Walker JJ, Brewster DH, Gould A, Raab GM. Trends in incidence of and mortality from invasive cancer of the uterine cervix in Scotland (1975-94). Public Health 1998; 112: 373-378.

Wright EP, Smith AB, Cull AM, Velikova G, Gould A, Forman D, Perren T, Brown J, Stead M, Selby PJ. Automated screening and recording of quality of life data: a comparison of data capture methods. Psycho-oncol 1998; 7: 67.

Northern Ireland Cancer Registry

Cancer Incidence in Northern Ireland 1993-1995 Editors: AT Gavin, J Reid, 1999

Patterns and Trends Northern Ireland Cancer Registry, 1995

Gavin AT, O'Reilly D. Cancer in Northern Ireland 2002 Ulster Medical Journal, 1996;65,2:106-112

Gavin AT, O'Reilly D, Reid J, Middleton R. Asbestos related mortality in Northern Ireland: 1985-1994 Journal of Public Health Medicine 1999;21(1):95-101.

Childhood Cancer Research Group

Committee on Medical Aspects of Radiation in the Environment (COMARE). Third Report. Report on the Incidence of Childhood Cancer in the West Berkshire and North Hampshire area, in which are situated the Atomic Weapons Research Establishment, Aldermaston and the Royal Ordnance Factory, Burghfield. 1989. London, HMSO.

Committee on Medical Aspects of Radiation in the Environment (COMARE). Fourth Report. The incidence of cancer and leukaemia in young people in the vicinity of the Sellafield site, West Cumbria: Further studies and an update of the situation since the publication of the report of the Black Advisory Group in 1984. Bridges, B. A. 1996. Wetherby, Department of Health.

Committee on Medical Aspects of Radiation in the Environment (COMARE). Fifth Report. The incidence of cancer and leukaemia in the area around the former Greenham Common Airbase. An investigation of a possible association with measured environmental radiation levels. 1998. Chilton, NRPB.

Stiller CA, Bunch KJ, Lewis IJ. Ethnic group and survival from childhood cancer: report from the UK Children's Cancer Study Group. *Br J Cancer* 2000;82:1339-43.

Alexander FE, Boyle P, Carli PM, Coebergh JW, Ekbom A, Levi F et al. Population density and childhood leukaemia: Results of the EUROCLUS study. Eur J Cancer 1999;35:439-44.

Wray NR, Alexander FE, Muirhead CR, Pukkala E, Schmidtmann I, Stiller CA. A comparison of some simple methods to identify geographical areas with excess incidence of a rare disease such as childhood leukaemia. Statistics in Medicine 1999;18:1501-16.

Stiller CA, Benjamin S, Cartwright RA, Clough JV, Gorst DW, Kroll ME et al. Patterns of care and survival for adolescents and young adults with acute leukaemia – a population based study. *Br J Cancer* 1999;79:658-65.

dos Santos Silva I, Swerdlow AJ, Stiller CA, Reid A. Incidence of testicular germ-cell malignancies in England and Wales: trends in children compared with adults. Int J Cancer 1999;83:634.

Bithell JF, Draper GJ. Uranium-235 and childhood leukaemia around Greenham Common airfield. J Radiol Prot 1999;19:253-9.

Stiller CA, Eatock EM. Patterns of care and survival for children with acute lymphoblastic leukaemia diagnosed between 1980-94. Arch Dis Child 1999;81:202-8.

Passmore SJ, Draper GJ, Brownbill PA, Kroll ME. Case-control studies of relation between childhood cancer and neonatal vitamin K administration. BMJ 1998;316:178-84.

Parkin DM, Kramárová E, Draper GJ, Masuyer E, Michaelis J, Neglia J et al. International Incidence of Childhood Cancer: Volume 2. Lyon: IARC Scientific Publications No 144, 1998.

Alexander FE, Boyle P, Carli PM, Coebergh JW, Draper GJ, Ekbom A et al. Spatial and temporal patterns in childhood leukaemia: further evidence for an infectious origin. *Br J Cancer* 1998;77:812-7.

Satgé D, Sasco AJ, Carlsen NLT, Stiller CA, Rubie H, Hero B et al. A lack of neuroblastoma in Down Syndrome: a study from 11 European countries. Cancer Res 1998;58:448-52.

Stiller CA, Allen MB, Bayne AM, Brownbill PA, Draper GJ, Eatock EM et al. United Kingdom: National Registry of Childhood Tumours, England and Wales, 1981-1990. In Parkin DM, Kramárová E, Draper GJ, Masuyer E, Michaelis J, Qureshi S et al, eds. International Incidence of Childhood Cancer: Volume 2, pp 365-7. Lyon: IARC Scientific Publications No 144, 1998.

Powell JE, Esteve J, Mann JR, Parker L, Frappaz D, Michaelis J et al. Neuroblastoma in Europe: differences in the pattern of disease in the UK. Lancet 1998;352:682-7.

Doyle P, Bunch KJ, Beral V, Draper GJ. Cancer incidence in children conceived with assisted reproduction technology. (research letter). Lancet 1998;352:452-3.

Narod SA, Hawkins MM, Robertson CM, Stiller CA. Congenital anomalies and childhood cancer in Great Britain. Am J Hum Genet 1997;60:474-85.

Stiller CA. Reliability of cancer registration data (editorial). Eur J Cancer 1997;33: 812-4.

Alexander FE, Boyle P, Carli PM, Coebergh JW, Draper GJ, Ekbom A et al. Spatial clustering of childhood leukaemia: summary results from the EUROCLUS project. *Br J Cancer* 1997;77:818-24.

Cartwright RA, McNally RJQ, Rowland DJ, Thomas J, Stiller C. Introduction to the Data Collection Study. In Cartwright RA, McNally RJQ, Thomas J, Staines A, Stiller C,

eds. The Descriptive Epidemiology of Luekaemia and Related Conditions in Parts of the United Kingdom 1984-1993, pp 1-8. London: Leukaemia Research Fund, 1997.

Cartwright RA, McNally RJQ, Rowland DJ, Staines A, Stiller C. Malignancies in the adolescent. In Cartwright RA, McNally RJQ, Rowland DJ, Thomas J, Staines A, Stiller C, eds. The Descriptive Epidemiology of Leukaemia and Related Conditions in Parts of the United Kingdom 1984-1993, pp 46-55 and CD-Rom. London: Leukaemia Research Fund, 1997.

Mott MG, Mann JR, Stiller CA. The United Kingdom Children's Cancer Study Group – the first 20 years of growth and development. Eur J Cancer 1997;33:1448-52.

Arrundale J, Bain M, Botting B, Brewster D, Cartwright R, Chalmers J et al. Handbook And Guide To The Investigation Of Clusters Of Diseases. University of Leeds: Leukaemia Research Fund Centre for Clinical Epidemiology, 1997.

Parkin DM, Clayton D, Black RJ, Masuyer E, Friedl HP, Ivanov E et al. Childhood leukaemia in Europe after Chernobyl: 5 year follow-up. Br J Cancer 1996;73:1006-12.

Alexander FE, Wray N, Boyle P, Bring J, Coebergh JW, Draper GJ et al. Clustering of childhood leukaemia: a European study in progress (review). J Epidemiol Biostats 1996;1:13-24.

Little MP, Muirhead CR, Stiller CA. Modelling lymphocytic leukaemia incidence in England and Wales using generalisations of the two-mutation model of carcinogenesis of Moolgavkar, Venzon and Knudson. Statistics in Medicine 1996;15:1003-22.

Levitt G, Bunch KJ, Rogers CA, Whitehead B. Cardiac transplantation in childhood cancer survivors in Britain. Eur J Cancer 1996;32:826-30.

Stiller CA, Parkin DM. Geographic and ethnic variations in the incidence of childhood cancer. Br Med Bull 1996;52:682-703.

Kramárová E, Stiller CA, Ferlay J, Parkin DM, Draper GJ, Michaelis J et al. The International Classification of Childhood Cancer. Lyon: IARC, 1996.

Kramárová E, Stiller CA. The international classification of childhood cancer. Int J Cancer 1996;68:759-65.

Draper GJ, Sanders BM, Lennox EL, Brownbill PA. Patterns of childhood cancer among siblings. *Br J Cancer* 1996;74:152-8.

Stiller CA, Boyle PJ. Effect of population mixing and socioeconomic status in England and Wales, 1979-85, on lymphoblastic leukaemia in children. BMJ 1996;313:1297-300.

Draper GJ. Cancer. In Botting B, ed. The Health of Our Children, pp 135-47. London: HMSO, 1995.

Lyons RA, Monaghan SP, Heaven M, Littlepage BNC, Vincent TJ, Draper GJ. Incidence of leukaemia and lymphoma in young people in the vicinity of the petrochemical plant at Baglan Bay, South Wales, 1974-1991. Occup Environ Med 1995;52:225-8.

Stiller CA, Allen MB, Eatock EM. Childhood cancer in Britain; The national registry of childhood tumours and incidence rates 1978-1987. Eur J Cancer 1995;31:2028-34.

Draper GJ. Childhood cancer: trends in incidence survival and mortality (editorial). Eur J Cancer 1995;31:653-4.

Stiller CA. International variations in the incidence of childhood carcinomas. Cancer Epid Bio Prev 1994;3:305-10.

Stiller CA, Nectoux J. International incidence of childhood brain and spinal tumours. Int J Epidemiol 1994;23:458-64.

Robertson CM, Hawkins MM, Kingston JE. Late deaths and survival after childhood cancer: implications for cure. BMJ 1994;309:162-6.

Stiller CA, Chessells JM, Fitchett M. Neurofibromatosis and childhood leukaemia/lymphoma: a population-based UKCCSG study. *Br J Cancer* 1994;70: 969-72.

Draper GJ, Kroll ME, Stiller CA. Childhood Cancer. Cancer Surv 1994;19/20:493-517.

Stiller CA, Eatock EM. Survival from acute non-lymphocytic leukaemia, 1971-88: a population based study. Arch Dis Child 1994;70:219-23.

Mann JR, Stiller CA. Changing pattern of incidence and survival in children with germ cell tumours (GCTs). Advances in Biosciences 1994;91:59-64.

Stiller CA, Parkin DM. International variations in the incidence of childhood soft-tissue sarcomas. Paediatr Perinat Epidemiol 1994;8:107-19.

Stiller CA. Population based survival rates for childhood cancer in Britain, 1980-91. BMJ 1994;309:1612-6.

Bithell JF, Dutton SJ, Draper GJ, Neary NM. The distribution of childhood leukaemias and non-Hodgkin lymphomas near nuclear installations in England and Wales. BMJ 1994;309:501-5.

Parkin DM, Cardis E, Masuyer E, Friedl HP, Hanslukwa H, Bobev D et al. Childhood leukaemia following the Chernobyl accident: The European Childhood Leukaemia-Lymphoma Incidence Study (ECLIS). Eur J Cancer 1993;29:87-95.

Stiller CA. Cancer registration: its uses in research, and confidentiality in the EC. J Epidemiol Community Health 1993;47:342-4.

Stiller CA. Trends in neuroblastoma in Great Britain: incidence and mortality, 1971-1990. Eur J Cancer 1993;29:1008-12.

LIST OF PUBLICATIONS USING CANCER REGISTRATION DATA *Appendix I*

Parkin DM, Stiller CA, Nectoux J. International variations in the incidence of childhood bone tumours. Int J Cancer 1993;53:371-6.

Draper GJ, Stiller CA, Cartwright RA, Craft AW, Vincent TJ. Cancer in Cumbria and in the vicinity of the Sellafield nuclear installation, 1963-90. BMJ 1993;306:89-94.

Black RJ, Urquhart JD, Kendrick SW, Bunch KJ, Warner J, Adam Jones D. Incidence of leukaemia and other cancers in birth and schools cohorts in the Dounreay area. BMJ 1992;304:1401-5.

Muirhead CR, Butland BK, Green BMR, Draper GJ. An analysis of childhood leukaemia and natural radiation in Britain. Radiat Prot Dosim 1992;45:657-60.

Stiller CA. Survival of patients in clinical trials and at specialist centres. In Williams CJ, ed. New Treatments for Cancer: Practical, Ethical and Legal Problems, pp 120-36. John Wiley & Sons Ltd, 1992.

Hawkins MM, Swerdlow AJ. Completeness of cancer and death follow-up obtained through the national health service central register for England and Wales. *Br J Cancer* 1992;66:408-13.

Stiller CA, Parkin DM. International variations in the incidence of neuroblastoma. Int J Cancer 1992;52:538-43.

Stiller CA, Bunch KJ. Brain and spinal tumours in children aged under two years: incidence and survival in Britain, 1971-85. *Br J Cancer* 1992;66:S50-S53.

Draper GJ, Sanders BM, Brownbill PA, Hawkins MM. Patterns of risk of hereditary retinoblastoma and applications to genetic counselling. *Br J Cancer* 1992;66:211-9.

Robertson CM, Stiller CA, Kingston JE. Causes of death in children diagnosed with non-Hodgkin's lymphoma between 1974 and 1985. Arch Dis Child 1992;67:1378-83.

Narod SA, Stiller CA, Lenoir GM. An estimate of the heritable fraction of childhood cancer. *Br J Cancer* 1991;63:993-9.

Muirhead CR, Butland BK, Green BMR, Draper GJ. Childhood leukaemia and natural radiation (corres). Lancet 1991;337:503-4.

Stiller CA, McKinney PA, Bunch KJ, Bailey CC, Lewis IJ. Childhood cancer and ethnic group in Britain: a United Kingdom children's cancer study group (UKCCSG) study. *Br J Cancer* 1991;64:543-8.

Draper GJ, Stiller CA, O'Connor CM, Vincent TJ, Elliott P, McGale P et al. The Geographical Epidemiology of Childhood Leukaemia and Non-Hodgkin Lymphomas in Great Britain, 1966-83, OPCS Studies on Medical and Population Subjects No.53. London: HMSO, 1991.

Kinlen LJ, Hudson CM, Stiller CA. Contacts between adults as evidence for an infective origin of childhood leukaemia: an explanation for the excess near nuclear establishments in West Berkshire? *Br J Cancer* 1991;64:549-54.

Hawkins MM. Risks of myeloid leukaemia in children treated for solid tumours (corres). Lancet 1990;336:887-.

Hawkins MM, Kingston JE, Kinnier Wilson LM. Late deaths after treatment for childhood cancer. Arch Dis Child 1990;65:1356-63.

Hawkins MM. Second primary tumours following radiotherapy for childhood cancer. Int J Radiol Oncol Biol Phys 1990;19:1297-301.

Stiller CA, Parkin DM. International variations in the incidence of childhood renal tumours. *Br J Cancer* 1990;62:1026-30.

Stiller CA, Bunch KJ. Trends in survival for childhood cancer in Britain diagnosed 1971-85. *Br J Cancer* 1990;62:806-15.

Levitt GA, Stiller CA, Chessells JM. Prognosis of Down's syndrome with acute leukaemia. Arch Dis Child 1990;65:212-6.

Stiller CA, Parkin DM. International variations in incidence of childhood lymphomas. Paediatr Perinat Epidemiol 1990;4:303-24.

United Kingdom Association of Cancer Registries

Davies TW, Williams EMI (eds). The Cancer Registry Handbook. Cambridge: United Kingdom Association of Cancer Registries, 1994.

Bell CMJ (ed). Reducing risk, improving outcome in cancer. London: United Kingdom Association of Cancer Registries, 1998.

Small Area Health Statistics Unit (SASHU)

Elliott P. Small area health statistics in Europe. Information Services and Use 1990; 10: 39-45.

Draper GJ, Elliott P. Variations in incidence rates and factors affecting them – Summary. In: Draper G (Ed). The Geographical Epidemiology of Childhood Leukaemia and Non-Hodgkin Lymphomas in Great Britain, 1966-83. Studies in Medical and Population Subjects No. 53. London: HMSO, 1991, pp 57-59.

Elliott P, McGale P, Vincent TJ. Population data and description of areas. In: Draper G (Ed). The Geographical Epidemiology of Childhood Leukaemia and Non-Hodgkin Lymphomas in Great Britain, 1966-83. Studies in Medical and Population Subjects No. 53. London: HMSO, 1991, pp 17-23.

Rodrigues L, Hills M, McGale P, Elliott P. Socioeconomic factors in relation to childhood leukaemia and non-Hodgkin lymphomas: An analysis based on small area statistics for census tracts. In: Draper G (Ed). The Geographical Epidemiology of Childhood Leukaemia and Non-Hodgkin Lymphomas in Great Britain, 1966-83. Studies in Medical and Population Subjects No. 53. London: HMSO, 1991, pp 47-56.

Elliott P, Cuzick J, English D, Stern R (eds). Geographical and Environmental Epidemiology: Methods for Small Area Studies. Oxford: Oxford University Press, 1992.

Elliott P, Kleinschmidt I, Westlake AJ. Use of routine data in studies of point sources of environmental pollution. In: Elliott P, Cuzick J, English D, Stern R, eds. Geographical and Environmental Epidemiology: Methods for Small Area Studies. Oxford: Oxford University Press, 1992, pp 106-114.

Elliott P, Beresford J, Jolley DJ, Pattenden S, Hills M. Cancer of the larynx and lung around incinerators of waste solvents and oils in Britain. In: Elliott P, Cuzick J, English D, Stern R, eds. Geographical and Environmental Epidemiology: Methods for Small Area Studies. Oxford: Oxford University Press, 1992, pp 359-367.

Elliott P, Hills M, Beresford J, Kleinschmidt I, Pattenden S, Jolley D, Rodrigues L, Westlake AJ, Rose G. Incidence of cancer of the larynx and lung near incinerators of waste solvents and oils in Great Britain. Lancet, 1992; 339: 854-858.

Elliott P, Westlake, AJ, Kleinschmidt I, Rodrigues L, Hills M, McGale P, Marshall K, Rose G. The Small Area Health Statistics Unit: a national facility for investigating health around point sources of environmental pollution in the United Kingdom. J Epidemiol Community Health, 1992; 46: 345-349.

Westlake AJ, ed. Geographical methods in small area health studies. Proceedings of a workshop held at the London School of Hygiene and Tropical Medicine on 22 June 1990. London, Small Area Health Statistics Unit, 1992

Quinn MJ, Elliott P. Small Area Health Statistics in the UK. In: Proceedings of an East European-COST meeting on air pollution epidemiology, held in Budapest, Hungary, May 22-25, 1991. Brussels: Commission of the European Communities, COST 631/2, 1993.

Sans S, Elliott P, Kleinschmidt I, Shaddick G, Pattenden S, Walls P, Grundy C, Dolk H. Cancer incidence and mortality near the Baglan Bay petrochemical works, South Wales. Occup Environ Med, 1995; 52: 217-224.

Martuzzi M, Hills M. Estimating the degree of heterogeneity between event rates using likelihood. Am J Epidemiol, 1995; 141: 369-74.

Elliott P, Martuzzi M, Shaddick G. Spatial statistical methods in environmental epidemiology: a critique. Stat Methods Med Res 1995; 4: 139-161.

Briggs D, Elliott P. The use of geographic information systems in studies of environment and health data. W Health Stats Q 1995; 48: 85-94.

Elliott P. Investigation of disease risk in small areas. Occup Environ Med 1995; 52: 785-89.

Small Area Health Statistics Unit. Proceedings of a Meeting on The Use of Deprivation Indices in Studies of Environment and Health. London School of Hygiene and Tropical Medicine, 1994. J Epidemiol Community Health 1995; 49 (Suppl. 2).

Diggle PJ, Elliott P. Disease risk near point sources: statistical issues for analyses using individual or spatially aggregated data. J Epidemiol Community Health 1995; 49(Suppl.2): S20-S27.

Kleinschmidt I, Hills M, Elliott P. Smoking behaviour can be predicted by neighbourhood deprivation measures. J Epidemiol Community Health 1995; 49(Suppl.2): S72-S77.

Dolk H, Mertens B, Kleinschmidt I, Walls P, Shaddick G, Elliott P. A standardisation approach to the control of socioeconomic confounding in small area studies of environment and health. J Epidemiol Community Health 1995; 49(Suppl.2): S9-S14.

Elliott P. Small-area studies. In: Environmental Epidemiology: Exposure and Disease. R Bertolini, M D Lebowitz, R Saracci, D A Savitz (eds). World Health Organisation-Lewis Publications (CRC Press), Florida, USA, 1996, pp 187-199.

Elliott P, Shaddick G, Kleinschmidt I, Jolley D, Walls P, Beresford J, Grundy C. Cancer incidence near municipal solid waste incinerators in Great Britain Br J Cancer 1996; 73: 702-710.

Olsen SF, Martuzzi M, Elliott P. Cluster analysis and disease mapping – why, when, and how? A step by step guide. Br Med J 1996; 313: 863-6.

Shaddick G, Elliott P. Use of Stone's method in studies of disease risk around point sources of environmental pollution. Stat Med 1996; 15: 1927-1934.

Diggle P, Elliott P. Use of deprivation indices in small area studies. J Epidemiol Community Health 1996; 50: 690. (Letter to the Editor).

Dolk H, Shaddick G, Walls P, Grundy C, Thakrar B, Kleinschmidt I, Elliott P. Cancer incidence near radio and television transmitters in Great Britain. I. Sutton Coldfield transmitter. Am J Epidemiol 1997; 145:1-9.

Dolk H, Elliott P, Shaddick G, Walls P, Thakrar B. Cancer incidence near radio and television transmitters in Great Britain. II. All high power transmitters. *Am J Epidemiol* 1997; 145:10-17.

Elliott P, Kleinschmidt I. Angiosarcoma of the liver in Great Britain in proximity to vinyl chloride sites. Occup Environ Med 1997; 54: 14-18.

Wilkinson P, Thakrar B, Shaddick G, Stevenson S, Pattenden S, Landon M, Grundy C, Elliott P. Cancer incidence and mortality around the Pan Britannica Industries pesticide factory, Waltham Abbey. Occup Environ Med 1997; 54: 101-107.

Eaton N, Shaddick G, Dolk H, Elliott P of the Small Area Health Statistics Unit. Small-area study of the incidence of neoplasms of the brain and central nervous system among adults in the West Midlands Region, 1974-86. *Br J Cancer* 1997;75:1080-83.

Diggle P, Elliott P, Morris S, Shaddick G. Regression modelling of disease risk in relation to point sources. J R Statist Soc A 1997; 160(3): 491-505.

Elliott P, Briggs D J. Recent developments in the geographical analysis of small area health environmental data. In: Progress in Public Health. G Scally (ed). Royal Society of Medicine Press, London, UK, 1997, pp 101-25.

Aylin P, Mahewaran R, Cockings S, Elliott P. A national system for rapid initial assessment of apparent disease clusters. Parliamentary Information Technology Committee Journal (PITCOM), GIS Supplement: Geographic Information and Public Policy) 1998; April Supplement, 14-15.

Elliott P, Best N. Geographical patterns of disease. In: Encyclopaedia of Biostatistics. P Armitage, T Colton (eds). Vol. 2, John Wiley & Sons Ltd, UK, 1998, pp 1694-1701.

Wakefield JC. The United Kingdom Small Area Health Statistics Unit. Journal of Japan Society for Atmospheric Environment 1998; 33: A60-A66.

Dolk H, Thakrar B, Walls P, Landon M, Grundy C, Saez Lloret I, Wilkinson P, Elliott P. Mortality among residents near cokeworks in Great Britain. Occup Environ Med. 1999; 56: 34-40.

Morton Jones T, Diggle P, Elliott P. Investigation of excess environmental risk around putative sources: Stone's test with covariate adjustment. Stat Med 1999; 18: 189-197.

Elliott P, Wakefield JC. Small-area studies of environment and health. In: Statistics for the Environment 4: Statistical Aspects of Health and the Environment. V Barnett, A Stein, K F Turkman (eds). John Wiley & Sons Ltd, UK. 1999, pp. 3-27.

Aylin P, Maheswaran R, Wakefield J, Cockings S, Jarup L, Arnold R, Wheeler G, Elliott P. A national facility for small area disease mapping and rapid initial assessment of apparent disease clusters around a point source: the UK Small Area Health Statistics Unit. J Public Health Med 1999; 21: 289-298.

Maheswaran, R, Arnold RA, Jessop EG. Multiple myeloma in south Cumbria: prediction fulfilled. J Epidemiol Community Health 1999; 53: 255.

Wakefield JC, Elliott P. Issues in the statistical analysis of small-area health data. Stat Med 1999; 18: 2377-2399.

Wilkinson P, Thakrar B, Walls P, Landon M, Falconer S, Grundy C, Elliott P. Lympho-haematopoietic malignancy around all industrial complexes that include major oil refineries in Great Britain. Occup Environ Med 1999; 56: 577-80.

Best NG, Wakefield JC. Accounting for inaccuracies in population counts and case registration in cancer mapping studies. J R Statist Soc A 1999; 162: 363-382.

Best NG. Bayesian ecological modelling. In: Disease Mapping and Risk Assessment for Public Health. A Lawson, A Biggeri, D Böhning, E Lesaffre, J-F Viel and R Bertollini (eds), John Wiley and Sons, Chichester, UK, 1999, pp. 193-201.

Best NG, Arnold RA, Thomas A, Waller LA, Conlon EM. Bayesian models for spatially correlated disease and exposure data. In: Bayesian Statistics 6. J M Bernardo, J O Berger, A P Dawid, A F M Smith (eds). Oxford University Press, Oxford, UK, 1999, pp. 131-56.

Wakefield J, Morris S. Spatial dependence and errors-in variables in environmental epidemiology. In: Bayesian Statistics 6. J M Bernardo, J O Berger, A P Dawid, A F M Smith (eds). Oxford University Press, Oxford UK, 1999, pp. 657-84.

Arnold R, Elliott P, Wakefield J, Quinn M (eds). Population Counts in Small Area Studies: Implications for Studies of Environment and Health. Studies on Medical and Population Subjects No. 62: Office for National Statistics, UK, 1999, 85 pp.

Elliott P, Arnold R, Cockings S, Eaton N, Jarup L, Jones J, Quinn M, Rosato M, Thornton L, Toledano M, Tristan E, Wakefield J. Risk of mortality, cancer incidence and stroke in a population potentially exposed to cadmium. Occup Environ Med 2000; 57: 94-97.

Nieuwenhuijsen MJ, Toledano MB, Eaton NE, Fawell J, Elliott P. Chlorination disinfection byproducts in water and their association with adverse reproductive outcomes: a review. Occup Environ Med 2000; 57: 73-85.

Elliott P, Eaton N, Shaddick G, Carter R. Cancer incidence near municipal solid waste incinerators in Great Britain 2: Histopathological and case-note review of primary liver cancer cases. *Br J Cancer* 2000; 82(5): 1103-1106.

Pascutto C, Wakefield JC, Best NG, Richardson S, Bernardinelli L, Staines A, Elliott P. Statistical issues in the analysis of disease mapping data. Stat Med (in press).

International collaborative publications

Data from some of the cancer registries in England, from ONS (for England and Wales as a whole) and from ISD Scotland, are included in the database used to produce the global reference volume Cancer Incidence in Five Continents. Similar data are used in the EUROCARE studies of cancer survival. In addition to the two EUROCARE

monographs published by IARC, a large number of papers have been published on survival for specific cancer sites using the first EUROCARE database, for example Sant et al 1995. And in depth analyses of the EUROCARE II data were published in a special issue of the European Journal of Cancer. A paper (on colorectal cancer) giving the results of the first of a series of EUROCARE "high resolution" studies has been published in *Gut*. Data from the UK are also included in two major analyses of international cancer trends. Data on childhood cancer in the UK are included in an IARC monograph.

Parkin DM, Whelan SL, Ferlay J, Raymond L, Young J (eds). Cancer Incidence in Five Continents, Volume VII. IARC Scientific Publications No 143. Lyons: International Agency for Research on Cancer, 1997.

Ferlay J, Parkin DM, Pisani P. GLOBOCAN 1: Cancer incidence and mortality worldwide, (CD-ROM). Lyons: International Agency for Research on Cancer, 1998.

Berrino F, Sant M, Verdecchia A, Capocaccia R (eds). Survival of cancer patients in Europe – the EUROCARE study. IARC Scientific Publications No.132. Lyons: International Agency for Research on Cancer, 1995.

Sant M, Capocaccia R, Verdecchia A, Estève J, Gatta G, Micheli A, Coleman MP, Berrino F. Survival of women with breast cancer in Europe: variation with age, year of diagnosis and country. *International Journal of Cancer* 1998; 77: 679-683.

Berrino F, Capocaccia R, Estève J, Gatta G, Hakulinen T, Micheli M, Sant M, Verdecchia A (eds). Survival of cancer patients in Europe – the EUROCARE study II. IARC Scientific Publications No.151. Lyons: International Agency for Research on Cancer, 1999

Coebergh JWW, Sant M, Verdecchia A, Capocaccia R, Hakulinen T, Estève J (eds). Survival of Adult Cancer Patients in Europe diagnosed from 1978-1989. The EUROCARE II Study. *European Journal of Cancer* 1998; 34: 2137-2278.

Gatta G, Capocaccia R, Sant M, Bell CMJ, Coebergh JWW, Damhuis RAM, Faivre J, Martinez-Garcia C, Pawlega J, Ponz de Leon M, Pottier D, Raverdy N, Williams EMI, Berrino F. Understanding variations in survival for colorectal cancer in Europe: a EUROCARE high-resolution study. *Gut* 2000; 47: 533-538.

Coleman MP, Estève J, Damiecki P, Arslan A, Renard H. Trends in cancer incidence and mortality. IARC Scientific Publications No.121. Lyons: International Agency for Research on Cancer, 1993.

Doll R, Fraumeni JF Jr, Muir CS (eds). Cancer Surveys Volume 19/20: Trends in Cancer Incidence and Mortality. Imperial Cancer Research Fund/Cold Spring Harbor Laboratory Press, 1994.

Parkin DM, Kramárová E, Draper GJ, Masuyer E, Michaelis J, Qureshi S ET AL. International Incidence of Childhood Cancer: Volume 2. IARC Scientific Publications No.144. Lyons: International Agency for Research on Cancer, 1998.

Office for National Statistics

Quinn MJ, Babb PJ, Kirby EA, Brock A. Report: Registrations of cancer diagnosed in 1994-1997, England and Wales. Health Statistics Quarterly 2000; 7: 71-82.

Office for National Statistics. Cancer statistics – registrations, England & Wales, 1994. Series MB1 no.27. London: The Stationery Office, 2000.

Quinn MJ, Babb PJ, Jones J, Baker A, Ault C. CD-ROM – Cancer 1971-1997: Registrations of cancer cases and deaths in England and Wales by sex, age, year, health region and type of cancer. London: ONS, 1999.

Coleman MP, Babb PJ, Harris S, Quinn MJ, Sloggett A, De Stavola BL. Report: Cancer survival in England and Wales, 1991-1998. Health Statistics Quarterly 2000; 6: 54-63.

Coleman MP, De Stavola BL, Harris S, Sloggett A, Quinn MJ, Babb PJ. Cancer survival in the Health Authorities of England up to 1998. Patients diagnosed 1991-93 and followed up to 31st December 1998. (A report prepared for the National Health Service Executive under contract with the National Centre for Health Outcomes Development). London: London School of Hygiene & Tropical Medicine, 2000.

Coleman MP, Babb PJ, Damiecki P, Grosclaude P, Honjo S, Jones J, Knerer G, Pitard A, Quinn MJ, Sloggett A, De Stavola B. Cancer Survival Trends in England and Wales, 1971-1995: deprivation and NHS Region. Studies in Medical and Population Subjects No. 61. London: The Stationery Office, 1999.

Coleman MP, Babb PJ, Mayer D, Quinn MJ, Sloggett A. CD-ROM – Cancer survival trends in England and Wales, 1971-1995: deprivation and NHS region. London: ONS, 1999.

Office for National Statistics. Report: Death registrations 1999: cause, England and Wales. Health Statistics Quarterly 2000; 6: 64-70.

Office for National Statistics. Mortality statistics 1998: cause, England and Wales. Series DH2 no.25. London: The Stationery Office,1999.

Office for National Statistics. 20th Century Mortality (England & Wales 1901-1995) CD-ROM. 1997. London: ONS.

Roman E, Carpenter L. Cancer incidence in England, 1981-87. In Drever F (ed): Occupational health. Series DS no.10. Ch 7. The Stationery Office: London. 1995.

Drever F, Whitehead M. Health Inequalities DS No.15 London: The Stationery Office. 1997.

Office for National Statistics. Social Trends 34. 1999 Edition. London: The Stationery Office. 1999.

Office for National Statistics. Regional Trends 34. 1999 Edition. London: The Stationery Office. 1999.

Quinn MJ, Elliott PE. Small Area Statistics in the UK. Proc Int Conf Air Pollution Methodology (COST 613/2) (Budapest). Brussels: Commission of the European Communities, 1991.

Quinn MJ. Confidentiality. In: Elliott P, Cuzick J, English D, Stern R (eds) Geographical and Environmental Epidemiology – Methods for Small-Area Studies. Oxford: Oxford University Press, 1992: 132-140.

Schrijvers CTM, Mackenbach JP, Lutz J-M, Quinn MJ, Coleman MP. Deprivation and survival from breast cancer. *British Journal of Cancer* 1995; 72: 738-743.

Quinn MJ, Allen EJ, on behalf of the United Kingdom Association of Cancer Registries. Changes in incidence of and mortality from breast cancer in England and Wales since introduction of screening. *British Medical Journal* 1995; 311: 1391-1395.

Schrijvers CTM, Mackenbach JP, Lutz J-M, Quinn MJ, Coleman MP. Deprivation, stage at diagnosis and cancer survival in south east England. *International Journal of Cancer* 1995; 63: 324-329.

Quinn MJ, Charlton JRH. Person-based follow up of cancer cases: an example of linkage using fuzzy matching. Proc Int Conf IASS/IAOS Satellite Meeting on Longitudinal Studies, Jerusalem, 1997.

Arrundale J, Bain M, Botting B, Brewster D, Cartwright R A, Chalmers J, Coggan D, Elliott PE, Jackson I, McKinney P, McNally R, Miles DPB, Quinn MJ, Sharp L, Staines A, Stiller C, Wilkinson P. Handbook and Guide to the Identification of Clusters of Diseases. Leukaemia Research Fund Centre for Clinical Epidemiology, University of Leeds, 1997.

Quinn MJ. Cancer registration: UK, England and Wales. In: Parkin DM, Whelan SL, Ferlay J, Raymond L, Young J (eds). Cancer Incidence in Five Continents, Volume VII. IARC Scientific Publications No 143. Lyons: IARC, 1997.

Quinn MJ. Making it all add up. In: NHS Breast Screening Programme 98 Review. Sheffield: NHS BSP, 1998: 10-11.

Quinn MJ, Martinez-Garcia C, Berrino F. Variations in survival from breast cancer in Europe by age and country, 1978-1989. *European Journal of Cancer* 1998; 34: 2204-2211.

Reeves GK, Beral V, Bull D, Quinn MJ. Estimating relative survival among people registered with cancer in England and Wales. *British Journal of Cancer* 1999; 79(1): 18-22.

Quinn MJ, Babb PJ, Jones J, Allen E, on behalf of the United Kingdom Association of Cancer Registries. The effect of screening on the incidence of and mortality from cancer of the cervix in England: evaluation based on routinely collected statistics. *British Medical Journal* 1999; 318; 904-908.

Edwards DM & Jones J. Incidence of and survival from upper aerodigestive tract cancers in the UK: the influence of deprivation. *European Journal of Cancer* 1999; 35: 968-972.

Schillinger JA, Grosclaude PC, Honjo S, Quinn MJ, Sloggett A, Coleman MP. Survival from acute lymphocytic leukaemia: Socio-economic status and regi on. *Archives of Disease in Childhood* 1999; 80: 311-317.

Quinn MJ. Cancer registration – an example for other disease registers? *British Journal of Healthcare Computing & Information Management* 1999; 16(10): 18-22.

Arnold R, Elliott P, Wakefield J, Quinn MJ (eds). Population counts in small areas: implications for studies of environment and health. Studies on Medical and Population Subjects No.62. London: ONS, 1999.

Elliott P, Arnold A, Cockings S, Eaton M, Jarup L, Jones J, Quinn MJ, Rosato M, Thornton I, Toledo M, Tristan E, Wakefield J. Risk of mortality, cancer incidence and stroke in a population potentially exposed to cadmium. *Journal of Occupational and Environmental Medicine* 2000; 57: 94-97.

Moss SM, Blanks RG, Quinn MJ. Screening mammography re-evaluated. *Lancet* 2000; 355: 748 (letter).

Coleman MP, Babb PJ, Quinn MJ, Sloggett A, De Stavola B. Socio-economic inequalities in cancer survival. Proc 7th Biennial Symposium on Minorities, the Medically Underserved & Cancer "Addressing the Unequal Burden of Cancer". Washington DC, USA, February 2000.

Richards MA, Stockton D, Babb PJ, Coleman MP. How many deaths have been avoided through improvements in cancer survival? *British Medical Journal* 2000; 320: 895-898.

Grosclaude P, Coleman MP, Sloggett A, Quinn MJ. Problèmes posés par les catégories imprécis dans l'analyse des données de survie: l'exemple de la morphologie tumorale. Proc Annual Conference of the Latin Language Cancer Registries. San Sebastian, June 2000.

Majeed A, Babb PJ, Jones J, Quinn MJ. Prostate cancer in England and Wales 1971-1998. *British Journal of Urology International* 2000; 85: 1058-1062.

Quinn MJ, Botting B, Foote D, Read A.. Registration of ovarian cancer in England and Wales. *British Journal of Cancer* 2000; 83: 278-279 (letter).

Quinn MJ. Progress on flagging cancers at NHSCR. *The Researcher* 2000; 14: 4-5.

Coleman MP, Babb PJ, Stockton D, Forman D, Møller H. Trends in breast cancer incidence, survival and mortality in England and Wales. *Lancet* 2000; 356: 590-591 (letter)

Blanks RG, Moss SM, McGahan CE, Quinn MJ, Babb PJ. Effect of NHS breast screening programme on mortality from breast cancer in England and Wales, 1990-8: comparison of observed with predicted mortality. *British Medical Journal* 2000; 321: 665-669.

Quinn MJ, Babb PJ. Cancer trends in England and Wales, 1950-1999. *Health Statistics Quarterly* 2000; 8:5-19.

Coleman MP, Babb PJ, Sloggett A, Quinn MJ, De Stavola BL. Socio-economic inequalities in cancer survival in England and Wales. *Cancer* (in press).

Kaye JA, Derby LE, del Mar Melero-Montes M, Quinn MJ, Jick H. The incidence of breast cancer in the General Practice Research Database compared with national cancer registration data. *British Journal of Cancer* (in press).

Longitudinal Study

Davey-Smith G, Leon D, Shipley M, Rose G. Socio-economic differentials in cancer among men. *International Journal of Epidemiology* 1991;20:2:339-345.

Dolin P, A descriptive study of occupation and bladder cancer in England and Wales. *British Journal of Cancer* 1992;65:3:476-478.

Dolin PJ, Cook-Mozaffari P. Occupation and bladder cancer: a death certificate study. *British Journal of Cancer* 1992;66:3:568-579.

Grulich A, Swerdolw A, Dos Santos Silva I, Beral V. Is the apparent rise in cancer mortality in the elderly real? Analysis of changes in certification and coding of cause of death in England and Wales, 1970-1990. *International Journal of Cancer* 1995;63:164-168.

Harding S, Allen EJ. Sources and uses of data on cancer among the ethnic groups. *British Journal of Cancer* 1996;74:Supplement XXIX:S17-S21.

Harding S. The incidence of cancers among second generation Irish living in England and Wales. *British Journal of Cancer* 1998;78(7):958-961.

Harding S, Rosato M. Incidence of cancers in migrant groups living in England and Wales. Ethnicity and Health 1998.

Kogevinas M. Longitudinal Study: Socio-demographic differences in cancer survival. OPCS Series LS 1990 No5 London:HMSO.

Kogevinas M. Social Inequalities and Cancers, In: Preventing Cancers. Heller T, Bailey L, Patterson S, Milton Keynes: The Open University Press:5-17.

Kogevinas M, Marmot MG, Fox AJ, Goldblatt PO. Socio-economic differences in cancer survival. Journal of Epidemiology and Community Health 1991;45:216-219.

Murphy M, Goldblatt P, Thornton-Jones H, Silcocks P. Survival amongst women with cancer of the uterine cervix; the influence of marital status and social class. *Journal of Epidemiology and Community Health* 1990;44:293-296.

Pugh H, Power C, Goldblatt P, Arber S. Women's Lung Cancer Mortality, Socio-Economic Status and Changing Smoking Patterns. Social Science and Medicine 1991;32:10:1105-1110.

GLOSSARY

Adenocarcinoma – a malignant epithelial tumour derived from glandular tissue or where the tumour cells form recognisable glandular structures

Adenoma – a benign epithelial tumour derived from glandular tissue or which exhibits clearly defined glandular structures. Some adenomas can progress to become malignant

Age-specific rate – an incidence or mortality rate for a specific (usually five year) age group, e.g. 40-44. The numerator is usually the number of cases or deaths in a calendar year; and the denominator is the mid-year population for the relevant age group

Age standardisation – controlling for differences in the age structure of populations between geographical areas or overtime to give unbiased comparisons of incidence rates or mortality

Age-standardised rate – see appendix H

Basal cell – regarded as malignant but does not show all the characteristics of malignant tumours, eg does not tend to metastasise

Benign – a tumour which is usually slow growing, retaining many of the structural and functional features of its tissues of origin and not invading surrounding tissue or metastasising to distant organs

Carcinoma – a malignant tumour derived from epithelial tissue

Cohort – a group of people selected by their year of birth. Their characteristics can be followed as they enter successive age and time periods

Cohort effects – variations in health status that arise from the different causal factors affecting each birth cohort in the population exposed as environment and society change over time

Cohort incidence/mortality ratios – the ratio of observed to expected cases/deaths. The expected cases/deaths are derived by applying the method of indirect standardisation to the observed rates, using 1926 as the base cohort for incidence and 1916 for deaths

Confidence interval – a range of values for a variable (eg a rate), constructed so that this range has a specified probability of including the true value of the variable

Crude incidence/death rate – the total number of cases/deaths in specified period divided by the total mid-year population for that period

Crude survival – the proportion of a cohort of subjects alive at the end of a specified time interval since diagnosis (irrespective of the cause of death)

Cumulative risk – the lifetime chance of developing/dying from the specified cancer. Usually expressed as a percentage

Deprivation – socio-economic deprivation was measured using the Carstairs index. This combines several variables from the census into a score for each enumeration district (ED). All individuals living in a particular ED are assigned the same deprivation score. (See Appendix H for further details.)

Epithelium – A closely packed sheet of cells arranged in one or more layers, the component cells of which usually adhere to each other along their edges and surfaces

Frequency distribution – the percentage distribution of cases per defined age group

Grade – tumours originating from a specific cell type may be of varying degrees of differentiation and of malignancy. Grade I has the least degree of malignancy and grade IV has the greatest

Incidence – the rate at which new cases occur in the population, ie the number of new cases of a disease in a specified period, divided by the population at risk of getting the disease during the specified period

Indirect standardisation – compares study populations where the age-specific rates are unknown. The age-specific rates for the standard population are averaged using the distribution of the study population as a weight. The ratio of the crude rate for the study population to the weighted average obtained is the standardised incidence/mortality ratio

In situ – the tumour, which is not malignant, remains at the site of origin and has not spread to other parts of the body

Latent period – the interval between disease initiation to onset of clinical symptoms and signs

Lead time bias – if an individual participates in a screening programme and the disease is detected earlier than it would have been in the absence of screening, the amount of time by which diagnosis is advanced as a result of screening is the lead time. As the point of diagnosis is then advanced in time, survival as measured from diagnosis is automatically lengthened, even if total length of life is not increased

Leukaemia – cancers of the blood forming organs, characterised by abnormal proliferation and development of leucocytes and their precursors in the bone marrow, blood, lymph and lymph glands

Lymphoma – a cancer of cells of the immune system confined to lymph glands and related tissues

GLOSSARY

Malignant – a tumour with the potential for invading neighbouring tissue and/or metastasising to distant body sites, or one that has already done so

Metastases – the spread or transfer of cancer from its original site to another place in the body where the disease process starts up. This usually occurs by way of the bloodstream or the lymphatic system

Mortality:incidence ratios – the number of deaths from a cancer in a particular period (usually a calendar year) divided by the number of new cases in the same period. [N.B. For cancers with good survival, those dying in a given year may have been diagnosed many years previously.]

Mortality – the number of deaths from a disease in a given population during a specified time period (usually a calendar year), divided by the mid-year population

Neoplasm – a collection of cells, derived from a common origin, that is increasing in number and expanding or spreading, either locally or to remote sites ie a new growth

Papilloma – a benign tumour of skin or mucous membrane in which epithelial cells grow outward from a surface around a connective tissue core containing blood vessels

Prevalence – the number of people diagnosed with cancer in a specified period who are still alive at the end of that period

Relative survival – the ratio of the observed survival in the group being studied and the survival that would have been expected had they been subject only to the mortality rates of the general population. (See Appendix H for further details.)

Risk – the proportion of people in a population who develop the disease within a specified time period

Risk factors – an aspect of personal behaviour or lifestyle, an environmental exposure, or an inherited characteristic which is known to be associated with a health related condition

Sarcoma – a tumour, often highly malignant, developing in the connective tissue of bones, muscles, blood vessels, cartilage etc.

Screening – the routine examination of numbers of apparently healthy people to identify those with a particular disease at an early stage

Squamous cell – common skin tumours can arise from squamous epithelium e.g. tongue, larynx, bronchus

Stage – describes the tumour with respect to its size, extent and how widespread, if any, the metastases are

Subsite – the anatomical site as specified by the fourth digit of the ICD code

Transitional cell – a type of tumour that arises anywhere in the urinary tract particularly the bladder. There may be multiple lesions, often fronded

Tumour – a mass of abnormal tissue, the growth of which exceeds and is uncoordinated with the normal tissue from which it originates, and which persists in the same excessive manner after the stimuli which evoked the change have ceased

REFERENCES

Chapter 1

1. Department of Health. *Saving Lives: Our Healthier Nation.* London: Department of Health, 1999.

2. Department of Health. *The NHS Plan: A plan for investment: A plan for reform.* CM4818–I. London: Department of Health, 2000.

3. NHS Executive. *Expenditure on cancer research 1995/96.* Leeds: NHS Executive, 1996.

4. Quinn MJ, Babb PJ, Kirby EA, Brock A. Report: Registrations of cancer diagnosed in 1994–1997, England and Wales. *Health Statistics Quarterly* 2000; 7: 71–82.

5. Coleman MP, Babb PJ, Damiecki P, Grosclaude P, Honjo S, Jones J, Knerer G, Pitard A, Quinn MJ, Sloggett A, De Stavola BL. *Cancer survival trends in England and Wales, 1971–1995: deprivation and NHS region.* Studies in Medical and Population Subjects No.61. London: The Stationery Office, 1999.

6. Berrino F, Sant M, Verdecchia A, Capocaccia R (eds). *Survival of cancer patients in Europe – the EUROCARE study.* IARC Scientific Publications No.132. Lyons: International Agency for Research on Cancer, 1995.

7. Berrino F, Capocaccia R, Estève J, Gatta G, Hakulinen T, Micheli M, Sant M, Verdecchia A (eds). *Survival of cancer patients in Europe – the EUROCARE study II.* IARC Scientific Publications No.151. Lyons: International Agency for Research on Cancer, 1999.

8. National Cancer Institute. *SEER Stat – cancer incidence public use database 1973–95. Release 1.1.* Bethesda MD: National Cancer Institute, 1998.

9. Department of Health. *National Cancer Plan.* London: Department of Health, 2000.

10. dos Santos Silva I. *Cancer Epidemiology: Principles and Methods.* Lyons: International Agency for Research on Cancer, 1999.

11. Parkin DM, Chen VW, Ferlay J, Galceran J, Storm HH, Whelan SL. *Comparability and Quality Control in Cancer Registration.* IARC Technical Report No.19. Lyons: International Agency for Research on Cancer, 1994.

12. Jensen OM, Parkin DM, Maclennan R, Muir CS, Skeet RG (eds). *Cancer Registration: Principles and Methods.* IARC Scientific Publications No.95. Lyons: International Agency for Research on Cancer, 1991.

13. Swerdlow AJ. Cancer registration in England and Wales: some aspects relevant to interpretation of the data. *Journal of the Royal Statistical Society Series A* 1986; 149: 146–160.

14. Office for National Statistics. *Mortality statistics 1998: cause, England and Wales.* Series DH2 no.25. London: The Stationery Office, 1999.

15. Rooney C, Devis T. *Mortality trends by cause of death in England and Wales 1980–94: the impact of introducing automated cause coding and related changes in 1993.* Population Trends 1996; 86: 29–35.

16. Doll R, Peto R. *The Causes of Cancer.* Oxford: Oxford University Press, 1981.

17. Doll R, Fraumeni JF Jr, Muir CS (eds). *Cancer Surveys Volume 19/20: Trends in Cancer Incidence and Mortality.* Imperial Cancer Research Fund/Cold Spring Harbor Laboratory Press, 1994.

18. Coleman MP, Babb PJ, Stockton D, Forman D, Møller H. Trends in breast cancer incidence, survival and mortality in England and Wales. *Lancet* 2000; 356: 590–591 (letter).

Chapter 2

1. Quinn MJ, Babb PJ, Kirby EA, Brock A. Report: Registrations of cancer diagnosed in 1994–1997, England and Wales. *Health Statistics Quarterly* 2000; 7: 71–82.

2. Office for National Statistics. *Cancer statistics 1994 – registrations, England and Wales.* Series MB1 no.27. London: The Stationery Office, 2000.

3. World Health Organisation. *International Statistical Classification of Diseases and Related Health Problems – Tenth Revision.* Geneva: WHO, 1992.

4. Department of Health & Social Security. *Breast cancer screening: report to the health ministers of England, Wales, Scotland and Northern Ireland.* London: HMSO, 1986. (Forrest report.)

5. Doll R, Peto R. *The Causes of Cancer.* Oxford: Oxford University Press, 1981.

6. Coleman MP, Babb PJ, Damiecki P, Grosclaude P, Honjo S, Jones J, Knerer G, Pitard A, *Quinn MJ, Sloggett A, De Stavola BL. Cancer survival trends in England and Wales, 1971–1995: deprivation and NHS region.* Studies in Medical and Population Subjects No.61. London: The Stationery Office, 1999.

7. Quinn MJ, Allen E, on behalf of the United Kingdom Association of Cancer Registries. Changes in incidence of and mortality from breast cancer in England and Wales since introduction of screening. *British Medical Journal*

1995; 311: 1391–1395.

8. Chouillet A, Bell CMJ, Hiscox J. Management of breast cancer in southeast England. *British Medical Journal* 1994; 308: 168–171.

9. Department of Health & Welsh Office. *A policy framework for commissioning cancer services. A report by the Expert Advisory Group on Cancer to the Chief Medical Officers of England and Wales.* London: Department of Health, 1995. (Calman–Hine report.)

10. Blanks RG, Moss SM, McGahan CE, Quinn MJ, Babb PJ. Effect of NHS breast screening programme on mortality from breast cancer in England and Wales, 1990–8: comparison of observed with predicted mortality. *British Medical Journal* 2000; 321: 665–669.

11. Swerdlow AJ, dos Santos Silva I, Reid A, Qiao Z, Brewster DH, Arrundale J. Trends in cancer incidence and mortality in Scotland: description and possible explanations. *British Journal of Cancer* 1998; 77 (supplement 3): 1–16.

12. Brewster DH, Fraser LA, Harris V, Black RJ. Rising incidence of prostate cancer in Scotland: increased risk or increased detection? *British Journal of Urology International* 2000; 85: 463–473.

13. Brawley OW. Prostate carcinoma incidence and patient mortality. The effects of screening and early detection. *Cancer* 1997; 80: 1857-1863.

14. Majeed A, Babb PJ, Jones J, Quinn MJ. Prostate cancer in England and Wales 1971–1998. *British Journal of Urology International* 2000; 85: 1058–1062.

15. Beral V. Cancer of the cervix: a sexually transmitted infection? *Lancet* 1974; i: 1037–1040.

16. Quinn MJ, Babb PJ, Jones J, Allen E, on behalf of the United Kingdom Association of Cancer Registries. Effect of screening on incidence of and mortality from cancer of the cervix in England: evaluation based on routinely collected statistics. *British Medical Journal* 1999; 318: 904–908.

17. Sasieni P, Adams J. Effect of screening on cervical cancer mortality in England and Wales: analysis of trends with an age–cohort model. *British Medical Journal* 1999; 318: 1244–1245.

18. Office for National Statistics. *Regional Trends 34.* London: The Stationery Office, 1999.

19. Parkin DM, Pisani P, Ferlay J. Estimates of the world–wide incidence of 25 major cancers in 1990. *International Journal of Cancer* 1999; 80: 827–841.

20. Coleman MP, Babb PJ, Harris S, Quinn MJ, Sloggett A, De Stavola BL. Report: Cancer survival in England and Wales, 1991–1998. *Health Statistics Quarterly* 2000; 6: 54–63.

21. Langlands AO, Pocock SJ, Kerr GK, Gore M. Long–term survival of patients with breast cancer: a study of the curability of the disease. *British Medical Journal* 1979; 2: 1247–1251.

22. Zahl P–H, Tretli S. Long–term survival of breast cancer in Norway by age and clinical stage. *Statistics in Medicine* 1997; 16: 1435–1449.

23. Stiller CA. Population based survival rates for childhood cancer in Britain, 1980–91. *British Medical Journal* 1994; 309: 1612–1616.

24. Coleman MP, De Stavola B, Harris S, Sloggett A, Quinn MJ, Babb PJ. *Cancer survival in the Health Authorities of England up to 1998. Patients diagnosed 1991–93 and followed up to 31st December 1998.* (A report prepared for the National Health Service Executive under contract with the National Centre for Health Outcomes Development.) London: London School of Hygiene and Tropical Medicine, 2000.

25. Tomatis L. Socio–economic factors and human cancer. *International Journal of Cancer* 1995; 62: 121–125.

26. Drever F, Whitehead M (eds). *Health inequalities.* Office for National Statistics Series DS no.15. London: The Stationery Office, 1997.

27. Department of Health. *Independent inquiry into inequalities in heath.* London: The Stationery Office, 1998. (Acheson report.)

28. Auvinen A, Karjalainen S. Possible explanations for social class differences in cancer patient survival. In: Kogenvinas M, Pearce N, Susser M, Boffetta P (eds) *Social inequalities and cancer.* IARC Scientific Publications No.138. Lyons: International Agency for Research on Cancer, 1999.

29. Leon DA, Wilkinson RG. Inequalities in prognosis: socio–economic differences in cancer and heart disease survival. In: Fox J (ed) *Health inequalities in European countries.* Aldershot: Gower, 1989.

30. Schrijvers CTM, Mackenbach J, Lutz J–M, Quinn MJ, Coleman MP. Deprivation and survival from breast cancer. *British Journal of Cancer* 1995; 72: 738–743.

31. Schrijvers CTM, Mackenbach J, Lutz J–M, Quinn MJ, Coleman MP. Deprivation, stage at diagnosis and cancer survival. *International Journal of Cancer* 1995; 63: 324–329.

REFERENCES

32. Scottish Cancer Intelligence Unit. *Trends in cancer survival in Scotland 1971–1995*. Edinburgh: Information and Statistics Division, 2000.

33. Berrino F, Sant M, Verdecchia A, Capocaccia R (eds). *Survival of cancer patients in Europe – the EUROCARE study*. IARC Scientific Publications No.132. Lyons: International Agency for Research on Cancer, 1995.

34. Coebergh JWW, Sant M, Verdecchia A, Capocaccia R, Hakulinen T, Estève J (eds). Survival of Adult Cancer Patients in Europe diagnosed from 1978–1989. The EUROCARE II Study. *European Journal of Cancer* 1998; 34: 2137–2278.

35. Berrino F, Capocaccia R, Estève J, Gatta G, Hakulinen T, Micheli M, Sant M, Verdecchia A (eds). *Survival of cancer patients in Europe – the EUROCARE study II*. IARC Scientific Publications No.151. Lyons: International Agency for Research on Cancer, 1999.

36. National Cancer Institute. *SEER Stat – cancer incidence public use database 1973–95*. Release 1.1. Bethesda MD: National Cancer Institute, 1998.

37. Sant M, Capocaccia R, Verdecchia A, Estève J, Gatta G, Micheli M, Coleman MP, Berrino F. Survival of women with breast cancer in Europe: variation with age, year of diagnosis and country. *International Journal of Cancer* 1998; 77: 679–683.

38. Sant M, Capocaccia R, Verdecchia A, Gatta G, Micheli M, Mariotto A, Hakulinen T, Berrino F. Comparisons of colon cancer survival among European countries: the EUROCARE study. *International Journal of Cancer* 1995; 63: 43–48.

39. Quinn MJ, Martinez–Garcia C, Berrino F. Variations in survival from breast cancer in Europe by age and country, 1978–1989. *European Journal of Cancer* 1998; 34: 2204–2211.

40. Gatta G, Capocaccia R, Sant M, Bell CMJ, Coebergh JWW, Damhuis RAM, Faivre J, Martinez–Garcia C, Pawlega J, Ponz de Leon M, Pottier D, Raverdy N, Williams EMI, Berrino F. Understanding variations in survival for colorectal cancer in Europe: a EUROCARE high–resolution study. *Gut* 2000; 47:533-578.

41. Richards M. International comparisons. In: Cancer Action 2. London: NHS Executive, 2000.

Chapter 3

1. Coleman MP, Estève J, Damiecki P, Arslan A, Renard H. *Trends in cancer incidence and mortality*. IARC Scientific Publications No.121. Lyons: International Agency for Research on Cancer, 1993.

2. Silverman DT, Morrison AS, Devesa SS. Bladder cancer. In: Schottenfeld D, Fraumeni JF Jr (eds) *Cancer Epidemiology and Prevention*. 2nd Edition, Chapter 54. New York: Oxford University Press, 1996.

Chapter 4

1. Neal A, Hoskins P. CNS tumours. *Clinical oncology*. Chapter 12. London: Edward Arnold, 1994.

2. Preston–Martin S, Mack WJ. Neoplasms of the nervous system. In: Schottenfeld D, Fraumeni JF Jr (eds). *Cancer Epidemiology and Prevention*. 2nd Edition, Chapter 58. New York: Oxford University Press, 1996.

3. Muir CS, Storm HH, Polednak A. Brain and other nervous system tumours. In: Doll R, Fraumeni JF Jr, Muir CS (eds) *Cancer Surveys Volume 19/20: Trends in Cancer Incidence and Mortality*. Imperial Cancer Research Fund/Cold Spring Harbor Laboratory Press, 1994.

4. Swerdlow AJ, dos Santos Silva I, Reid A, Qiao Z, Brewster DH, Arrundale J. Trends in cancer incidence and mortality in Scotland: description and possible explanations. *British Journal of Cancer* 1998; 77 (supplement 3): 1–16.

Chapter 5

1. McPherson K, Steel CM, Dixon JM. Breast cancer – epidemiology, risk factors, and genetics. *British Medical Journal* 1994; 309: 1003–1007.

2. Evans DGR, Fentiman IS, McPherson K, Asbury D, Ponder BAJ, Howell A. Familial breast cancer. *British Medical Journal* 1994; 308: 183–187.

3. Department of Health & Social Security. *Breast cancer screening: report to the health ministers of England, Wales, Scotland and Northern Ireland*. London: HMSO, 1986. (Forrest report.)

4. Quinn MJ, Allen E, on behalf of the United Kingdom Association of Cancer Registries. Changes in incidence of and mortality from breast cancer in England and Wales since introduction of screening. *British Medical Journal* 1995; 311: 1391–1395.

5. Blanks RG, Moss SM, McGahan CE, Quinn MJ, Babb PJ. Effect of NHS breast screening programme on mortality from breast cancer in England and Wales, 1990–8: comparison of observed with predicted mortality. *British Medical Journal* 2000; 321: 665–669.

6. Karjalainen S, Aareleid T, Hakulinen T, Pukkala E, Rahu M, Tekkel M. Survival of female breast cancer patients in Finland and Estonia: stage at diagnosis an important determinant of the difference between countries. *Social Science Medicine* 1989; 28: 233–238.

7. Scottish Cancer Intelligence Unit. *Trends in cancer survival in Scotland 1971–1995*. Edinburgh: Information and Statistics Division. 2000.

8. Schrijvers CTM, Mackenbach JP, Lutz JM, Quinn MJ, Coleman MP. Deprivation and survival from breast cancer. *British Journal of Cancer* 1995; 72: 738–743.

9. Quinn MJ, Martinez–Garcia C, Berrino F. Variations in survival from breast cancer in Europe by age and country, 1978–1989. *European Journal of Cancer* 1998; 34: 2204–2211.

10. Sant M, Capocaccia R, Verdecchia A, Estève J, Gatta G, Micheli A, Coleman MP, Berrino F. Survival of women with breast cancer in Europe: variation with age, year of diagnosis and country. *International Journal of Cancer* 1998; 77: 679–683.

Chapter 6

1. Parkin DM, Pisani P, Ferlay J. Estimates of the worldwide incidence of 25 major cancers in 1990. *International Journal of Cancer* 1999; 80: 827–841.

2. Beral V. Cancer of the cervix: a sexually transmitted infection? *Lancet* 1974; i: 1037–1040.

3. Brinton LA. Epidemiology of cervical cancer – overview. In: Muñoz N, Bosch FX, Shah KV, Meheus A (eds) *The epidemiology of human papillomavirus and cervical cancer*. IARC Scientific Publications No.119. Lyons: International Agency for Research on Cancer, 1992.

4. Day NE. Screening for cancer of the cervix. *Journal of Epidemiology and Community Health* 1989; 43: 103–106.

5. Farmery E, Gray JAM. *Report of the First Five Years of the NHS Cervical Screening Programme*. Oxford: National Co–ordinating Network, 1994.

6. Murphy MFG, Campbell MJ, Goldblatt PO. Twenty years' screening for cancer of the uterine cervix in Great Britain 1964–84: further evidence for its ineffectiveness. *Journal of Epidemiology and Community Health* 1987; 42: 49–53.

7. Quinn MJ, Babb PJ, Jones J, Allen E, on behalf of the United Kingdom Association of Cancer Registries. Effect of screening on incidence of and mortality from cancer of the cervix in England: evaluation based on routinely collected statistics. *British Medical Journal* 1999; 318: 904–908.

8. Woodman CBJ, Rollason T, Ellis J, Tierney R, Wilson S, Young L. Human papillomavirus infection and risk of progression of epithelial abnormalities of the cervix. *British Journal of Cancer* 1996; 73: 553–556.

9. Sasieni P, Adams J. Effect of screening on cervical cancer mortality in England and Wales: analysis of trends with an age–cohort model. *British Medical Journal* 1999; 318: 1244–1245.

10. Walker JJ, Brewster D, Gould A, Raab GM. Trends in incidence of and mortality from invasive cancer of the uterine cervix in Scotland (1975–1994). *Public Health* 1998; 112: 373–378.

11. Herbert A, Breen C, Bryant TN, Hitchcock A, Macdonald H, Millward–Sadler GH, Smith J. Invasive cervical cancer in Southampton and South West Hampshire: effect of introducing a comprehensive screening programme. *Journal of Medical Screening* 1996; 3: 23–28.

Chapter 7

1. Krongborg O, Fenger C, Olsen J, Jorgensen OD, Sondergaard O. Randomised study of screening for colorectal cancer with faecal–occult–blood test. *Lancet* 1996; 348: 1467–1471.

2. Hardcastle JD, Chamberlain JO, Robinson MHE, Moss SM, Amar SS, Balfour TW, James PD, Mangham CM. Randomised controlled trial of faecal–occult–blood screening for colorectal cancer. *Lancet* 1996; 348: 1472–1477.

Chapter 8

1. McCredie M. Bladder and kidney cancers. In: Doll R, Fraumeni JF Jr, Muir CS (eds) *Cancer Surveys Volume 19/20: Trends in Cancer Incidence and Mortality*. Imperial Cancer Research Fund/Cold Spring Harbor Laboratory Press, 1994.

2. McLaughlin JK, Blot WJ, Devesa SS, Fraumeni JF Jr. Renal Cancer. In: Schottenfeld D, Fraumeni JF Jr (eds) *Cancer Epidemiology and Prevention*. 2nd Edition, Chapter 53. New York: Oxford University Press, 1996.

Chapter 9

1. Austin DF, Reynolds P. Neoplasms of the nervous system. In: Schottenfeld D, Fraumeni JF Jr (eds) *Cancer Epidemiology and Prevention*. 2nd Edition, Chapter 30. New York: Oxford University Press, 1996.

2.	Coleman MJ, Estève J, Damiecki P, Arslan A, Renard H. *Trends in cancer incidence and mortality*. IARC Scientific Publications No.121. Lyons: International Agency for Research on Cancer, 1993.

Chapter 10

1.	Linet MS, Cartwright RA. The leukemias. In: Schottenfeld D, Fraumeni JF Jr (eds) *Cancer Epidemiology and Prevention*. 2nd Edition, Chapter 40. New York: Oxford University Press, 1996.

2.	Kane EV, Roman E, Cartwright R, Parker J, Morgan G. Tobacco and the risk of acute leukaemia in adults. *British Journal of Cancer* 1999; 81: 1228–1233.

Chapter 11

1.	Coleman MP, Babb PJ, Damiecki P, Grosclaude P, Honjo S, Jones J, Knerer G, Pitard A, Quinn MJ, Sloggett A, De Stavola BL. *Cancer survival trends in England and Wales 1971–1995: deprivation and NHS Region*. Studies in Medical and Population Subjects No.61. London: The Stationery Office; 1999.

Chapter 12

1.	Cancer Research Campaign. *Lung Cancer and Smoking*, CRC Factsheet No. 11. London: CRC, 1996.

2.	Koop C Everett. *Journal of the National Cancer Institute* (US) 1989; 81(21): 1610.

3.	Action on Smoking and Health. *Fact Sheet No.19*. London: ASH, 1999.

4.	Schrijvers CTM, Mackenbach J, Lutz J–M, Quinn MJ, Coleman MP. Deprivation, stage at diagnosis and cancer survival. *International Journal of Cancer* 1995; 63: 324–329.

Chapter 13

1.	Cancer Research Campaign. *Malignant Melanoma – UK*, CRC Factsheet No. 4. London: CRC, 1995.

2.	UK Skin Cancer Prevention Working Party. *Consensus statement*. London: British Association of Dermatologists, 1994.

3.	MacKie R, Hole DJ. Incidence and thickness of primary tumours and survival or patients with cutaneous malignant melanoma in relation to socio–economic status. *British Medical Journal* 1996; 312: 1125–1128.

4.	Schrijvers CTM, Mackenbach J, Lutz J–M, Quinn MJ, Coleman MP. Deprivation, stage at diagnosis and cancer survival. *International Journal of Cancer* 1995; 63: 324–329.

Chapter 14

1.	Herrington LJ, Weiss NS, Olshan AF. In: Schottenfeld D, Fraumeni JF Jr (eds) *Cancer Epidemiology and Prevention* 2nd Edition, Chapter 43. New York: Oxford University Press, 1996.

2.	Coleman MP, Estève J, Damiecki P, Arslan A, Renard H. *Trends in cancer incidence and mortality*. IARC Scientific Publications No.121. International Agency for Research on Cancer: Lyons, 1993.

3.	Berrino F, Sant M, Verdecchia A, Capocaccia R (eds). *Survival of cancer patients in Europe – the EUROCARE study*. IARC Scientific Publications No.132. Lyons: International Agency for Research on Cancer, 1995.

4.	Berrino F, Capocaccia R, Estève J, Gatta G, Hakulinen T, Micheli M, Sant M, Verdecchia A (eds). *Survival of cancer patients in Europe – the EUROCARE study II*. IARC Scientific Publications No.151. Lyons: International Agency for Research on Cancer, 1999.

Chapter 15

1.	Coleman MP, Estève J, Damiecki P, Arslan A, Renard H. *Trends in cancer incidence and mortality*. IARC Scientific Publications No.121. International Agency for Research on Cancer: Lyons, 1993.

2.	Hartge P, Devesa SS, Fraumeni JF Jr. Hodgkin's and non–Hodgkin's lymphomas. In: Doll R, Fraumeni JF Jr, Muir CS (eds) *Cancer Surveys Volume 19/20: Trends in Cancer Incidence and Mortality*. Imperial Cancer Research Fund/Cold Spring Harbor Laboratory Press, 1994.

Chapter 16

1.	Swerdlow AJ, dos Santos Silva I, Reid A, Qiao Z, Brewster DH, Arrundale J. Trends in cancer incidence and mortality in Scotland: description and possible explanations. *British Journal of Cancer* 1998; 77 (supplement 3): 1–16.

2.	Dolan K, Sutton R, Walker SJ, Morris AI, Campbell F, Williams EMI. New classification of oesophageal and gastric carcinomas derived from changing patterns in epidemiology. *British Journal of Cancer* 1999; 80: 834–842.

3.	Blot WJ, Devesa SS, Hermanek P, Fraumeni Jr JF. Rising incidence of adenocarcinoma of the esophagus and gastric cardia. *Journal of American Medical Association* 1991; 265: 1287–1289.

4.	Ricaurte O, Flejou JF, Vissauzaine C, Goldfain D, Rotenburg D, Cadiot G, Potet P. Helicobacter pylori infection in patients with Barrett's oesophagus: a prospective immunohistochemical study. *Journal of Clinical Pathology* 1996; 49: 176–177.

5.	Chow WH, Finkle WD, McLaughlin JK, Frankl H, Ziel HK, Fraumeni Jr JF. The relation of gastrooesophageal disease and its treatment to adenocarcinomas of the oesophagus and gastric cardia. *Journal of American Medical Association* 1995; 274: 474–477.

Chapter 17

1.	Booth M. Aetiology and epidemiology of ovarian cancer. In Blackledge GRP, Jordan JA, Shingleton HM (eds) *Textbook of Gynaecologic Oncology*. London: WB Saunders, 1991.

2.	Hildreth NG, Kelsey JL, Livolsi VA, Fischer DB, Holford TR, Mostow ED, Schwartz PE, White C. An epidemiologic study of epithelial carcinoma of the ovary. *American Journal of Epidemiology* 1981; 114: 398–405.

3.	Swerdlow AJ, dos Santos Silva I. Recent trends in incidence of and mortality from breast, ovarian and endometrial cancers in England and Wales and their relation to changing fertility and oral contraceptive use. *British Journal of Cancer* 1995; 72: 485–492.

4.	Booth M, Beral V, Smith P. Risk factors for ovarian cancer: a case–control study. *British Journal of Cancer* 1989; 60: 592–598.

5.	Cancer Research Campaign. *Ovarian Cancer – UK*. CRC Factsheet No.17. London: CRC, 1997.

6.	Crawford RAF, Jacobs IJ. Molecular genetics of ovarian cancer. *Contemporary Reviews in Obstetrics & Gynaecology* 1996, 8: 44–49.

7.	Schrijvers CTM, Mackenbach J, Lutz J–M, Quinn MJ, Coleman MP. Deprivation, stage at diagnosis and cancer survival. *International Journal of Cancer* 1995; 63: 324–329.

8.	Folke Petersson (ed). *Annual report on the results of treatment in gynaecological cancer*. 22nd Vol International Federation of Gynaecology and Obstetrics, 1994.

9.	Jacobs IJ, Skates SJ, MacDonald N, Menon U, Rosenthal AN, Prys Davies A, Woolas R, Jeyarajah AR, Sibley K, Lowe DG, Oram DH. Screening for ovarian cancer: a pilot randomised controlled trial. *Lancet* 1999; 353: 1207–1210.

Chapter 18

1.	International Agency for Research on Cancer. *Tobacco smoking*. Monographs on the evaluation of the carcinogenic risk of chemicals to humans. Vol. 38. Lyons: International Agency for Research on Cancer. 1986.

2.	Anderson KE, Potter JD, Mack TM. Pancreatic cancer. In: Schottenfeld D, Fraumeni Jr JF (eds) *Cancer Epidemiology and Prevention*. 2nd Edition, Chapter 35. New York: Oxford University Press, 1996.

Chapter 19

1.	Breslow N, Chan CW, Dhom G et al. Latent carcinoma of prostate at autopsy in seven areas. *International Journal of Cancer* 1977; 20: 680–688.

2.	Potosky AL, Kessler LG, Gridley G, Brown CC, Horm JW. Rise in prostatic cancer incidence associated with increased use of transurethral resection. *Journal of the National Cancer Institute* 1990; 82: 1624–1628.

3.	Potosky AL, Miller BA, Albertsen PC, Kramer BS. The role of increasing detection in the rising incidence of prostate cancer. *Journal of the American Medical Association* 1995; 273: 548–552.

4.	Berrino F, Estève J, Coleman MP. Basic issues in the estimation and comparison of cancer patient survival. In: Berrino F, Sant M, Verdecchia A, Capocaccia R, Hakulinen T, Estève J (eds) *Survival of cancer patients in Europe: the EUROCARE study*. IARC Scientific Publications No. 132. Lyons: International Agency for Research on Cancer, 1995.

5.	Doll R, Peto R. *The causes of cancer*. Oxford: Oxford University Press. 1981.

6.	Swerdlow AJ, dos Santos Silva I, Reid A, Qiao Z, Brewster DH, Arrundale J. Trends in cancer incidence and mortality in Scotland: description and possible explanations. *British Journal of Cancer* 1998; 77 (supplement 3): 1–16.

7.	Brawley OW. Prostate carcinoma incidence and patient mortality. The effects of screening and early detection. *Cancer* 1997; 80: 1857-1863.

8.	Thames Cancer Registry. *Cancer in South East England* 1992. Sutton: Thames Cancer Registry, 1995.

9.	Selley S, Donovan J, Faulkner A, Coast J, Gillatt D. Diagnosis, management and screening of early localised prostate cancer. *Health Technology Assessment* 1 (2), 1997.

REFERENCES

Chapter 20

1. Parkin DM, Pisani P, Ferlay J. Estimates of the worldwide incidence of 25 major cancers in 1990. *International Journal of Cancer* 1999; 80: 827-841.

Chapter 21

1. Cancer Research Campaign. *Testicular Cancer – UK.* CRC Factsheet No.16. London: CRC, 1998.

2. Horwich A (ed). *Testicular cancer: investigation and management.* London: Chapman Hall, 1992.

3. Thames Cancer Registry. *Cancer in South East England 1992.* Sutton: Thames Cancer Registry, 1996.

4. Medical Research Council Working Party on Testicular Tumours. Prognostic factors in advanced non– seminomatous germ–cell testicular tumours: results of a multicentre study. *Lancet* 1985, i: 8–11.

Chapter 22

1. Elwood JM, Cole P, Rothman KJ, Kaplan SD. Epidemiology of endometrial cancer. *Journal of the National Cancer Institute* 1977; 59: 1055–1060.

2. Henderson BE, Casagrande JT, Pike MC, Mack T, Rosario I, Duke A. The epidemiology of endometrial cancer in young women. *British Journal of Cancer* 1983; 47: 749–756.

3. Swerdlow AJ, dos Santos Silva I. Recent trends in incidence of and mortality from breast, ovarian and endometrial cancers in England and Wales and their relation to changing fertility and oral contraceptive use. *British Journal of Cancer* 1995; 72: 485–492.

4. Jensen OM, Parkin DM, MacLennan R, Muir CS, Skeet RG (eds). *Cancer registration: principles and methods.* IARC Scientific Publications No.95. Lyons: International Agency for Research on Cancer, 1991.

Chapter 23

1. Chow W–H, Linet MS, Liff JM, Greenberg RS. Cancers in children. In: Schottenfeld D, Fraumeni JF Jr (eds) *Cancer Epidemiology and Prevention.* 2nd Edition, Chapter 54. New York: Oxford University Press, 1996.

2. Kinlen LJ. Epidemiological evidence for an infective basis in childhood leukaemia. *British Journal of Cancer* 1995; 71: 1–5.

3. Kinlen LJ. High–contact paternal occupations, infection and childhood leukaemia: five studies of unusual population–mixing of adults. *British Journal of Cancer* 1997; 76: 1539–1545.

4. Scherr PA, Mueller NE. Non–Hodgkin's lymphoma. In: Schottenfeld D, Fraumeni JF Jr (eds) *Cancer Epidemiology and Prevention.* 2nd Edition, Chapter 42. New York: Oxford University Press, 1996.

5. Draper GJ, Kroll ME, Stiller CA. Childhood cancer. In: Doll R, Fraumeni JF Jr, Muir CS (eds) *Cancer Surveys Volume 19/20: Trends in Cancer Incidence and Mortality.* Imperial Cancer Research Fund/Cold Spring Harbor Laboratory Press, 1994.

6. Storm HH. Cancers of the soft tissues. In: Doll R, Fraumeni J F Jr, Muir CS (eds) *Cancer Surveys Volume 19/20: Trends in Cancer Incidence and Mortality.* Imperial Cancer Research Fund/Cold Spring Harbor Laboratory Press, 1994.

7. Neal AJ, Hoskin PJ. *Clinical oncology: a textbook for students.* London: Edward Arnold, 1994.

8. Stiller CA. Population based survival rates for childhood cancer in Britain, 1980–91. *British Medical Journal* 1994; 309: 1612–1616.

9. Yeazel MW, Ross JA, Buckley JD, Woods WG, Ruccione K, Robison LL. High birth weight and risk of specific childhood cancers: a report from the Children's Cancer Group. *Journal of Pediatrics* 1997; 131: 671–677.

Appendix G

1. Parkin DM, Chen VW, Ferlay J, Galceran J, Storm HH, Whelan SL. *Comparability and Quality Control in Cancer Registration.* IARC Technical Report No. 19. Lyons: International Agency for Research on Cancer, 1994.

2. Jensen OM, Parkin DM, Maclennan R, Muir CS, Skeet RG (eds). *Cancer Registration: Principles and Methods.* IARC Scientific Publications No.95. Lyons: International Agency for Research on Cancer, 1991.

3. Office of Population Censuses and Surveys. *Report of the Advisory Committee on Cancer Registration.* London: OPCS, 1970.

4. Office of Population Censuses and Surveys. *Report of the Advisory Committee on Cancer Registration, 1980: Cancer registration in the 1980s.* Series MB1 no.6. London: HMSO, 1981.

5. Department of Health. *Working for Patients. The Health Service.* Caring for the 1990s. CM 555. London: HMSO, 1989.

6. Office of Population Censuses and Surveys. *A Review of the National Cancer Registration System in England and Wales.* Series MB1 no.17. London: HMSO, 1990.

7. NHS Management Executive. *Minimum data set for the National Cancer Registration System.* EL(92)95. London: Department of Health, 1995.

8. Office for National Statistics. *Cancer statistics 1994 – registrations, England & Wales.* Series MB1 no.27. London: The Stationery Office, 2000.

9. Quinn MJ, Charlton JRH. Person–based follow–up of cancer cases: linkage using fuzzy matching. Proc Int Conf IASS/IAOS Satellite Meeting on Longitudinal Studies, Jerusalem, 1997.

10. Gill LE, Simmons HM. *An algorithm for fixed–length proper name compression: Oxford Name Compression Algorithm (ONCA).* University of Oxford, Unit of Clinical Epidemiology, 1992.

11. Newcombe HB. *Handbook of Record Linkage – Methods for Health and Statistical Studies, Administration and Business.* Oxford: Oxford Medical Publications, Oxford University Press, 1988.

12. Kendrick S, Clarke J. The Scottish Record Linkage System. Health Bulletin (Edinburgh) 51 No.2, March 1993.

13. Quinn MJ, Babb PJ, Jones J, Brock A. Report: Registrations of cancer diagnosed in 1993–1996, England and Wales. *Health Statistics Quarterly* 1999; 4: 59–70.

14. Quinn MJ, Babb PJ, Kirby EA, Brock A. Report: Registrations of cancer diagnosed in 1994–1997, England and Wales. *Health Statistics Quarterly* 7; 2000: 71–82.

15. NHS Executive. *Core contract for purchasing Cancer Registration.* EL(96)7. London: NHS Executive, 1996.

16. Winyard G. EL(96)7: *Core contract for purchasing cancer registration* (letter). Leeds: NHS Executive, 1998.

17. Department of Health & Welsh Office. *A policy framework for commissioning cancer services. A report by the Expert Advisory Group on Cancer to the Chief Medical Officers of England and Wales.* London: Department of Health, 1995. (Calman–Hine report.)

18. Parkin DM, Whelan SL, Ferlay J, Raymond L, Young J. *Cancer Incidence in Five Continents, Volume VII.* IARC Scientific Publications No.143. Lyons: International Agency for Research on Cancer, 1997.

19. Davies TW, Williams EMI (eds). The Cancer Registry Handbook. Cambridge: United Kingdom Association of Cancer Registries, 1994.

20. Bell CMJ (ed). *Reducing risk, improving outcome in cancer.* London: United Kingdom Association of Cancer Registries, 1998.

Appendix H

1. Parkin DM, Whelan S, Ferlay J, Raymond L, Young J (eds). *Cancer Incidence in Five Continents, Vol. VII.* IARC Scientific Publications No.143. Lyons: International Agency for Research on Cancer, 1997.

2. Quinn MJ, Babb PJ, Kirby EA, Brock A. Report: Registrations of cancer diagnosed in 1994–1997, England and Wales. *Health Statistics Quarterly* 2000; 7: 71–82.

3. World Health Organisation. *International Statistical Classification of Diseases and Related Health Problems – Tenth Revision.* Geneva: WHO, 1992.

4. Swerdlow AJ. Cancer registration in England and Wales: Some aspects relevant to interpretation of the data. *Journal of the Royal Statistical Society Series A* 1986; 149: 146–160.

5. Hawkins MM, Swerdlow AJ. Completeness of cancer and death follow–up obtained through the National Health Service Central Register for England and Wales. *British Journal of Cancer* 1992; 66: 408–413.

6. Villard–Mackintosh L, Coleman MP, Vessey MP. The completeness of cancer registration in England: an assessment from Oxford FPA study. *British Journal of Cancer* 1988; 58: 507–511.

7. Office for National Statistics. *Cancer statistics 1993 – registrations, England and Wales.* Series MB1 no.26. London: The Stationery Office, 1999.

8. Parkin DM, Chen VW, Ferlay J, Galceran J, Storm HH, Whelan SL. *Comparability and Quality Control in Cancer Registration.* IARC Technical Report No.19. Lyons: International Agency for Research on Cancer, 1994.

9. Office of Population Censuses and Surveys. *Registry/ONS Interface Document. National Cancer Registration System, England and Wales.* London: OPCS, 1994 (subsequently revised).

10. NHS Executive. *Core contract for purchasing Cancer Registration.* EL(96)7. London: NHS Executive, 1996.

11. Winyard G. EL(96)7: *Core contract for purchasing cancer registration* (letter). Leeds: NHS Executive, 1998.

12. Huggett C. *Review of the Quality and Comparability of Data held by Regional Cancer Registries.* Bristol: Bristol Cancer Epidemiology Unit incorporating the South West Cancer Registry, 1995.

13. Seddon DJ, Williams EMI. Data quality in population based cancer registration: an assessment of the Merseyside and Cheshire Cancer Registry. *British Journal of Cancer* 1997; 76: 667–674.

14. Brewster D, Crichton J, Muir C. How accurate are Scottish cancer registration data? *British Journal of Cancer* 1994; 70: 954–959.

15. Office of Population Censuses and Surveys. *Cancer statistics 1988 – registrations, England and Wales.* Series MB1 no.21. London: HMSO, 1994.

16. Quinn MJ, Babb PJ, Jones J, Baker A, Ault C. *Cancer 1971–1997.* Registrations of cancer cases and deaths in England and Wales by sex, age, year, health region and type of cancer (CD–ROM). London: Office for National Statistics, 1999.

17. American Cancer Society. *Manual of tumour nomenclature and coding (MOTNAC).* Washington DC: American Cancer Society, 1951.

18. World Health Organisation. *International Classification of Diseases – Ninth Revision.* Geneva: WHO, 1977.

19. Office of Population Censuses and Surveys. *Mortality statistics – comparison of 8th and 9th Revision of the International Classification of Diseases.* Series DH1 no.10. London: HMSO, 1983.

20. Quinn MJ. Progress on flagging cancers at NHSCR. *The Researcher* 2000; 14: 4–5

21. Office of Population Censuses and Surveys and General Register Office (Scotland). *1991 Census: Definitions, Great Britain.* London: HMSO, 1992.

22. Office for National Statistics. *Mortality statistics 1998: cause, England and Wales.* Series DH2 no.25. London: The Stationery Office, 1999.

23. Office of Population Censuses and Surveys. *Mortality statistics 1984: cause, England and Wales.* Series DH2 no.11. London: HMSO, 1985.

24. Office for National Statistics. *Mortality statistics 1993 (revised) and 1994: cause, England and Wales.* Series DH2 no.21. London: The Stationery Office, 1996.

25. Rooney C, Devis T. *Mortality trends by cause of death in England and Wales 1980–94: the impact of introducing automated cause coding and related changes in 1993.* Population Trends 1996; 86: 29–35.

26. Doll R, Peto R. *The Causes of Cancer.* Oxford: Oxford University Press, 1981.

27. Coleman MP, Babb P, Damiecki P, Grosclaude P, Honjo S, Jones J, Knerer G, Pitard A, Quinn MJ, Sloggett A, De Stavola B. *Cancer Survival Trends in England and Wales, 1971–1995: deprivation and NHS Region.* Studies in Medical and Population Subjects No. 61. London: The Stationery Office, 1999.

28. Coleman MP, Babb PJ, Harris S, Quinn MJ, Sloggett A, De Stavola BL. Report: Cancer survival in England and Wales, 1991–1998. *Health Statistics Quarterly* 2000; 6: 54–63.

29. Office for National Statistics. Report: Death registrations 1999: cause, England and Wales. *Health Statistics Quarterly* 2000; 6: 64–70.

30. Quinn MJ, Allen EJ, on behalf of the United Kingdom Association of Cancer Registries. Changes in incidence of and mortality from breast cancer in England and Wales since introduction of screening. *British Medical Journal* 1995; 311: 1391–1395.

31. Blanks RG, Moss SM, McGahan CE, Quinn MJ, Babb PJ. Effect of NHS breast screening programme on mortality from breast cancer in England and Wales, 1990–8: comparison of observed with predicted mortality. *British Medical Journal* 2000; 321: 665–669.

32. Quinn MJ, Babb PJ, Jones J, Allen E, on behalf of the United Kingdom Association of Cancer Registries. The effect of screening on the incidence of and mortality from cancer of the cervix in England: evaluation based on routinely collected statistics. *British Medical Journal* 1999; 318: 904–908.

33. Sasieni P, Adams J. Effect of screening on cervical cancer mortality in England and Wales: analysis of trends with an age–cohort model. *British Medical Journal* 1999; 318: 1244–1245.

34. Defoe D. *A journal of the plague year.* London, 1722.

35. Heasman MA, Lipworth L. *Accuracy of certification of cause of death.* Studies in Medical and Population Subjects no.20. London: HMSO, 1966.

36. Alderson MR, Meade TW. Accuracy of diagnosis on death certificates compared with that in hospital records. *British Journal of Preventive and Social Medicine* 1967; 21: 22–29.

37. Grulich AE, Swerdlow AJ, dos Santos Silva I, Beral V. Is the apparent rise in cancer mortality in the elderly real? Analysis of changes in certification and coding of cause of death in England and Wales, 1970–1990. *International Journal of Cancer* 1995; 63: 164–168.

38. Percy CL, Muir CS. The international comparability of cancer mortality data: results of an international death certificate study. *American Journal of Epidemiology* 1989; 129: 934–946.

39. Ashworth TG. Inadequacy of death certification: proposal for change. *Journal of Clinical Pathology* 1991; 44: 265–268.

40. Percy CL, Dolman AB. Comparison of the coding of death certificates related to cancer in seven countries. *Public Health Reports* 1978; 93: 335–350.

41. Percy CL, Stanek E, Gloeckler Ries LA. Accuracy of cancer death certificates and its effect on cancer mortality statistics. *American Journal of Public Health* 1981; 71: 242–250.

42. Percy CL, Miller BA, Gloeckler Ries LA. Effect of changes in cancer classification and the accuracy of cancer death certificates on trends in cancer mortality. *Annals of the New York Academy of Sciences* 1990; 609: 87–97.

43. Hoel DG, Ron E, Carter R, Mabuchi K. Influence of death certificate errors on cancer mortality trends. *Journal of the National Cancer Institute* 1993; 85: 1063–1068.

44. Lindahl BI, Johansson LA. Multiple cause–of–death data as a tool for detecting artificial trends in the underlying cause statistics: a methodological study. *Scandinavian Journal of Social Medicine* 1994; 22: 145–158.

45. Garne JP, Aspegren K, Balldin G. Breast cancer as cause of death – a study over the validity of the officially registered cause of death in 2631 breast cancer patients dying in Malmö, Sweden 1964–1992. *Acta Oncologica* 1996; 35: 671–675.

46. Coleman MP & Aylin P (eds). *Death certification and mortality statistics: an international perspective.* Studies on Medical and Population Subjects No.64. London: TSO, 2000.

47. Coleman MP, Babb PJ, Stockton D, Forman D, Møller H. Trends in breast cancer incidence, survival and mortality in England and Wales. *Lancet* 2000; 356: 590–591 (letter).

48. Parkin DM, Kramarova E, Draper GJ, Masuyer E, Michaelis J, Neglia J, Qureshi S, Stiller CA. *International Incidence of Childhood Cancer Volume II.* IARC Scientific Publications No.144. Lyons: International Agency for Research on Cancer, 1998.

49. dos Santos Silva I. *Cancer Epidemiology: Principles and Methods.* Lyons: International Agency for Research on Cancer, 1999.

50. Breslow NE, Day NE. *Statistical Methods in Cancer Research, Volume II – The Design and Analysis of Cohort Studies.* IARC Scientific Publications No.82. Lyons: International Agency for Research on Cancer, 1987.

51. Beral V. Cancer of the cervix: a sexually transmitted infection? *Lancet* 1974; i: 1037–40.

52. Swerdlow AJ, dos Santos Silva I, Reid A, Qiao Z, Brewster DH & Arrundale J. Trends in cancer incidence and mortality in Scotland: description and possible explanations. *British Journal of Cancer* 1998, 77 (Supplement 3): 1–16.

53. Schouten LJ, Straatman H, Kiemeney LALM & Verbeek ALM. Cancer incidence: life table risk versus cumulative risk. *Journal of Epidemiology and Community Health* 1994; 48: 596–600.

54. Carstairs V, Morris R. Deprivation and mortality: an alternative to social class? *Community Medicine* 1989; 11: 210–219.

55. Zanetti R, Micheli A, Rosso S, Sant M. The prevalence of cancer: a review of the available data. *Tumori* 1999; 85: 408–413.

56. Estève J, Benhamou E, Croasdale M, Raymond L. Relative survival and the estimate of net survival: elements for further discussion. *Statistics in Medicine* 1990; 9: 529–538.